Civil Jet Aircraft Design

Lloyd R. Jenkinson

Senior Lecturer, Loughborough University, UK

Paul Simpkin

Formerly Head of Aircraft Performance, Rolls Royce, UK

Darren Rhodes

Civil Aviation Authority – NATS Department of Operational Research
and Analysis (DORA), UK

ARNOLD

A member of the Hodder Headline Group
LONDON • SYDNEY • AUCKLAND

First published in Great Britain in 1999 by
Arnold, a member of the Hodder Headline Group,
338 Euston Road, London NW1 3BH
http://www.arnoldpublishers.com

Co-published in North America by
American Institute of Aeronautics and Astronautics, Inc.,
1801 Alexander Bell Drive,
Reston, VA 20191-4344

British Library Cataloguing in Publication Data
A catalogue record for this book is available from the British Library

Library of Congress Cataloging-in-Publication Data
A catalog record for this book is available from the Library of Congress

ISBN 0 340 74152 X

1 2 3 4 5 6 7 8 9 10

Commissioning Editor: Matthew Flynn
Production Editor: James Rabson
Production Controller: Sarah Kett
Cover design: Terry Griffiths

Typeset in 10/12 pt Times by AFS Image Setters Ltd, Glasgow
Printed and bound in Great Britain by MPG Books Ltd, Bodmin, Cornwall

What do you think about this book? Or any other Arnold title?
Please send your comments to feedback.arnold@hodder.co.uk

This book is dedicated
to our nearest and dearest:
Marie, Dot, Ken and Marjorie

Contents

Preface

Excellence in design is one of the principal factors that enable a developed nation to stay competitive in a global economy. The ability of companies to 'add value' to a product or manufactured goods is highly regarded throughout the world. Within this context, aircraft manufacture and operation is regarded as a desirable commercial activity. Many countries who previously were not involved in aeronautics are now moving into the business. However, as aircraft are made technically more complex, involve increasing interdependence between component parts (airframe, engines and systems), and become more multi-national due to the large initial development costs, there is an increasing challenge to the industry. The aircraft design team is in the forefront of this challenge.

Set against this scenario are the conflicts between the various objectives and requirements for new aircraft projects. The designer is concerned about increasing performance and quality, meeting production deadlines, promoting total-life product support, and satisfying customer and infrastructure requirements. Above all these aspects the designer is expected to meet established safety and environmental requirements and to anticipate the sociological and political impact of the design. Balancing all these aspects within an acceptable cost and timescales is what makes aircraft design such a professionally challenging and ultimately satisfying activity.

One of the educational objectives of an aeronautical engineering curriculum is to introduce students to the procedures and practices of aircraft design as a means of illustrating the often conflicting requirements mentioned above. This textbook describes the initial project stages of civil transport aircraft design as an example of such practices. Obviously, the methods used have had to be simplified from industrial practice in order to match the knowledge, ability and timescales available to students. However, this simplification is made in the level of specialisation and detail design and not in the fundamental principles. An example of this approach is the substitution of general purpose spreadsheet methods in place of the specialist procedural-based mainframe programs commonly used for aircraft analysis in industry.

Apart from a general introduction to aircraft project design, this book provides an extension to the classical 'Flight Mechanics' courses. It also bridges the gap between specialist lectures in aerodynamics, propulsion, structures and systems, and aircraft project coursework. The scope of the book has purposely been limited to meet the objectives of undergraduate study. In order to illustrate the basic principles, a simplified approach has had to be made. Where appropriate, reference is made to other texts for more detailed study. Some prior knowledge of conventional theories in aerodynamics, propulsion, structures and control is necessary but where possible the analysis presented in the book can be used without further study. The terminology and the significance of various parameters is explained at the point of application and the main notation listed at the end of the book.

The book is arranged in two parts. In Chapters 1–14 each of the significant influences on aircraft project design is described. This part starts with a broad introduction to civil air transport, followed by a detailed description of the design process and a description of aircraft layout procedures. The next set of chapters are concerned with detailed descriptions of the design methods and an introduction to the principal aircraft components. The concluding chapters deal with the parametric methods used to refine the design configuration and a description of the formal presentation of the baseline design.

The second part of the book (Chapters 15–19) includes an introduction to the use of spreadsheet methods in aircraft design work and four separate design studies. The studies illustrate the application of such spreadsheet methods. Each study deals with a separate design topic. The first shows how a simple design specification is taken through the complete design process. As a contrast the second study deals with non-passenger aircraft design, considering a transport aircraft. As both the previous studies deal with conventional configurations the third study shows how to assess unorthodox layouts. Finally, the last design study shows how you may use the methods to analyse topics other than pure aircraft technical aspects.

No book alone will provide the key to good design. You can only achieve this through the acquisition of knowledge, hard methodical work, broad experience on many different aircraft projects and an open and creative mind. However, we hope that this book will eliminate some of the minor stumbling blocks that young engineers find annoying, confusing and time-wasting at the start of their design work.

Wherever possible System International (SI) units have been used. However, in aeronautics several parameters continue to be used and quoted in non-SI units (for example altitude is normally in feet). It is therefore necessary to have an understanding of different systems of units. To allow conversion between different systems of units a conversion table offers help to both aspiring young designers and older engineers who struggle to convert their past experience into the new system of units.

Finally, it is impossible to make the book complete. The contents and data do not cover all the aspects of civil transport design required by every user. For example, we have not included anything on supersonic aircraft because it is still uncertain if we can solve the environmental problems (noise and emissions) associated with high and fast operations. By the same token we have not gone too far on the inclusion of advanced technologies and materials. Such developments will not affect the main design process and you could allow for them in future studies by establishing factors for use in the standard formulae (e.g. mass and drag reduction factors).

However, allowing for these omissions, we have made a genuine attempt to produce a book that is a starting point for students who want to know more about the fascinating process of commercial turbofan aircraft design.

Acknowledgements

In writing this book we have received a great deal of assistance from many individuals, institutions and companies who have given information, effort and encouragement to us.

We are indebted to the Department of Aeronautical and Automotive Engineering and Transport Studies at Loughborough University who freely supported the preparation and development of the book. We express our thanks to the heads of department, Stan Stevens and Jim McGuirk, and all the members of the aeronautical staff. We are particularly grateful to secretaries Ann French and Mary Bateman for their fortitude in tackling the typing of the original manuscript.

Of all the companies who have provided help and information for this book we would like to thank Rolls Royce PLC and specifically John Hawkins of the advanced projects office, Bruce Astride and members of the aircraft project group.

We also recognise the assistance of Patrick Farmar with respect to the case studies and Bob Caves for his help with various operational issues.

Accompanying data

In association with this book is a series of data sets which are located on the publisher's web site (**www.arnoldpublishers.com/aerodata**). There are five separate sets of data:

- Data A contains technical information on over 70 civil jet aircraft.
- Data B contains details of over 40 turbofan engines.
- Data C includes geographical and site data for around 600 airports.
- Data D defines the International Standard Atmosphere (ISA) and various operational speeds.
- Data E includes definitions and conversions between different parameters and systems of units used in aeronautics.

Students will find the information in these data sets useful in conducting project design studies. Reference to these sets is made at the appropriate points throughout the book and particularly in the chapters concerned with the case studies.

1

Introduction

On the third of May 1952 a De Havilland Comet 1 aircraft (Fig. 1.1) took off to inaugurate the world's first scheduled jet airline service. This flight from London to Johannesburg was celebrated close to the fiftieth anniversary of the Wright brothers' historic first powered passenger-carrying flight and only 11 years after the first flight of the Whittle jet engine. In the 40-plus years since the Comet flight, jet aircraft have established a dominant position in the civil air transport market and continue to show steady increase in numbers as demand for business, leisure and private flying continues to grow.

Since the end of the Second World War annual air transport passenger numbers have risen from about 18 million to over a billion in recent years. This growth is expected to continue as more nations become industrialised and the world's population becomes more air-minded. This demand for air travel establishes a

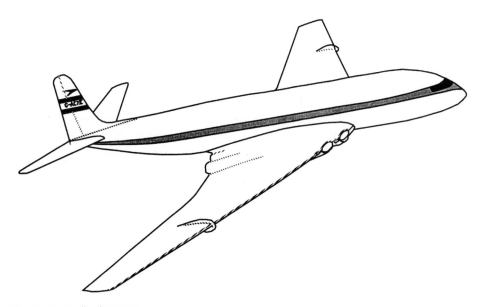

Fig. 1.1 De Havilland Comet 1.

requirement for more aircraft and a larger network of routes. Commercial opportunities arise for both aircraft manufacturers and airlines to meet the expected market.

Estimating traffic growth

To plan aircraft development programmes it is necessary to estimate future trends in air transport for both passenger and freight businesses. The number of aircraft movements is mainly related to the demand for passenger travel. Econometric analysis of historical data shows a strong correlation between world economic growth and the demand for air transport. This confirms that an expansion in business travel and cargo transport are linked to growth in commercial and world trading activity. The level of personal disposable income affects the demand for leisure travel. The standard of service provided has a direct influence on the customer's motivation to travel. All these issues are affected by such factors as the price of the air ticket, international currency exchange rates, availability and frequency of the service, expansion and development of routes and changes in the regulations governing airline operations.

World economic growth is measured by national and global gross national products (GNP). Although the expansion in air transport generally follows the variation in GNP, it has consistently shown a much larger growth rate. In the period 1960–1990 world GNP increased at an average annual rate of 3.8% in real terms whereas airline scheduled passenger traffic (measured in revenue passenger miles, RPM) increased at an annual rate of 9.5%. Over this period the rate of growth of both GNP and RPM have progressively decreased. In the last decade the average world GNP rose by only 2.4% and RPM by 5.7% per year. Even at this lower rate air travel doubles in a 12-year period. These high growth rates have resulted in congestion at the busiest airports and in the most frequently used flight corridors.

Apart from the capital required to purchase new aircraft to meet the demand, large investments are necessary in airports and the associated infra-structure to provide the service. The resultant expansion at existing airports (increased length and number of runways and new terminal facilities) attracts environmental objections and inevitable political interference with the economic model. These factors are difficult to predict and could have considerable effect on the natural expansion of traffic at the busiest city centre airports. Technological developments in air transport management and aircraft design may be the only way that more serious environmental and political restrictions can be overcome.

Factors outside the control of the air transport industry have been shown to affect the national growth in traffic. Sudden changes in fuel price by cartel trading was the cause of depression in demand for air travel in the mid to late 70s. Political unrest, (e.g. terrorism, or in extreme cases war) have always depressed air travel. In the future, the market for air transport may be affected by the expansion of new communication and information technology systems (e.g. tele-conferencing and the internet). This may either depress travel by reducing the need for business trips, or stimulate the market by generating more trade and a stronger demand for

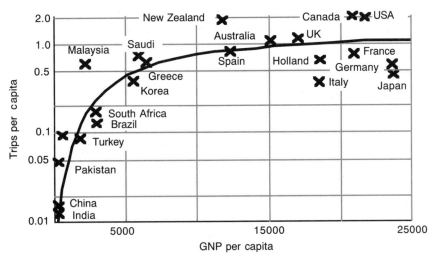

Fig. 1.2 Demand for air travel (source IMF).

holiday travel. Such influences make forecasts based on historical analysis very suspect but the consistency of demand for air travel over the past 30 years gives reassurance that the effects tend to be self balancing.

The influence of national GNP values, and the resulting personal disposable income associated with this, is clearly illustrated by plotting the number of trips per capita against national GNP. The distribution is shown in Fig. 1.2.

In developed countries the air transport market is regarded as mature with growth at about 3–4% per year. The latent demand for air travel is seen in the developing and under-developed industrial countries. The developing countries have air transport growth rates about double that of the developed countries. The poorer countries need the stimulation of industrialisation to provide growth in air transport. This is shown clearly in the potential for growth in China and in the countries of the original USSR.

The UK Office of Science and Technology in a 'Technology Forecast' in 1995 identified three main influences on the growth of air travel:

- gross domestic product (GDP)
- airfares
- propensity to travel

GDP is the dominant parameter and as such needs to be carefully considered in any forecast. The current value of 3% may be varied by ±0.9% in a best and worst case scenario. Air fares have historically reduced due largely to competitive trading and economies of scale but the future is somewhat confused as airline bankruptcies and amalgamations reduce the number of operators. Fares are also affected by the cost of fuel which may vary in the future due to changes in production rates and the addition of environmental taxes (e.g. carbon tax). All these effects need to be reflected in the methods used for forecasting.

The economic modelling of demand for air transport is based on the following two models.

$$(PK) = a \cdot (GDP)^b \cdot (PR)^c \qquad\qquad (FTK) = d \cdot (EX)^e \cdot (FR)^f$$

where:

(PK) = passenger–kilometres (FTK) = freight tonne–kilometres
(GDP) = gross domestic product (EX) = world exports
(PR) = passenger revenue per unit (PK) (FR) = freight revenue per unit (FTK)

Historical analysis provides values for the coefficients a, b, c, d, e, f. As passenger demand increases airlines can respond by scheduling extra flights, by using larger aircraft, or by increasing passenger load factors. The linking of passenger–kilometres to aircraft–kilometres involves estimates of passenger load factor (i.e. the ratio of the number of seats filled to the total available) and aircraft size. The following relationships apply:

aircraft–kilometres = (passenger–kilometres)/(passenger load factor)

where: passenger load factor = (passenger–kilometres)/(seat–kilometres) and

aircraft size = (seat–kilometres)/(aircraft–kilometres)

The historical trends in passenger load factor and aircraft size are shown in Fig. 1.3.

It can be seen that since the early 80s both aircraft size and passenger load factor have increased only modestly. This effect may have resulted from the introduction of several new medium-sized aircraft into the market at this time and the expansion of airlines due to deregulation. As a result, the number of city pairs linked by scheduled services nearly doubled in this period. These factors are likely to continue into the near future but may be curtailed by restrictions in air movement growth at some airports due to congestion.

Overcoming the operational problems at airports during busy times of the day has been cited as the main reason to develop new larger aircraft to meet future traffic growth. Larger aircraft would allow more passengers to be moved from

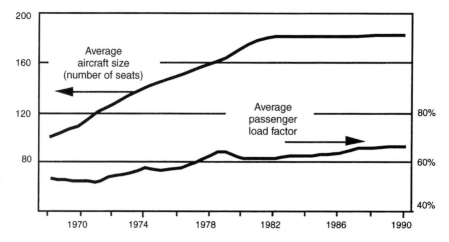

Fig. 1.3 Trend in aircraft size and passenger load factor (source ICAO).

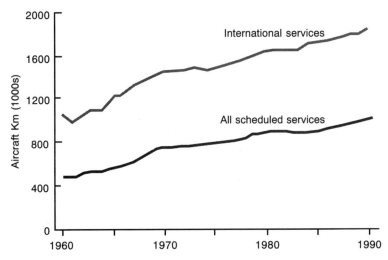

Fig. 1.4 Historical trend in average aircraft stage length (source ICAO).

existing runways and in the airspace near the airport. An estimate by the International Civil Aviation Organisation (ICAO) made in 1992 predicted average aircraft size to rise from 183 seats in 1990 to 220 in 2001.

Airport planning demands a knowledge of the number of aircraft departures. Aircraft stage length links aircraft departures to aircraft–kilometres:

$$\text{aircraft departures} = (\text{aircraft–km})/(\text{stage length})$$

where: stage length = (aircraft–km)/(aircraft departures).

The historical trend in average stage length is shown in Fig. 1.4.

In the past 20 years the growth in average stage length has been between 1 and 2% per annum. The figure above shows that the largest growth has been in the long-haul routes. This has been made possible mainly by the increased range capability of new aircraft types and the development of new markets, particularly in the Pacific region. This trend is expected to continue and result in annual growth of 1% in average stage length.

Forecasting growth can be achieved by considering the average annual rates of change in each of the main variables:

$$\% \text{ (aircraft–km)} = \% \text{ (passenger–km)} - \% \text{ (load factor)} - \% \text{ (average size)}$$

$$\% \text{ (departures)} = \% \text{ (aircraft–km)} - \% \text{ (stage length)}$$

ICAO made the following predictions in 1992 (Table 1.1).

The world's commercial activity associated with aerospace is huge (amounting to over $100B in the mid-90s). This is shared almost equally between companies specialising in airframes, engines, airframe systems, airport systems and airport facility.

There are currently about 350 000 civil aircraft registered in the world but most of these are in the light/personal aircraft category. Only 50 000 are registered in the civil commercial category (hire and reward), and only about 10 000 of these are

Table 1.1 Historical traffic data (source ICAO)

	1970	1980	1990	2000	Average annual growth rate (%)		
					70–80	80–90	90–00
Passenger–km (billions)	382	929	1654	2830	9.3	5.9	5.0
Passenger load factor (%)	52	61	66	68	1.6	0.8	0.3
Passenger aircraft size (seats)	109	171	183	220	4.6	0.7	1.7
Aircraft stage length (km)	738	875	983	1100	1.7	1.2	1.0
Aircraft–km (millions)	7004	9350	14 307	19 800	2.9	4.3	3.0
Aircraft departures (thousands)	9486	10 691	14 553	18 000	1.2	3.1	2.0

The table above includes all-freight movements but excludes operations of aircraft registered in the Russian Federation.

turbofan airliners. Although civil turbo-powered airliners represent a small percentage of the total aircraft population they account for 75% of the total value of all aircraft. The dominance of turbofan-powered aircraft is illustrated by the historical trend of commercial aircraft by type as shown in Fig. 1.5. In the early days most scheduled flights were made by piston engined aircraft but as the turbojet and later the turbofan engines were developed these dominated the market and superseded the piston types. At the present time few new piston-powered

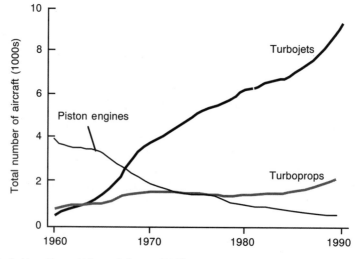

Fig. 1.5 Historical trend in world fleet mix (source ICAO).

Table 1.2 World fleet mix forecast (source UK Department of Trade)

Category	Current fleet	Forecast deliveries	Deliveries value ($B)	% of total
Turbofan airliners	9680	12 100	725	75
Turboprop airliners	5000	3900	50	5
Business jets	8000	7200	75	8
Business turboprops	9500	6000	30	3
Turbine helicopters	12 000	9000	80	8
Light aircraft	300 000	20 000	3	<1
Civil aircraft (total)	344 180	58 200	963	100%

aircraft are used by airlines. For the short-haul market the fuel efficiency of the turboprop aircraft still attracts operators.

The current and projected fleet of civil aircraft together with the projected cost data for new aircraft, up to 2015, are shown in Table 1.2.

The projected traffic growth over the next 20 years (Fig. 1.6) averages 5% per year but this figure is not expected to be constant across all regions. For developed airline networks (mostly North America and Europe) the growth is predicted to be 4%, whereas for the Asia/Pacific Ring area a faster growth of 7% is expected. The projected figures do not include the potential for air travel from undeveloped economic areas like China, Russia and Africa. The projected growth forecasts could be greatly under-estimated if these areas develop faster than expected.

The demand for air travel is affected by the cost of air fares which over the past 30 years have progressively reduced in real terms. On this basis the transAtlantic return ticket is currently only about 40% of the fare charged in 1960. This reduction has been achieved by the development of a strong market with several

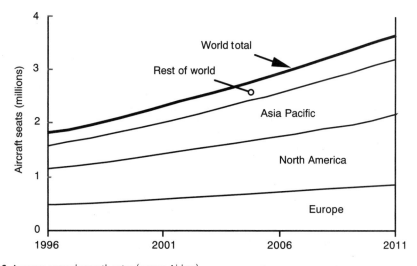

Fig. 1.6 Average annual growth rates (source Airbus).

airlines competing for business. Over the past few years this fierce competition has resulted in reduced revenue and poor commercial returns for some airlines. If a more commercially sensible trading environment is adopted in the future this may result in more expensive airfares which in turn will reduce demand and therefore the potential revenue. This dilemma illustrates the precarious nature of airline business.

Modal choice

The travelling public has available a wide choice of modes of transport including car, bus, train, ship and aircraft. By far the most significant advantage of air travel is the time saved by the fast cruising speed. Professor Bouladon of the Geneva Institute aptly described this in his analysis of transport gaps in 1967.

The total trip time shown in Fig. 1.7 is a combination of delay caused by the infrequency of the service, the speed of travel and the wasted time due to the interconnection of services.

With only minor alterations his hypothesis remains valid today. Of the three 'gaps' identified, the short- and long-haul ones are directly targeted by the air transport industry. Reducing each of the component times contributing to the overall trip time presents opportunities for both operational and technical improvements in new air transport and continues to challenge aircraft designers, airline managers and airport operators. For short stages it is no longer acceptable to have long reporting times prior to boarding. The most successful local air services attempt to copy bus and train operations in which tickets are bought at the boarding gate at the time of embarking. For longer journeys the pre-loading of luggage and cabin supplies means that earlier reporting is necessary.

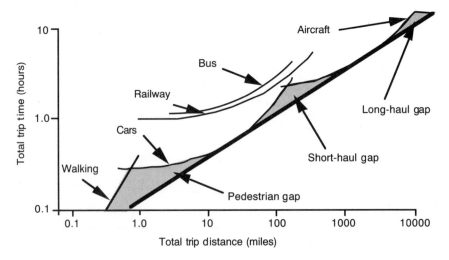

Fig. 1.7 Transport gaps (source Bouladon).

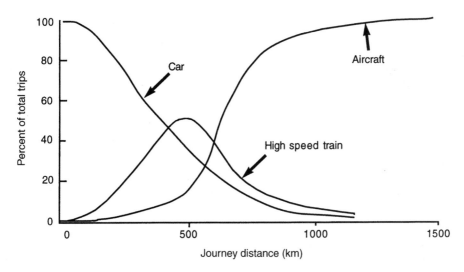

Fig. 1.8 Modal traffic split (source Airbus).

The influence of time saving is shown by the modal split (for business travel) between the three major transport forms of travel (Fig. 1.8).

As we all know, for shorter journeys and where a suitable public transport system is not available the private car is the natural choice of travel. For journeys less than 400 km/250 miles the car is the dominant mode of transport. In this market the train and bus are seen to be disadvantaged by the infrequency of service, by the out-of-pocket cost and the slow journey times (especially for distances greater than 250 km/150 miles. As public transport services are developed into a frequent, fast and comfortable option (e.g. by the introduction of high speed trains), the competition to air becomes stronger in the mid-range, (250–600 km/150–350 miles). Over about 600 km/350 miles the time saving of air travel becomes attractive and air dominates the market. As shown by Bouladon, the total journey time is affected by delays and transfers between modes. The links to the airport (road, rail and public services) is a major influence on the success of the service. For leisure travel the choice of mode is strongly influenced by ticket price and airport convenience. This has led to the development of the non-scheduled/charter air transport sector. As personal disposable income increases and more holiday resorts are developed this sector will become increasingly significant.

The aircraft market

Traffic growth and modal split are the major factors in the demand for new aircraft but these are not the only parameters. Aircraft have only a finite life and this means that there is a considerable market for replacements. It is estimated that only about 60% of the total new aircraft will be required to meet the expected increase in traffic. The rest will be needed to replace existing aircraft which have reached the end of their useful life. Some of the replacement aircraft will be introduced because of environmental and regulatory changes (e.g. older types not meeting local noise regulations) and others will be needed to replace aircraft which are less efficient and more costly to operate than new types.

The number of aircraft required is also affected by the trend to larger aircraft, longer-range non-stop flights, and the passenger's preference for the wide-body configuration. In the next 20 years it is estimated that about 13 500 new aircraft will be needed to match the increase in demand for air travel and to replace older types. More than half of these aircraft will be sized at less than 200 seats. This represents between $B950 and $B1500 turnover to cover the purchase of new equipment and the associated after-sales trade. As the economies of the developing countries grow, new manufacturing companies will be attracted to the civil aircraft market opportunities. It is likely that these new companies will be started as new joint ventures with established aircraft manufacturers. The number of competing companies may therefore increase, adding further commercial pressure to the current manufacturing industry.

The market for new aircraft is dominated by very few airlines. About 50% of all turbofan airliners are held by only 18 airlines. The top 100 airlines operate 81% of all jet airliners. This centralisation may change in the future due to trade and traffic liberalisation and the resulting challenges to existing monopolistic business practices.

Although aircraft design deals mainly with the technical/engineering aspects it must be remembered that many non-technical influences are important. It is necessary to understand these influences at the onset of the design so that they can be fully considered and due allowance be given to them in the development of the aircraft specification. For example, analysis of demand for civil aircraft shows how the air transport sector is only one option available to the travelling public. For most journeys a mixture of transport modes is necessary; therefore each mode is not independent. Within this context the availability, cost and efficiency of the road and rail networks to and from an airport are known to be major components in the choice of air travel.

The increase in demand for air travel will lead to more aircraft movements. This will result in higher aircraft utilisation and the requirement for more aircraft to be produced. The operational problems from such developments are now becoming a major concern. Congestion at airports and in the main air routes, environmental damage and the consumption of scarce resources will increasingly be factors to be considered by the designers and operators.

The commercial transport market may be roughly divided into four sections depending on aircraft size and capability (large, wide-bodied, narrow-bodied and regional). The largest aircraft (often referred to as jumbos) are wide-bodied multi-deck configurations of which the Boeing 747 is typical. Passenger capacities exceeding 400 with range capability exceeding 7000 nm and take-off length of about 11 000 ft are typical. With the expected growth in passenger traffic and the shortage of available take-off slots at convenient times there is a renewed interest in expanding the large aircraft sector. Competitors to the B747 will include new super-jumbos (new large aircraft) with passenger capacity between 500 and 700 with a range of 7500 nm.

The next largest sector includes wide-bodied turbofan aircraft which are smaller than the jumbos. These have capacities between 200 and 400 passengers and a range capability of over 6000 nm. It is this sector that has seen most competition over the past decade with Airbus, Boeing and the original McDonnell–Douglas companies offering a range of aircraft types to the airlines.

The narrow-body sector includes aircraft with capacities from about 100 to just under 200 passengers, flying up to 5000 nm but aimed mainly at the 2000–3000 nm maximum stage length market. Several smaller aircraft manufacturers have aircraft types to compete with the big manufacturers but the market is mainly satisfied by Airbus and Boeing types.

The smallest commercial aircraft sector is referred to as the 'regionals' with passenger capacities ranging from 30 to 100 and a maximum range of typically 1500 nm. This sector is dominated by turbopropeller types which are speed limited due to propeller noise and aerodynamic factors. To offset the disadvantage caused by the slow speed (increased travel time) several aircraft manufacturers have introduced small turbofan-powered designs with capacity starting at 50 seats. The '-jet versus -prop' issue is one which will continue to dominate the top end of the regional sector. The increased speed and comfort of the jet types are likely to win market share in the higher capacity, longer range part of the sector.

The commercial transport market is now experiencing the development of a fifth sector, namely 'air freight/cargo'. The development of several world-wide parcel companies offering fast and secure delivery of mail, high value commodities and perishable goods has led to the conversion of passenger transport aircraft for freight or combined freight/passenger (known as 'combi') configurations. The freight sector covers each of the traditional sectors mentioned above with freight capacities of over 70 tonnes, 35–70 tonnes, 30–60 tonnes and below 30 tonnes respectively for each of the four sectors. The influence of the freight operations on airport movements (many during the night) can be seen by the ranking of the busiest airports (see Data C). The main freight/cargo 'hub' airports now figure in the most active sites. Most manufacturers of new aircraft will now offer freight and combi versions of their passenger aircraft. Some interest also exists in the conversion of military transport aircraft (e.g. C-17, An 124) into civil freighters.

Business jets present yet another departure from the four market sectors of commercial transport described above. These aircraft are generally small (maximum 12 seats) but may have long range capabilities with reduced occupancy. This sector is a specialised part of the aircraft market which, although using similar techniques and methods to the larger aircraft, is not covered in this book. However, the methods and data described can be adapted to assist in the preliminary design of such aircraft for student work.

Civil aircraft design to meet the market described above is clearly a challenging occupation which involves a unique mixture of technical excellence, engineering management and business acumen. In an ever changing social and political environment aeronautical engineering design presents opportunities to use the experience and skills associated with advanced technologies to produce new aircraft designs. The following chapters offer a background to these possibilities and show how preliminary design studies are undertaken.

2

Project design process

Objectives

Aircraft design is a compromise between many competing factors and constraints. It is important to recognise these and to understand the influence of each on the aircraft configuration. This chapter describes the process by which the design is conceived to meet the many constraints and regulations that exist. Such a process involves the coordination of many different specialist departments. Each of these work on the overall design but have a divided responsibility between the effectiveness of the aircraft and the professional objectives of their speciality. It is necessary to understand the way in which each of the main departments interacts on the definition of the design. Some of the subsequent chapters in this book show the detailed estimations that are made in the early stages of this process (e.g. mass estimations, aerodynamic assessments and performance predictions).

The specialist departments provide the input data to the technical and economic evaluators (sometimes called project managers). These designers coordinate a systematic search to find the 'optimum' configuration and settle disputes between conflicting specialist opinions.

Understanding the conflicting pressures placed on the overall design will allow you to appreciate the process and to take such influences into account when conducting your own studies. After reading this chapter you will be aware of the organisational framework in which the aircraft is designed and the compromises that have to be taken into account to reach a sensible and acceptable layout.

It has been said, ungenerously, that a camel is a horse designed by a committee – it is the job of the project designer/manager to avoid the aircraft turning out like a camel unless the main requirement is to travel for days across a desert without water supplies!

Project design

The project design process is the means by which the competing factors and constraints which affect the design are synthesised with the specialist analytical inputs to produce the overall configuration. The process may be considered in three different parts:

- conceptual design studies
- preliminary design studies
- detail design studies

The distinction between the second and third parts is somewhat blurred due to the individualistic definition of level of detail design in the project stage. The two parts are sometimes linked and termed the 'preliminary design phase'.

The project design activity ends when the configuration is 'frozen' and a decision is taken to proceed to the 'detail design phase'. It is in this phase that detailed component geometry is specified and the manufacturing processes planned.

The project design phase is followed by the detail design and manufacturing phases. During the detail design phase all the significant technical decisions are finalised and the aircraft committed to production. Throughout the activity, drawings are progressively released for production. At the beginning of the manufacturing phase some parts are built for test purposes and the first complete aircraft is used for the initial flight tests. After all the tests have been completed the aircraft type will be granted a Certificate of Airworthiness by the national (or federal) aviation authority. The aircraft type can then be introduced into airline service.

Figure 2.1 shows how the design and manufacturing activities are scheduled.

Note how the various phases are not sequential but tend to overlap. The curved arrow shows the build-up of cost expended in the project. The importance of the project design phase is shown by this cost escalation line. Less than 3% of the total

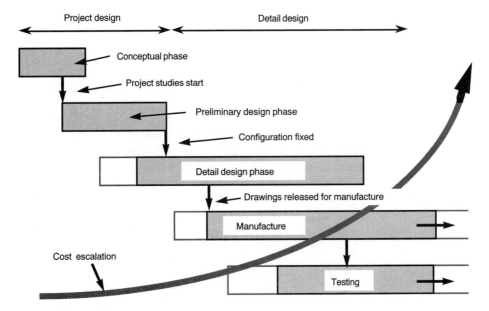

Fig. 2.1 Design and manufacturing schedule.

pre-production cost is attributable to the project design activity yet it is in this period that irrevocable decisions on the design of aircraft are taken. Modifications to the basic configuration after this period are difficult and costly, making it necessary to 'get the aircraft right'. This places a large responsibility on the project design team. To this end, they need to understand not only the detail design, manufacturing process and operation of the proposed design but also the sensitivity of the design to possible future changes. This makes it necessary to analyse many variants of the aircraft around a selected preferred (baseline) design. Computerised methods of analysis are currently used to conduct these studies which are sometimes inappropriately called 'optimised designs'.

Conceptual design studies

The first activity in the project design process is the 'conceptual design study'. In this phase, conventional and novel configurations are considered to determine layouts which are technically feasible and commercially viable. At the start of the phase all options are considered. For example, the aircraft overall geometrical configuration, the engine (type, number and position), the internal cabin arrangements, the aircraft systems and control, manufacturing methods and materials, the level of advanced technology to be included and the overall operational procedures are all to be decided. Each choice needs to be investigated as completely as the level of detail and time that is available will allow. It may be necessary to conduct research studies into some of the novel concepts and technologies to quantify the effectiveness of the idea on the aircraft layout. In some of the research areas it will not be possible to quantify the effectiveness with reasonable accuracy and an intelligent 'guesstimate' will be required.

During the concept design phase the quantity of data generated on each design will be relatively limited and the manpower expended small. The outcome of the study is a knowledge of the feasibility of the various concepts and an estimate of the rough size of the most likely configurations. A detailed analysis of each design at this stage is not possible and may be of limited value as most of the concepts will be discarded. For conventional layouts and for developments of existing aircraft, experience from previous designs (and competitor aircraft) will provide a good approximation to the aircraft size. If the layout is novel then available methods which have been developed from conventional designs will give only crude approximations. In these cases it is necessary to develop new assessment models. Throughout the project phase all the models will be improved and refined to reduce the technical and commercial risks associated with the novel layouts and technologies.

Preliminary design studies

At the end of the conceptual design phase all the design layouts will have been analysed. Those which were regarded as unfeasible or too commercially risky will be eliminated. The remainder will be compared after careful consideration of a suitable selection criterion. The 'best' one, or possibly two, designs will be identified and taken into the preliminary design phase. It is important not to carry too many options forward to the next stage as this will dissipate the available effort and slow down the detailed definition of the preferred design. However, care must be taken to avoid discarding design layouts too quickly as some may lead to evolutionary configurations which could give the aircraft a competitive advantage over aircraft from other companies.

In the preliminary design phase the preferred configuration(s) from the conceptual study is subjected to a more rigorous technical analysis. The objective of this phase is to find the best ('optimum') geometry for the aircraft with regard to the commercial prospects and in comparison with competitor aircraft. All the principal aircraft parameters are considered to be variable during this analysis. This allows sensitivity analyses to be conducted. It is common practice to establish a 'baseline' configuration and to perform a series of parametric studies around this layout. These parametric studies are done in parallel with the development of the baseline design but must be performed quickly enough to allow any desirable changes to the layout to be made before the design is 'frozen'. It is therefore necessary to conduct such studies at a level of detail that will provide quick but reasonably accurate answers.

At the same time as the parametric studies are being considered the design team will analyse competitive aircraft, perform trade-off studies in detailed technical areas, and test the sensitivity of the baseline design to changes in the constraints imposed by the specification. Again, many of these studies are done at a level of detail that produces quick results which are reasonably accurate.

Detail design studies

The detail design phase is started towards the end of the parametric analysis. In this part of the design process the layout is refined to a greater level of detail. With the external shape fixed, the structural framework will be defined. In areas of doubt, the theoretical calculations will be done in finer detail and validated by component tests. In this phase there will be an increasing reluctance to make radical geometric changes to the overall layout of the aircraft as this may invalidate work done on the baseline configuration by other departments. Detailed optimisation studies will be limited to parts which will not affect the overall configuration of the aircraft. Throughout this phase the aircraft weight and performance estimates will be continuously updated as more details of the aircraft layout become available. At the end of this phase the aircraft is 'released for production'.

The design process

Each part of the process is shown in Fig. 2.2 and briefly described in the following sections.

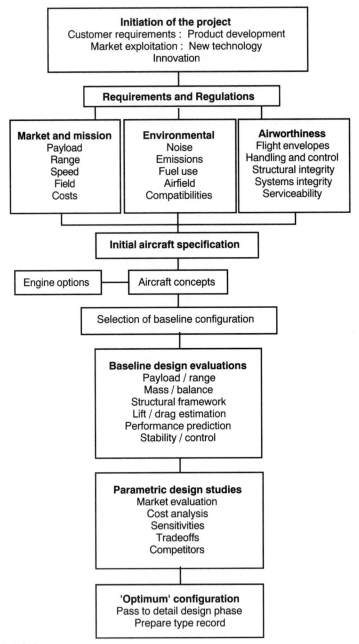

Fig. 2.2 Project design process.

Initiation

The start of the design process requires the recognition of a 'need'. This may come from established or potential customers, an analysis of the market and the trends for demands, the development of an existing product line (e.g. aircraft stretch), the introduction and exploitation of new technologies (e.g. composite structures) and products (e.g. new engines) and the application of innovation in research and development (e.g. laminar flow). For any of these possibilities it is necessary to fully understand the factors which will affect the design. For each new design the relative significance of the various features will be unique. It is essential that the most influential features are identified and understood. From such investigations it is possible to clearly state the specification that will have to be met by the design. This must contain clear and unambiguous statements which can be quantified, or at least form the basis on which decisions can be taken in the design management process.

Specification

During the early part of the project design phase, one of the main tasks is to clarify and test the specification upon which all later stages of the design process will be based. Identifying the role (or roles) of the aircraft in terms of the required performance and payload are the first considerations. In the early stages the aircraft specification cannot be rigidly set out; it is therefore necessary to investigate the effect of changes to the requirements. The preliminary design studies may be used to provide evidence on which to base decisions on aircraft role and performance.

In some cases the specification will be more tightly set by the customer and as a result the design process can be started earlier. Having obtained the specification for the aircraft but before getting too involved with the technical design, it is necessary to define the criteria to be used in making design decisions on various options.

Criteria

If the designers, manufacturers and customers can agree how to assess the effectiveness of an aircraft, the design team will have an easier task. Even if no agreement is possible the designers will need to define the criterion or criteria on which to judge competing designs. It is important that this definition is done carefully as it will be used throughout the whole design process as the standard on which to make design decisions.

The most sensible criterion for any company to use is 'return on investment'. Equally, the purchaser will use a similar criterion when comparing competing products. Hence this criterion will affect the sales potential of the aircraft and thereby the aircraft manufacturer's criterion. Unfortunately 'return on investment' is not the only criterion to be considered by the purchaser and producer of the aircraft. Political, economic, technological and sociological (environmental) aspects must also be considered when analysing the design but at the early design stage there will be insufficient knowledge of the future situation to make accurate prediction of the effects of these on the aircraft design. Also a generalised return on investment criterion is difficult to model due to the variability of economic performance of different airlines and manufacturing companies. In the past, this had led the design team to adopt a simpler alternative criterion. Up to about the mid-50s the only criterion which could be easily quantified was weight (e.g. minimum all-up weight or minimum zero–fuel weight) or performance criteria (e.g. minimum drag, maximum speed). With the introduction of standardised cost methods (e.g. Association of European Airlines–Direct Operating Cost Method) it was possible to consider non-weight/performance aspects in the design (e.g. maintenance costs, crew salaries, depreciation, finance, etc.) Nowadays the development of computer methods in project design permits a more complex multi-parameter criterion to be investigated. Such computer models are only applicable to the early stages of project work. As the design studies become more detailed the traditional criterion of minimum weight is still used in conjunction with a knowledge of operating, manufacturing and material costs. Over the past few years some effort has been concentrated on cost elimination by the introduction of 'value analysis' in the detail design stages. Consideration is now given to the 'whole life' costing process which includes operational aspects not covered in the direct operating cost models (e.g. fleet mix and aircraft and engine spares commonality). This may lead to the consideration of a 'family' approach to aircraft development in which similar aircraft layouts (airframe, engines and systems) are aimed at different payload/range specifications.

Constraints

Each of the proposed layouts will be subject to a number of active design constraints. These may arise from airworthiness requirements (e.g. second segment climb gradients), performance specifications (e.g. take-off field length), operational parameters (e.g. turn-round time), or be imposed from external influences (e.g. environmental noise rules). There will also be a number of constraints set by the design/manufacturing management team. These may include the definition of available materials and manufacturing methods, limits to the inclusion of advanced technologies, the exploitation of aerodynamic and structural innovations and the specification of particular components (e.g. engines and avionics). These constraints may be regarded as defining the technical and commercial risk that the company is prepared to accept in the new design.

It must be appreciated that it is possible to impose so many constraints to the design that a feasible solution does not exist. Also, the inclusion of unnecessary, or over-stringent constraints will inevitably lead to commercial penalties on the design. One of the main studies to be undertaken in the project design phase is to assess the sensitivity of all of the design constraints. Those which are shown to be most critical to the effectiveness of the aircraft will be carefully re-assessed to try to minimise the influence of the constraints on the overall design.

Defining the aircraft specification, determining the criteria to be used for judging the aircraft effectiveness, and setting the constraints, represent a substantial technical and managerial effort which must be undertaken prior to the start of any design work. If any of these areas is not considered carefully the final aircraft design may be commercially flawed. The reason for many of the past failures in aircraft design can be traced to a poor understanding of one or more of these basic factors.

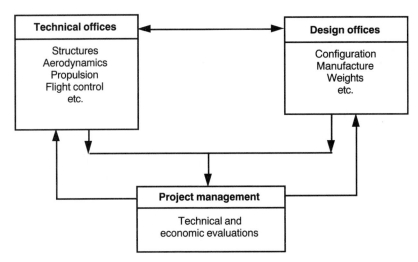

Fig. 2.3 Project management.

Project analysis

With several, but not enough, highly inter-related equations governing the design of the aircraft it is not surprising that an iterative approach is the traditional method of conducting the initial project design process. Initial 'guesses' based upon past experience or on data from similar types of aircraft form the starting point for the analysis. The iterative process progressively determines more accurate and 'acceptable' output which refines the design. The iterations are performed within specialisations but are all integrated through the aircraft configuration and performance analysis. Before considering each of the main specialisations in the design process it is worth seeing how each is inter-related into the total design activity.

Figure 2.3 illustrates the way in which data, or information, is passed between specialist groups during the design process. The main technologies are linked to the configuration design via the evaluation of the aircraft performance and economics. It is within this evaluation process that checks are made on the aircraft constraints and the overall criterion (sometimes called the objectives function) is determined. The specialist groups must consider the level of 'advanced technology' to be adopted (e.g. the use of new materials, micro-minimisation of systems, aerodynamic features, aerofoil and control innovations) together with all the other active constraints on the design. The data flow lines shown in the diagram are specific and indicate how the technology areas influence the configuration of the aircraft through the economic and performance evaluation.

The diagram shows clearly that the design process is not a single-step sequential process. It is a cyclic process without a natural starting position. Design solutions can only be achieved by an iterative process. Experience suggests that the best way of starting the process is to make an estimate of the configuration and use this 'cock-shy' to make the detailed estimation in each of the technological areas.

Previous analysis methods suggest that the equations governing the design are reasonably well behaved and will rapidly converge to a feasible solution (if one exists). Obviously, the closer the initial guess is to the final design configuration, the quicker will be the convergence to the best feasible design. As a guide to the selection of the initial configuration it is useful to analyse previous aircraft designs of similar specification, preferably using statistical correlations on the main parameters. This avoids introducing unwanted personal preferences into the choice of aircraft parameters.

Each of the specialist areas will now be described starting with the three principal sections (i.e. aerodynamics, propulsion and weights). The original cartoons in this section have been variously attributed to Bruhn and Miller in the early 'forties but their sentiments still have relevance over fifty years later!

Aerodynamics

The aerodynamicist's dream is of a super-smooth streamlined aircraft in which the wing forms the principal feature (Fig. 2.4). The fuselage, engines and undercarriage are all regarded as inconveniences which reduce aerodynamic efficiency.

The aerodynamics office data flow is shown in Fig. 2.5.

The fundamental technologies are contained in the upper left-hand box. These are based on the theories and methods covered in textbooks on aerodynamics, on empirical data from past designs, from wind tunnel tests, or by computational fluid dynamic analysis. Input data is shown in the upper right-hand box. This comes mainly from the configuration group which provides the geometry of the aircraft (e.g. cabin size). The improvement in aerodynamic efficiency, by controlling the profile shapes and integration of components, is the main objective for the aerodynamicists. For a given technology level, the geometric description of the wing, tail and fuselage is determined. Conflicting requirements from different mission stages must be resolved by the specification of compromised layouts (e.g.

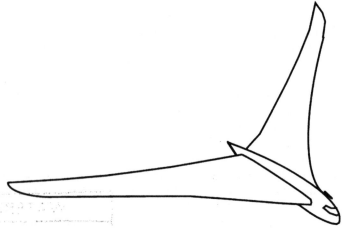

Fig. 2.4 The aerodynamicist's dream aircraft.

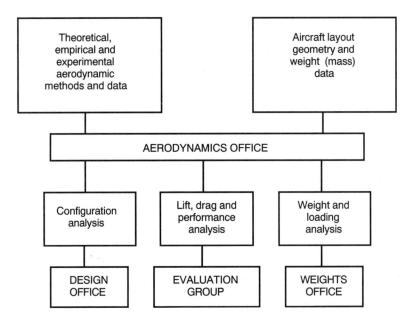

Fig. 2.5 The aerodynamics office.

wing area, flap complexity, etc.). Output from the aerodynamic analysis is fed as geometrical data to the configuration and weight sections and as aerodynamic functions/charts to the performance group (e.g. drag polars, flight envelope, manoeuvre loadings, aerodynamics forces and moments).

Propulsion

The engine experts see the aircraft as a means of carrying their masterpiece (Fig. 2.6)!

The data flow follows a similar pattern to the aerodynamics office and is shown in Fig. 2.7.

The enabling technology is shown in the left-hand box. This is associated with accepted theories, methods and data from engine tests. The technology level to be adopted is of particular concern to the engine designers because the choice may adversely affect engine reliability and maintenance costs. Much will be expected by the airframe manufacturer from engine improvements as these will directly influence the operational effectiveness of the aircraft (fuel-use, noise levels, emissions and engine life). Commercially binding assurances will be required by the aircraft manufacturing and airline companies with regard to the minimum engine performance.

Input data to the propulsion group (shown in the right-hand box) includes nacelle geometry, operating conditions, engine thrust de-rates, power off-takes, etc.

Output from the propulsion analysis is fed as geometry and weight data to the configuration and weight groups, and as performance data (fuel flow/thrust charts, power off-takes) to the performance section.

Fig. 2.6 The engine designer's dream aircraft.

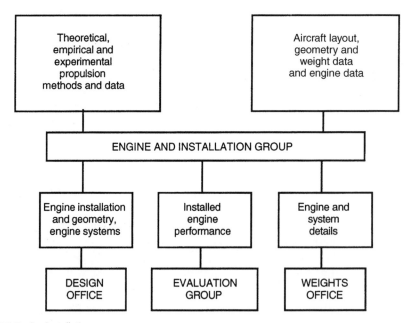

Fig. 2.7 Engine installation group.

The designer has three choices with regard to the engines to be specified:

- to use a current or proposed up-rated existing engine;
- to use a new engine which will be available in the same timeframe as the aircraft project;
- to use a hypothetical 'scaled' (rubber) engine based on a proposed new design.

The aircraft designer will be reluctant to exclusively specify an engine from one manufacturer. They would prefer to be in the stronger commercial position of specifying a range of engines from competing manufacturers. This may lead to the duplication of work for different engine types. The choice between new and existing engines presents a difficult decision. The potential advantages of a new product with improved technology levels are set against the risk of technical problems which may delay or even penalise the aircraft efficiency. Project studies with theoretical 'rubber' engines are sometimes used to identify the best choice of engine type and performance.

The installation of the engines on the aircraft offers a difficult decision for the configuration designer. How many engines should be used (2, 3 or 4)? Should the engines be wing or fuselage mounted? Should they be podded or buried? What type (turboprop, propfan or turbofan)? Such decisions will impact directly on the success or otherwise of the whole project.

Weights

The weight engineers are always fighting a battle against increased weight from the technical and design offices. The weight engineers' dream aircraft would be made as small as possible and would resemble a light model aircraft structure (Fig. 2.8).

Weight is regarded by the designers as one of the principal engineering groups. The data flow for this section is seen to be more complex than for the previous two specialisations (Fig. 2.9).

The fundamental principles on which the weight analysis is made are based on statistical data from previous aircraft designs. Input comes from each of the specialist groups plus the configuration design team. This information is in terms

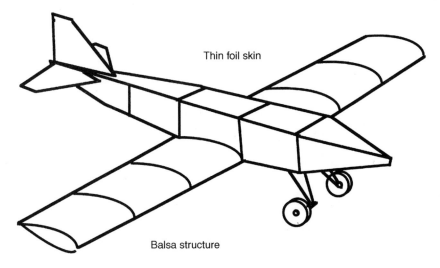

Thin foil skin

Balsa structure

Fig. 2.8 The weight engineer's dream aircraft.

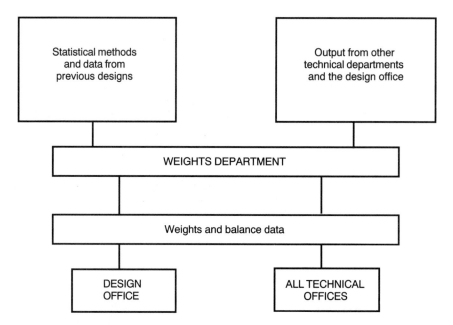

Fig. 2.9 The weights department.

of geometric data, aircraft loading and system definitions. Output is sent directly to the performance and economics group in the form of mass tables and centre of gravity distributions. The influence of weight growth on aircraft performance is known to be highly detrimental to the efficient operation of the aircraft. Therefore the group constantly predicts the aircraft weight. This acts as a barometer of potential aircraft performance problems. If weight growth is detected weight saving programmes will be instigated.

There are several other technical departments involved in the design process including the control and structures groups.

Control group

The flight control group have a difficult job. They are asked to predict the flying qualities of the aircraft, sometimes before the true shape is finalised. They would see their dream aircraft as one in which several interchangeable surfaces are available for different flight conditions and designs, or in which the control surfaces are so large they will not risk any unknown instability (Fig. 2.10).

The data flow in this department is similar to that of the aerodynamicists with whom they work closely (Fig. 2.11).

The basic concepts on which the analysis is based concern the stability and control (handling) assessment and system analysis. The technology level has been substantially affected by modern computer technology (flight management systems), transmission (fly by light), and information display ('glass' cockpits, new system design, lasers, synthetic vision, etc.). Input data comes from the aircraft con-

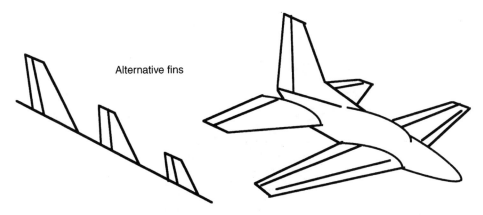

Fig. 2.10 The control group's dream aircraft.

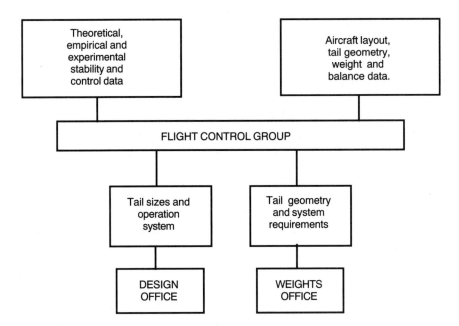

Fig. 2.11 The flight control department.

figuration [tail sizes, flight envelope, centre of gravity (c.g.), range]. Output is sent to the geometry and weight groups in the form of system descriptions and control surface sizes.

Structures group

The stress office would like to see straight and simple load paths and the avoidance of any localised twisting of the structure. Their dream aircraft would have no

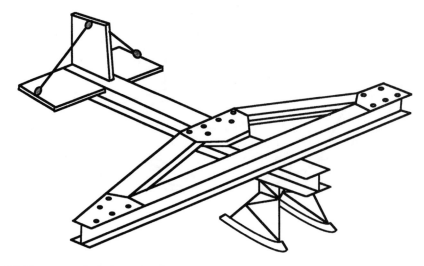

Fig. 2.12 The stress analyst's dream aircraft.

moving parts and no cut-outs and all the joints, where they cannot be avoided, would be positioned away from highly loaded areas and easily visible to allow them to be inspected regularly. Although not 'aesthetic' their design would be structurally simple as shown in Fig. 2.12.

The data flow in their office is shown in Fig. 2.13.

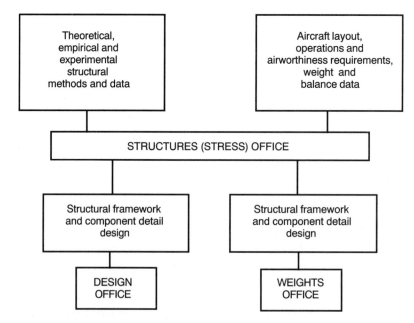

Fig. 2.13 The structures office.

The structural analysis of the aircraft represents another major technological section in the design organisation. Assurance of the airworthiness of the aircraft in terms of strength and stiffness is their main area of responsibility. The structural strength (stressing) and stiffness (aeroelasticity) of the aircraft are related to other technical areas (aerodynamics, propulsion and flight control as shown above).

The fundamental concepts on which the analysis is based include structural analysis methods, structural testing, material properties and manufacturing methods. The geometric and weight groups provide the input to the analysis in terms of shape, size and masses. Aerodynamic loads are either determined by the aerodynamics department or by a specialised loading group. Output to the configuration and weight groups consists of the geometry of the structural framework, material specifications and constructional details.

Computer-aided design and concurrent engineering

The advent of computer-aided drawing (CAD) programs and their universal adoption in the manufacturing industry has led to the ability of each technical department to access the current aircraft geometrical definition. This ability allows each department to be working on the design concurrently. This reduces the time necessary to analyse and develop the design compared to the older methods which relied on a sequencing of the different specialist departments. The process also allows more detailed investigations to be conducted on the design at the early stages, for example structural finite element analysis, computational fluid dynamics and dynamic flight simulation are performed in the knowledge of the same aircraft layout.

The integration of the aircraft design process requires careful management of the computer information system to avoid the use of different standards yet to allow changes to be incorporated as the design develops. The aircraft project manager has to control the total system design. This requires different knowledge and skills to that of the traditional chief designer. Several different software suites have been developed to meet the requirements of 'computer-aided design'. Each of the major manufacturing companies has incorporated a specific CAD system which it standardises for use in all the sections of the company; and they require subcontractors to be able to use a similar system.

The new design and development process has become known as concurrent engineering to represent the total integration of all the technical departments working on the aircraft project at the same time.

Configuration and performance/economic aspects

To understand the 'configurational' and 'performance and economic' analysis work it is necessary to study the individual aspects of aircraft design in more detail. The following chapters in this book are intended to give an introduction to each of these topics. These concentrate on the aeronautical engineering aspects (con-figuration, design, performance, etc.) but some important influences have had to be put into the background. Such aspects include aircraft and engine manufacturing processes, maintenance and reliability, operational and environmental features, social and political issues. Although it has not been possible to cover these aspects, designers must be aware of them as they may be critical to decisions made during the project development phase.

3

Overall configuration and systems

Objectives

This chapter describes how the overall configuration of the aircraft is decided from the many options available to the designers. From a brief historical perspective the chapter then looks at the unconventional layouts that have been considered in the recent past. Although these have not yet found favour with designers it is worth keeping them in mind as conditions and constraints may change in the future. Such changes could make current layouts less attractive or even not feasible. New designers should start a new project with a completely open mind and this means a careful consideration of all configurations particularly in a historical context.

Conventional layouts

The development of civil jet transport configurations has largely been an evolutionary process. Despite the many technical improvements that have been introduced over the past 40 years the wing, fuselage and control surface arrangement of modern aircraft appear little different to the original Comet design (Fig. 3.1). Only the engine installation has changed, from the original buried layout to the podded configurations used today. Even this change has its origins in the successful Boeing 707 (Fig. 3.2, not to the same scale as Fig. 3.1) layout which entered service only six years later than the Comet.

It is a tribute to the skill and foresight of the early designers that they selected an overall configuration which has not been beaten by the powerful computer technology now available in modern design offices. This review of the configuration disguises significant advances in aerodynamic drag reduction, improved lift generation, more efficient and lighter engines, changes to the structural materials and the methods of manufacture, the integration of systems, increased reliability, and the adoption of modern avionics and computer control technologies. The inaugural Comet flight from London to Johannesburg was slow by present day standards and it required refuelling stops at Rome, Beirut, Khartoum, Entebbe and Livingston. Although it took nearly 25 hours to complete the flight this halved

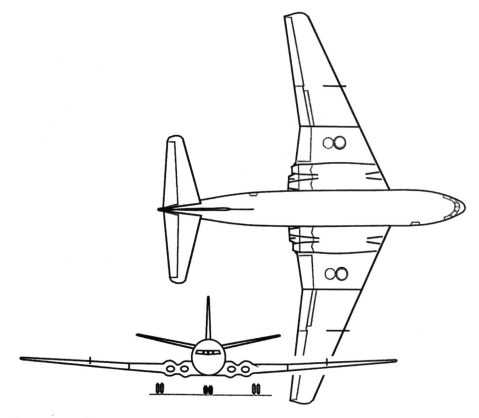

Fig. 3.1 Comet 1 layout.

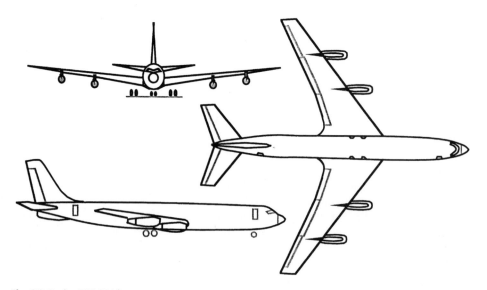

Fig. 3.2 Boeing 707-320 layout.

the previous scheduled flying time; a feat only equalled when the Concorde entered service 25 years later on the transatlantic route. The original aircraft carried only 36 passengers over a range of 2800 km. This specification would now class it as a small regional/domestic operation. Since 1952 aircraft designers have increased passenger capacity to more than 500 and extended the range capability on some aircraft to over 15 000 km. Engine designers have developed the original pure turbojet engine into highly fuel efficient turbofans with thrust exceeding 400 kN. All these technical improvements are impressive; redesigning the Comet with modern engines, aerodynamic improvements and modern materials would double the range for the same take-off weight.

Novel layouts

Aircraft designers have constantly searched for more efficient configurations. This has led to the study of many unconventional design concepts.[1] Although unusual

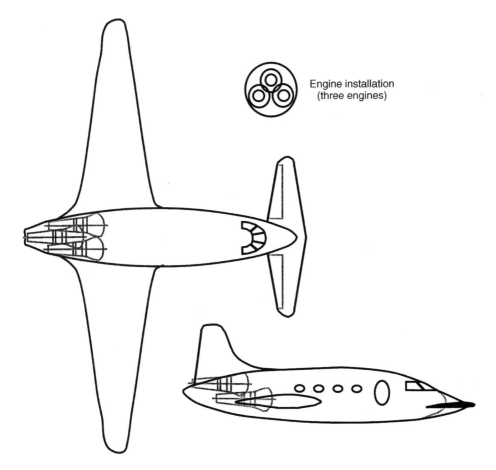

Engine installation
(three engines)

Fig. 3.3 Early Comet design layouts.

layouts have not yet superseded conventional designs, they have been studied seriously as each concept offers potential operational and technical advantages.

To the present time, the commercial risks involved in developing new layouts have been assessed as unacceptable. As air transport requirements evolve, unconventional designs may become more viable. It is therefore worth studying some of the most promising design concepts that have been investigated over the recent past.

All airliners are symmetrical about the vertical centreplane: the left-hand side is a mirror-image of the right-hand side. This is not surprising as all birds are like this and nature has had a longer evolutionary path! Most of the novel design layouts still retain this symmetry.

Even at the time of the first Comet, design teams studied unusual layouts.

Figure 3.3 shows the canard design layout studied by De Havilland in response to a UK government initiative in 1943–44. This shows how the early designers of commercial aircraft were striving to reduce the adverse effect of trim drag. This led the design team to consider the canard configuration. Compared to the previous propeller engined installations, the new compact jet engines could be positioned within the aircraft structure. On early engine designs intake geometry was not as critical as on later engines. Three engines were necessary on this design as larger thrust engines were not yet available.

The same design team also studied twin-boom and tail-less layouts for the initial Comet specification. It was felt at the time that the new jet engines offered the potential to introduce radical changes to the aircraft configuration but caution prevailed and a traditional layout was eventually selected for the new airliner.

The multi-hull layouts

In recent times the design of very large aircraft (>1000 passengers) has highlighted a number of problems associated with increasing the size of the conventional layout. One such problem relates to the emergency egress of passengers from large fuselage structures within the evacuation time specified in the airworthiness regulations. To overcome this problem and partially to take advantage of distributing the passenger load along the wing the multi-hull concept has been proposed[2,3] (shown in Fig. 3.4). The conventional fuselage layouts used in these multi-hull studies satisfy the evacuation requirements. Also, the positioning of the fuselages away from the aircraft centreline produces inertia relief loads to the wing lift. It is estimated that this relief load will substantially reduce wing and fuselage structure weight. Estimates on these designs predicted an 8% saving in aircraft empty weight.

In the twin-hull configuration the flight deck, positioned in the nose of one of the fuselages, is offset from the aircraft centreline. This has raised concern from pilots due to the unnatural 'feel' for the aircraft. Simulator tests have confirmed the difficulty of flying the aircraft in an emergency (e.g. with one engine failed) from the off-set pilot position. To avoid this problem the three-hull version has been proposed,

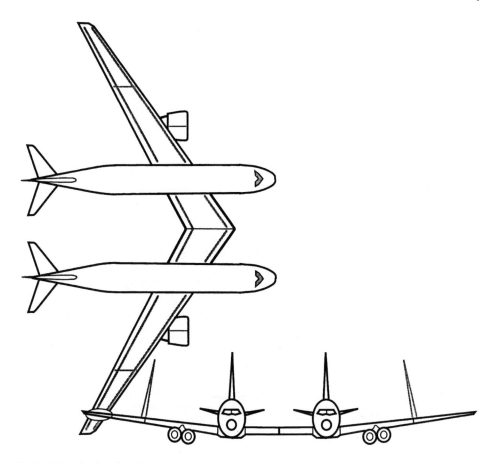

Fig. 3.4 Twin-fuselage layout.

see Fig. 3.5. This avoids the pilot control problem but still partially retains inertia relief on the wing from the outboard fuselages and satisfies escape criteria.

The main technical difficulties of multi-hull designs are concerned with the aerodynamic and structural analysis of the wing between the fuselages. The airflow conditions over the interconnecting surface(s) and the interference effects of the fuselage walls are unknown. The resulting airflow distributions raise concerns about the effectiveness of flaps and other controls and the consequential effects on aircraft dynamics. The structural analysis of the airframe bounded by the fuselages is complicated by the dynamic behaviour of the hulls. It is expected that the free motion of the hulls will generate substantial twisting and bending on the interconnecting structure. These effects are unknown and will undoubtedly cause a substantial increase in weight.

Due to the technical uncertainties described above the multi-hull configuration has not yet challenged conventional single-fuselage layouts; however, the potential of structural weight saving from directly coupling lift forces with inertia payload relief has resulted in a number of other unconventional design layouts.

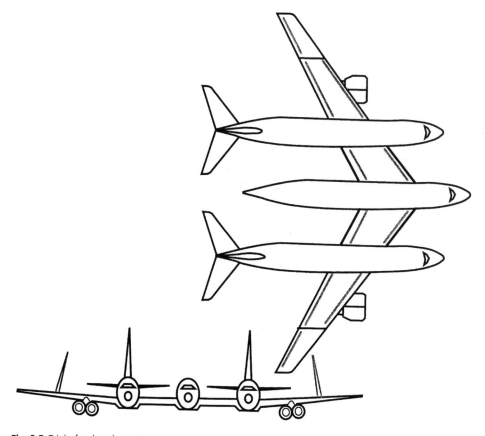

Fig. 3.5 Triple-fuselage layout.

The span-loader layout

In the span-loader aircraft shown in Fig. 3.6, the payload is held in the main wing-box structure. A small centrally positioned fuselage pod houses the flight deck and central services. In this configuration the wing structure is relieved of most of the bending loads because the aircraft operational weight is balanced directly by the lift from the wing sections. A saving of about 10% in aircraft take-off weight over an equivalent conventional layout has been claimed for the concept. Such a layout would require about the same overall dimensions (span and length) as a conventional design.

Boeing proposed a passenger variant of the span-loader as shown in Fig. 3.7. The central structure housed the engines (in this case nuclear power!) and the flight deck. The passenger cabin stretched the full span of the wings and had forward-facing windows positioned in the leading edge. Apart from the type of powerplants this concept was first proposed by Junkers in 1910!

The main disadvantages of the concept are the difficulties of loading the payload into the congested central wing space and the resulting structural layout implications. Concerns are also raised about egress routes and evacuation times in

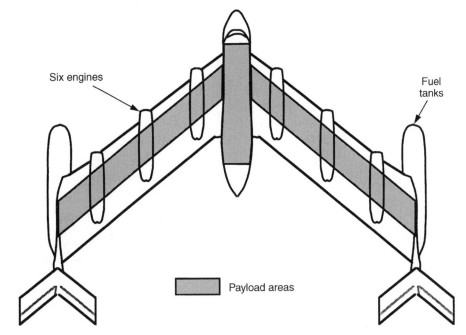

Six engines

Fuel tanks

Payload areas

Fig. 3.6 Span-loader layout.

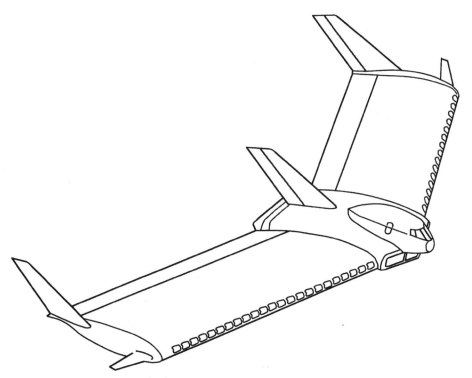

Fig. 3.7 Boeing's passenger span-loader.

emergency conditions. In the design layout shown in Fig. 3.6 the aircraft payload was specified as cargo only. The uncertainty of the design layout with respect to the unorthodox flight control system, and the reduced aircraft responsiveness in roll due to the increased aircraft moments of inertia, presents significant cause for concern.

The flying-wing layout

The prospect of distributing passengers in auditorium-type space has often been used to promote the concept of a flying-wing aircraft, now referred to as the 'blended-wing body' concept. Both Airbus and McDonnell–Douglas have recently shown such layouts, see Fig. 3.8.

The main advantage for this layout lies in the potential for increased passenger cabin volume and the associated improvement in comfort level. The designers also point out that the distributed payload will give increased inertia relief and therefore reduce structure weight.

There is less technical uncertainty attached to this configuration than the previous concepts but the need for increased wing thickness and the delta wing planform suggests that it will be less aerodynamically efficient than conventional layouts. The renewed interest in this layout is associated with improved aerodynamic efficiency from active boundary layer (laminar flow)

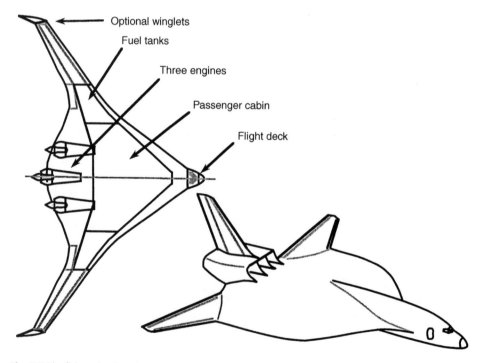

Fig. 3.8 The flying-wing layout.

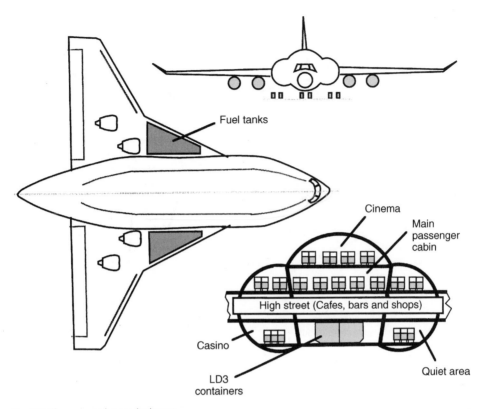

Fuel tanks

Cinema

Main passenger cabin

High street (Cafes, bars and shops)

Casino

LD3 containers

Quiet area

Fig. 3.9 The proposed mega-jet layout.

control systems and the need to build much higher capacity aircraft that will fit into existing airport facilities. The question to be answered is 'will passengers travel in an aircraft devoid of windows?' If so, the flying-wing may be seriously considered.

A hybrid between the flying wing and the distributed load has been suggested by the designer of the mega-jet.[4] The proposed layout is shown in Fig. 3.9.

This is a 1000+ passenger aircraft with about 50% more cabin space per passenger than conventional aircraft. This layout offers the prospect of providing ship-type passenger facilities (bars, casinos, cafes, shops and offices). Technically the concept appears feasible but the adoption of a radical concept in a very conservative industry seems unlikely unless other criteria (e.g. passenger comfort and facility provision) become important.

The canard layout

Apart from the original Comet projects mentioned earlier the canard layout has its ancestry in the original Wright Flyer which had the control surfaces ahead of the mainplane. This configurations avoids the tail-down forces required for trim on conventional layouts. Much research has been focused on a return to the canard

Fig. 3.10 The canard layout (source Airbus).

configuration for airlines as the trim drag in cruise will be reduced and useful fuel saving made.

Figure 3.10 shows a recent investigation into the advanced concept 220–260 seat airliner from Airbus.

In the past, the risk of nose-up overturning due to inadvertent mainplane stall whilst the canard surface is generating lift has stopped the adoption of the concept except for some military and home built designs. As aircraft computer control becomes more sophisticated and the reliability of such systems increases, the possibility of using the canard concept increases.

In some design proposals, a three surface (canard, mainplane and tailplane) layout has been suggested to split the balancing load and the control loads between front and rear surfaces. Although the aerodynamic stability and control features are simplified by this arrangement the addition of extra surfaces with the attendant structural and mechanical complication are serious disadvantages.

The tandem-wing layout

Extending the concept of dividing the lift generation between two surfaces to minimise the number of controls has led to the tandem-wing layout shown in Fig. 3.11.

It is interesting to note that the tandem wing configuration was studied in great detail in the early years of aircraft design.[5] Although there have always been strong proponents for this layout, as it provides more tolerance to centre of gravity movement, it has yet to be adopted for commercial designs.

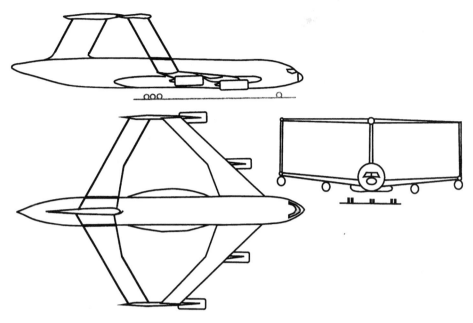

Fig. 3.11 The tandem-wing layout (source Lockheed).

To make the tandem-wing structures stiffer some designers have suggested joining the wing tips of the two surfaces. This idea produces the coupled biplane concept with a large longitudinal stagger between the two surfaces as shown above.

Each of the above configurations (canard, tandem and biplane) has the advantage of reducing overall aircraft wing span but retaining a reasonable aspect ratio for each surface. This, it is claimed, will give reduced cruise drag and improved aerodynamic efficiency. A structural weight reduction can also be claimed due to the integrated wing structure and the consequential reduction in fuselage bending stresses. The concepts have not yet been accepted because of the structural and aerodynamic uncertainty at the wing junctions and the risk of flight and structural instabilities.

The joined-wing

The tandem-wing layout has recently been revived for commercial aircraft layouts due to the difficulty of fitting larger span aircraft into existing airport facilities. Research[6] has shown that a narrower rear surface with a tip junction at about the 80% span position on the front surface is the optimum layout as shown in the projected 500 seat design in Fig. 3.12.

Although substantial structural and aerodynamic advantages are claimed for the joined-wing concept, the technical uncertainties surrounding aerodynamic interference effects and the novel structural framework have prevented its adoption.

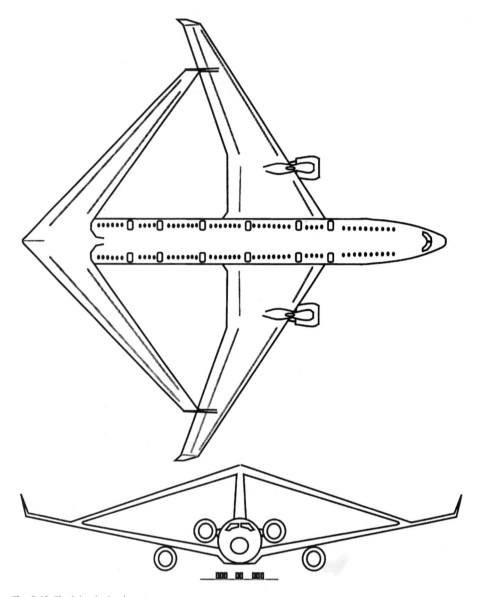

Fig. 3.12 The joined-wing layout.

The flatbed layout

The difficulties of loading a conventional fuselage with freight and passengers have led to the design of the modular aircraft. The 'flatbed' design shown in Fig. 3.13 can be quickly fitted with either a freight or passenger 'container' to reduce turn-round times for the aircraft and thereby increase aircraft utilisation and reduce airport stand occupancy.

There appears to be no technical reason why the concept should not be built

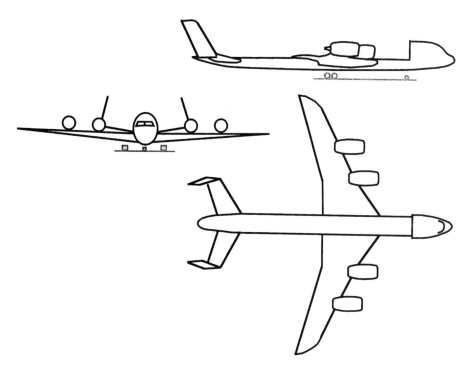

Fig. 3.13 The flatbed layout.

but operationally the layout is complicated and liable to misuse. Although modular designs may be adopted in the future they will need to be easier to use than the configuration shown above. This concept appears to offer plenty of scope for ingenuity to new designers.

Why not everything!

It appears that the imagination of designers knows no bounds. The combination of modular fuselage and three surface concepts has been proposed in the large Russian transporter shown in Fig. 3.14.

Back to the beginning

Until some unknown design or operational criteria come along to condition the aircraft layout to be different, it appears that the conventional configuration (shown in Fig. 3.15) is here to stay. The centrally positioned cylindrical fuselage set on a moderately swept trapezoidal low wing with podded engines and aft mounted control surfaces still offers aircraft designers ample opportunity to practice their skills.

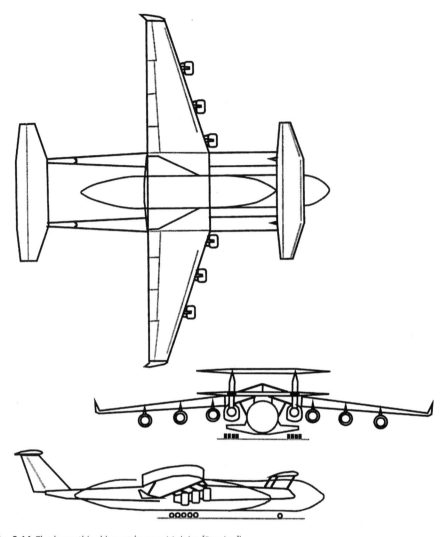

Fig. 3.14 The 'everything' layout (source Molniya [Russian]).

System considerations

The choice of aircraft configuration may be considered as an assembly of many sub-components. The airframe although significant represents only a small part of the total design effort. The inter-relationship between the airframe, engines and aircraft systems requires careful consideration when deciding the overall layout. For example, there are many moving parts on the aircraft and each must be controlled and instrumented so that the pilot and flight management system can make safe decisions. The structure must be able to react the loads that these parts impose and the airframe profile must enclose the units in an aerodynamic (low drag) shape.

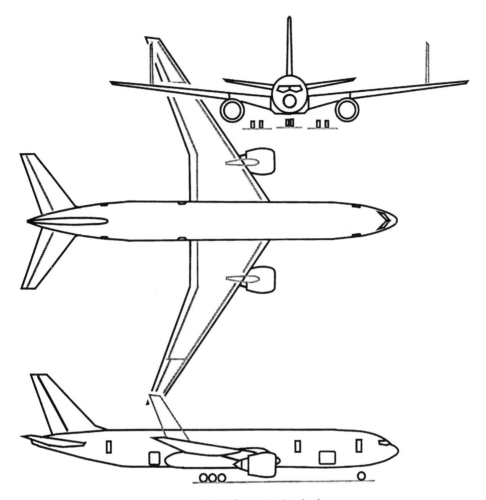

Fig. 3.15 The conventional layout: Boeing 777-200 (source Boeing data).

Internal space requirements for the configuration may also impose their own conditions. The fuel must be held in sealed tanks in the wing and (possibly) tail structures. Between a quarter and a half of the aircraft weight will be attributed to this fuel; this places a significant demand on the internal volume of the aircraft and in some cases limits the minimum size of the wing for the aircraft to fly a specified range.

The cabin air conditioning system also requires internal space provision for the ducting and flow control units. Although this is much less demanding in space than other requirements the system must run through the cabin taking up as little volume as possible and must be unobtrusive.

The aircraft flight management system although not physically demanding on space within the configuration has considerable influence on the size of control surfaces and the design of the associated operating systems. The design of the flight management system offers a dilemma to the designers; should the system be given

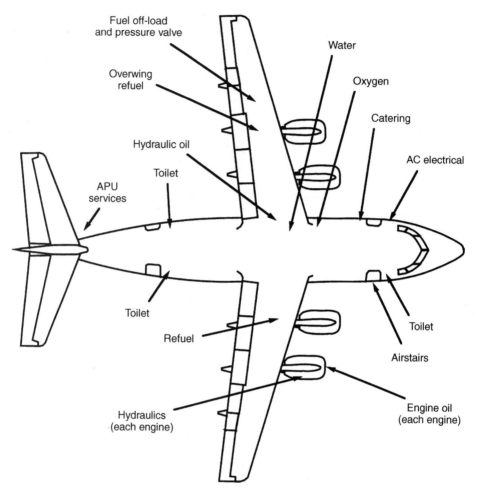

Fig. 3.16 Ground service requirements.

authority to over-ride unsafe decisions from the flight crew or should the captain retain the right to take full control of the aircraft.

Passengers are now expecting more facilities to be provided in the aircraft, particularly with regard to entertainment and business use. This places increased demands on internal volume, weight and power supply. Micro-miniaturisation of the electronics reduces these requirements but at the expense of increased aircraft purchase cost.

The aircraft designers must also consider layout requirements associated with servicing the aircraft (as shown in Fig. 3.16). This is particularly important for a quick turn-round on the airport apron. A full turn-round will involve refuelling, fresh water replenishment, re-supply of catering provisions, toilet servicing, cabin cleaning and cargo/baggage handling. Many of these actions are done con-currently. Space management around the aircraft during turn-round is a significant design consideration which may affect the overall layout of the aircraft components. For example, the positioning of the ground service vehicles around the aircraft during turn-round may dictate the position of service doors and other features. Such considerations may directly influence the aircraft layout.

Landing gear layout

One of the principal moving parts on the aircraft is the landing gear. This must be as light and small as possible. It must provide good ride dynamics during taxiing and safe energy absorption at touchdown. Retraction of the units is essential to reduce drag during flight. This places a strong demand for space to house the landing gear, usually at the wing fuselage junction. On many designs the landing gear bay requires separate fairing to enclose the wheels and shock absorber units.

A specialist section of the aerospace industry has been developed to deal with the design, manufacture and development of undercarriage components. Normally, but not always, the aircraft manufacturer will subcontract the landing gear design to one of these specialist companies. However, it is necessary for the aircraft project engineers to know something of the design parameters associated with landing gear design in order to produce an acceptable initial layout of the aircraft that will not compromise the design of the undercarriage and thereby increase the weight and cost.

The total landing gear system on an aircraft is a substantial contributor to weight and costs. Typically it represents from 3–6% of the aircraft take-off mass (representing 8–15% of the structure mass) and about 2% of the aircraft price.

For many years research has been conducted into methods of eliminating the units completely since they are only used at the beginning and the end of the flight and present a considerable weight penalty during cruise. Perhaps future development of vertical take-off and landing systems will provide an answer to this problem, although at present the landing gear on such aircraft tend to be more complicated than on conventional layouts!

In order that the design of the system can be properly understood it is necessary to carefully analyse the purpose of the landing gear. This can be listed as:

- to taxi to the take-off position and away from the runway at the end of the landing run;
- to allow the aircraft to accelerate without the use of special equipment (catapult, etc.) to allow rotation at the unstick speed;
- to allow a change of direction at the instance of touchdown from the flying altitude to the runway gradient;
- to assist in retarding the forward motion of the aircraft (braking) avoiding special equipment (arrester hooks).

Without the prospect of radical new ideas for landing gear design, the requirements listed above are met most conveniently by wheeled legs. For stability on the ground three contact points are required. The overall arrangement of these is at the designer's discretion. On some experimental designs two units (one forward and one aft of the aircraft centre of gravity) have been used. This layout is called the bicycle arrangement but as it is laterally unstable it needs wing mounted outriggers to stabilise the aircraft.

For the three unit landing gear there are basically two configurations. Each type has two main wheel units near the centre of gravity of the aircraft and the third either behind (tail-wheel) or in front (nose-wheel). The tail-wheel arrangement has largely gone out of favour because it produces an inclined passenger cabin floor, poor pilot visibility on the ground, is dynamically unstable (susceptible to ground looping) and presents the aircraft at a high drag attitude in the early part of the take-off run. The nose-wheel arrangement is now universally adopted for civil turbofan aircraft even though it is heavier and more expensive than the tail layout.

Detail layout (conventional landing gear)

The layout of the landing gear units is usually left until the overall aircraft configuration has been decided and an initial estimate made for the positions of the aircraft centre of gravity. The aircraft configuration may impose certain constraints on the landing gear layout due to retraction difficulties and structural aspects but in general the following procedures (as related to Fig. 3.17) can be used to fix the wheel positions.

1. Determine the height of the aircraft centre of gravity above the runway (distance h) accounting for shock strut length and travel (in static load condition), tyre size and retraction geometry.
2. Draw line AA through the main unit static ground position parallel to the fuselage centreline.
3. The longitudinal position of the nose-wheel attachment to the fuselage should be consistent with structural framework in the front fuselage, making sure that the flight deck floor line and the pressure bulkhead positions are not unduly compromised.

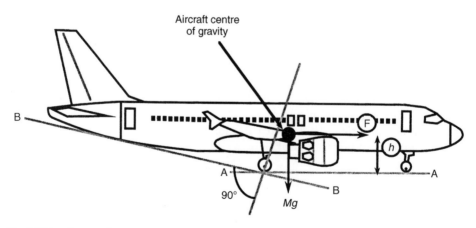

Fig. 3.17 Landing-gear layout.

4. The position of the main wheel behind the aircraft centre of gravity must satisfy the following criteria for the most adverse aircraft centre of gravity position:
   - provide an adequate reverse stabilising moment for backward towing and general stability (an estimate of braking force will have to be made);
   - provide a righting moment when the fuselage is pulled down onto its tail stop;
   - provide at least a static load of 8% W on the nose-wheel to give reasonable steering forces (where W is the aircraft take-off weight);
   - provide not more than 15% W static load on the nose-wheel, as more than this will make it difficult to rotate the aircraft at take-off without an excessive tail force;
   - provide a pitching frequency of the aircraft of about 100 cycles per minute (certainly greater than 30 cycles per minute) – this involves the ratio of the radius of gyration in pitch, the wheelbase and the undercarriage stiffness;
   - provide sufficient tail-down angle (angle between lines AA and BB) for rotation of the aircraft at take-off and in the landing attitude.

The required position of the three units in plan is a matter of simple geometrical calculation. For roll stability the main units should have a track (lateral distance between units) as large as possible with the general rule as shown in Fig. 3.18.

When more details are known a roll stiffness calculation should be attempted to show that the 'ride' is satisfactory. A track which is too wide should be avoided as manoeuvring along narrow taxi-ways may be difficult. The nose-wheel steering angle will be affected by track length for a given wheel base and this should be investigated (the aircraft must be able to turn about the centreline of either main unit).

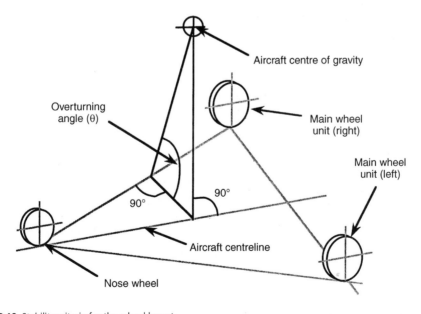

Fig. 3.18 Stability criteria for the wheel layout.

Energy absorption

The landing gear has to be capable of absorbing the energy due to the aircraft change of direction in landing from the final glide slope (usually about 3°) to the essentially horizontal runway. The energy to be absorbed is proportional to the vertical velocity of descent, v, squared, and the aircraft mass, M.

$$\text{Energy to be absorbed} = 0.5 \, Mv^2 \qquad [3.1]$$

This energy has to be absorbed by the shock strut and the tyre. Energy absorbed by the tyre is given by:

$$\delta_T \times \eta_T \times \lambda \times Mg \qquad [3.2]$$

where: δ_T = tyre deflection
η_T = tyre absorption efficiency
λ is the reaction factor = normal deceleration factor (i.e. number of 'g's)
M = aircraft landing mass
g = normal deceleration = $9.81 \, \text{ms}^{-2}$

Energy absorbed by the shock strut is given by:

$$S \times \eta_S \times \lambda \times Mg \qquad [3.3]$$

where: S = wheel travel
η_S = absorber efficiency

Equating [3.1] to [3.2] plus [3.3] and eliminating M we get:

$$\frac{v^2}{2g} = \lambda(\delta_T \eta_T + S \eta_S)$$

For a particular aircraft v and λ will be specified ($v = 3.5$ m/s and $\lambda = 2.0$ are typical values). For a particular tyre δ_T will be known and η_T can be assumed to equal 0.47.

The shock strut absorption efficiency (η_S) will depend upon the type of unit selected for the layout. For conventional oelo-pneumatic units η_S can be assumed to be 0.8. Hence the shock absorber travel can be determined.

Example calculation

Calculate the absorber travel required for a 60 seat regional jet aircraft with a vertical velocity of descent ($v = 3.5$ m/s) at a deceleration ($\lambda = 2$) with tyre characteristics ($\delta_T = 90$ mm, $\eta_T = 0.47$) and shock strut efficiency ($\eta_S = 0.8$). Putting the values into the equation above gives:

$$\frac{3.5^2}{2 \times 9.81} = \frac{2(0.47 \times 90 + S \times 0.8)}{1000}$$

Hence, absorber travel, $S = 270$ mm.

Tyre selection

The pneumatic tyre has several merits and is usually chosen for highly-loaded units. These advantages include the following.

- The contact surface stress can be chosen by simply selecting the inflation pressure. A solid wheel of either metal or rubber has contact stresses that may be too high for the runway surface.
- The energy storage capacity for an air spring is higher for a given weight than metal or solid rubber.
- During braking the rubber provides good adhesion to the runway surface.
- The elasticity of the tyre lowers rolling drag (due to the damping out of the irregularities of the surface).

In the initial project stage it will be sufficient to use tyre sizes similar to comparable existing designs. If space for retraction is restricted, the configuration is unorthodox, or the operation is unconventional (e.g. rough field requirements), the tyre size may be selected by the criteria below.

A pneumatic tyre is an interesting example of an inflated structure which reacts load by relief of internal pressure. Since the stiffness of the tyre is relatively low the ground load will be supported by the internal pressure on the ground contact area. Aircraft tyres usually operate at between one third and half the fully-squashed deflection. The load-carrying capability of a tyre can be expressed approximately as:

$$P = \text{constant} \cdot \delta \cdot p \cdot (DW)^{0.5}$$

where: δ = deflection
p = inflation pressure
D = diameter
W = width

The deflection will be a function of the dimensional properties of the tyre. As pressure increases the sizes of the tyre, DW, will reduce and thereby the weight, volume and frontal area. The designer will seek to use tyre pressure as high as permissible, consistent with the operational features of the aircraft and runway surface condition. Since the ground contact pressure is approximately equal to the inflation pressure, the landing surface must be capable of locally resisting the pressure. The maximum allowable inflation pressure is a function of the landing surface. Maximum tyre pressures for normal civil airport runways are around 120 psi. As the allowable pressure is a function of the runway foundation construction it could drop to below 90 psi for some small private airstrips.

For either flexible (asphalt or tarmacadam) or rigid (concrete slab) runways another limiting factor is the single wheel load. Each type of runway has a limiting strength and associated allowable tyre pressure. In the past several different types of index have been used to define runway performance but these have now been replaced by the internationally agreed (ICAO) 'Aircraft Classification Number – Pavement Classification Number (ACN–PCN)'. Values for ACN are quoted in ICAO Annex 14 which also includes the method of calculating the number. A good description of the method which also reproduces some typical values for ACN for several civil aircraft is given in the References.[7] Aircraft designed to be operated from runways of a particular value of PCN will require landing gear to be designed to an associated ACN value. This may dictate the use of more than one wheel per axle or restrict the maximum value for tyre pressure (refer to specialist texts on landing gear design and supporting airworthiness/operational calculations for more detail).

Once the tyre pressure has been decided, the main wheel tyre can be selected from the manufacturers' catalogues (considering the load on each of the main units as 45% of the aircraft gross weight). A check on the fully-squashed load should be made to ensure that the tyre does not 'bottom' before the full shock strut travel is reached. The aspect ratio, D/W, of the tyre is a design variable ranging from wide balloon tyres of less than 2.5 to narrow high pressure tyres at 5.0. The choice is dependent upon stowage, braking and inflation characteristics.

The load on the nose-wheel will be increased from the static case by braking. This must be considered in the selection of nose-wheel tyres. It is usual to take the design load as either the static case or 80% of the dynamic case whichever is the greater.

Mechanical design

The detailed design of the shock absorber and retraction mechanism is not usually considered in the project stage unless the overall aircraft configuration presents obvious difficulties in these areas (e.g. stowage space restrictions). To complete the aircraft general arrangement it will be sufficient to follow existing practice (e.g. adopting the geometry of the landing gear from a similar existing aircraft).

Future developments

Aircraft manufacture is an expensive and time consuming activity. The shortest period from concept initiation to aircraft operational flight is unlikely to be less than four years. Such timescales will only be possible for designs which are developments of an existing type (e.g. aircraft stretch). Typically, new designs take seven to ten years (or more) from conception to entry into service. During this period substantial pre-sale investment is required. Over the design period, operational requirements may change and so the design must be reasonably flexible. It is unlikely that a new aircraft will only be manufactured in the original configuration. New market opportunities, increases in productivity, development of new routes, etc. will result in 'stretched' designs. It is common for such stretches (plus re-engining and re-equipping of the original design) to produce a 'family' of aircraft as shown in Fig. 3.19. For example, the Boeing 777 airliner will be developed into a family of at least five (and possibly ten) aircraft types (see Table 3.1), each with a different payload/range capability aimed at different markets (routes) as demand for the aircraft dictates (see Fig. 3.20).

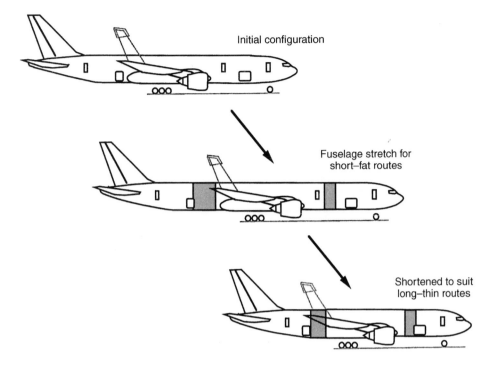

Fig. 3.19 Aircraft development.

Table 3.1 The proposed Boeing 777 family of aircraft

Boeing 777 (model)	MTOW (pounds)	Engine (lb-thrust)	Seats (see note*)	Range (nm)
-200A	545 000	77 000	300	4900
-200B	632 500	90 000	300	7400
-300	660 000	90 000	350	5700
-100	660 000	90 000	250	8700

* 3-class seating layouts.

If the 'family' approach to the design of a new aircraft is adopted the initial configuration may be a non-optimum design. For example it may carry a larger wing or the structural framework (e.g. landing gear) may be stronger than necessary for the initial design. The designers' task in the initial design phase is to balance the cost and complexity involved in future aircraft development (stretch) with the penalties imposed on the initial aircraft type. The marketing task is to sell commonality of the 'family' against the inherent but small inefficiency of the first design.

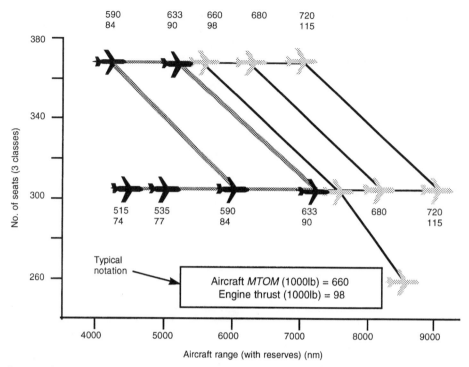

Fig. 3.20 The Boeing 777 development 'family' (source Boeing data).

In conclusion

To complete the overall configuration of the aircraft it is necessary to consider the detail design of each of the main components. The following chapters in this part of the book describe these issues and allow the layout of the aircraft to be finalised.

Before the design of components is considered in detail it is necessary to understand the safety and environmental issues that will affect the overall design. These aspects are described in the next chapter. For those who cannot wait to get into the main design topics, it is possible to move directly on to the detail design aspects (Chapter 5) with the intention of returning to Chapter 4 when you need to understand how the airworthiness and environmental issues influence the overall and detail design aspects of the aircraft.

References

1. Lange, R. H., Review of unconventional aircraft design concepts. *Journal of Aircraft.* **25**, 5: 385–92.
2. Lockheed-Georgia Co. *Multibody aircraft study – final contractors report.* May 1982.
3. Jenkinson, L. R. and Rhodes, D. P., Beyond future large transport aircraft, *AIAA paper 93-4791,* August 1993.

4. Ramsden, J. M., Towards the Megajet, *R. Ae. S Aerospace*, August 1994, 16–21.
5. Page, W. L., Further experiments on tandem aerofoils, *ARC Report 886*, May 1923.
6. Wolkovitch, J., The joined wing: an overview, *Journal of Aircraft*, 1986, 161.
7. Currey, N. S., Aircraft landing gear design, *AIAA Educ. Series*, 1988.

4

Safety and environmental issues

Objective

Some of the main constraints on the design of the aircraft result from mandatory and operational regulations. Before the aircraft is accepted into service it must be demonstrated by analysis, ground tests and flying tests that it meets all the airworthiness and environmental standards that currently apply to the type. When in service there is a continuing responsibility for the safe and efficient operation of the aircraft. At the early stages of the design of the aircraft it is essential to understand the nature of the regulations that must be met as subsequent modifications to comply with the legislation will be expensive and probably delay the project timescales.

The safety of the aircraft is assured by the airworthiness conditions laid down by national and international bodies. This chapter starts with a brief description of the framework in which these regulations are managed. The airworthiness regulations govern the ethos of the design. It is important to understand these aspects prior to the detailed consideration of aircraft component design as they may present constraints to the layout and performance of the aircraft.

Although airworthiness is described under the separate headings of structural integrity, system integrity, operation integrity and crashworthiness there is considerable interdependence involved in the overall aircraft configuration.

Environmental factors cover a wide range of aspects but noise and emissions are the main regulatory issues which affect the design of the aircraft. Both are largely associated with engine design aspects and aircraft flight profiles. With the expected growth in air traffic at the major airports the environmental pressure groups will become more vocal. This could result in more demanding criteria to be met by aircraft and engine manufacturers and operators.

For student designers it may be better to read this chapter after the ones on component design and analysis as these sections show the influence of airworthiness and environmental influences and so put the various regulations into context.

Airworthiness

Flying is potentially a dangerous form of transport. High speeds, a three-dimensional flight path, hostile ambient conditions (at cruise altitudes), a highly inflammable fuel mass positioned close to passengers; all these are combined with the vagaries of the weather, random occurrences of other natural hazards (e.g. ice and bird strikes) and, most significantly, human frailty. Within this scenario the travelling public demand an uneventful and comfortable journey unaware of any danger. Such expectations demand confidence in the mode of transport. This can only be achieved in a highly controlled operation leading to a public perception of safety. The fact that passengers seem more concerned about loss of luggage and long delays in airport terminals than the potential hazards is testament to the high quality of current airworthiness management.

The need to regulate the industry to achieve an acceptable standard of safety has been recognised since the early days of flight. The degree, or extent, of the regulation is complicated by the obvious international spectrum of the business. Since 1919 each nation has had absolute sovereignty over its airspace and been responsible for the execution of safety matters in its territory. This principle has been jealously protected and complete delegation of this responsibility to any international organisation (e.g. UN) has not been granted by any government. Nevertheless, several multi-national bodies have arisen in an attempt to coordinate safety policy and advise governments on the world-wide control of aircraft equipment and operations.

Set within the international framework, aircraft safety must encompass design, manufacture, operation and environmental aspects. As all these aspects are highly inter-dependent the regulations must reflect this inter-relationship. Safety is not totally a function of laws and decrees. It has always been recognised that an acceptable safety standard is dependent upon responsible attitudes on the part of all those who work and are involved in the air transport business. Thankfully, past experience has shown that this personal responsibility pervades the industry and we must ensure that this continues into the future.

The legal responsibility for airworthiness rests with:

- The airworthiness authority: for setting the standards and ensuring they are met. Each country has its own regulatory body (e.g. in the UK the Civil Aviation Authority [CAA] and in the USA the Federal Aviation Administration [FAA]). These vet the manufacturers and operators, issuing certificates of airworthiness, approving design and manufacturing companies, issuing operators' licences and controlling the standard of training for various personnel (e.g. flight crew and maintenance engineers).
- Aircraft manufacturers: to demonstrate that the aircraft meets the required design standard, to define the operating limitations, to define the maintenance schedules, to provide product support, and to take responsibility for product improvements during the life of the aircraft.
- Operators: to comply with airworthiness documentation, to comply with manufacturers' technical documents, to report defects and to assist the manufacturers to maintain airworthiness standards.

The airworthiness design requirements are principally aimed at the certification of new products but in recent years there has been a growing awareness of the significance of aircraft aging (i.e. the maintenance of airworthiness standards during the life of the product). The original manufacturer, irrespective of the operational life of the product, remains responsible for the design standards and all associated aspects of airworthiness of the vehicle throughout its operational life.

Design airworthiness requirements are contained in airworthiness regulations, for example, in the USA they are called Federal Airworthiness Regulations (FAR). Individual national European regulations are now being 'harmonised' and replaced by an international standard called Joint Airworthiness Requirements (JAR). For large civil transport aircraft with maximum take-off mass heavier than 12 500 lb (5669 kg), the appropriate document is FAR Part 25 for USA and JAR25 for Europe. These documents set out the design standard to be met and the method by which compliance to the standard can be demonstrated.

The main subject headings in JAR25 are shown below.

1. *Flight (analysed for normal and emergency thrust/power)*

   - Performance (stall, take-off, climb, en-route, landing, control speeds)
   - Controllability and manoeuvrability, trim, stability, stalling
   - Ground handling
   - Miscellaneous flight requirements (e.g. rough air, vibration and buffet, high speed)

2. *Structural*

   - Flight manoeuvre and gust conditions
   - Supplementary conditions (e.g. pressurisation, unsymmetrical loads)
   - Control surface and system loads
   - Ground loads, emergency landing conditions, fatigue evaluation, lightning protection

3. *Design and construction*

   - General (including materials, fabrication, fasteners, protection, fittings, flutter, bird strike)
   - Control surfaces and systems, landing gear, fuselage (including cockpit, windscreen, floors, etc.), emergency, heating and ventilation, pressurisation, fire protection
   - Miscellaneous (including electrical bondings, aircraft attitude levelling)

4. *Powerplant*

   - General (propellers, vibration, reverse thrust systems), fuel systems and components, oil systems, cooling, air intake systems, exhaust systems
   - Powerplant control and accessories, fire protection, auxiliary power unit

5. *Equipment*

- General (flight and navigation instruments), instrumentation, electrical systems and equipment

Safety

Safety is not an 'absolute' concept. The regulating authorities recognise acceptable 'levels of safety' which provide confidence to the travelling public and are also cost-effective.

The airworthiness of a particular operation is the status by which the aircraft (including engines and systems) is designed, maintained and operated to achieve an acceptable level of safety for passengers, crew and third parties. Within this context, design airworthiness is defined by a set of regulations and codes of practice. These provide an acceptable level of safety based on past experience.

The airworthiness regulations are continually under review and will be strengthened if incidents show an area of concern. Each nation controls its own airworthiness standard, usually through a government body or committee. For the United Kingdom, this control is exercised through the Civil Aviation Authority (CAA) and for the USA through the Federal Aviation Administration (FAA).

Table 4.1 Levels of safety (source: JAR)

**Levels of safety – Categories of effect
(Joint Airworthiness Requirements)**

Minor {Nuisance (10^{-2} to 10^{-3} per hour)}
(concerned with fleet service management and thereby occurring several times in the life of the aircraft)

- Operating limitations, routine changes to flight plan, emergency procedures (10^{-3} to 10^{-5})
- Physical effects but no injury to occupants (less than 10^{-5} per hour)

Major {Remote (10^{-5} to 10^{-7} per hour)}
(once in the operational life of an aircraft)

- Significant reductions in safety margins
- Difficulty for crew (adverse conditions which impair their efficiency)
- Passenger minor injuries

Hazardous {Extremely remote (10^{-7} to 10^{-9} per hour)}
(e.g. once in 20 years for a fleet)

- Large reductions in safety margins
- Crew extended due to increased workload or poor environment conditions (flight crew unable to perform their tasks accurately or completely)
- Serious injury
- Deaths of small number of occupants

Catastrophic {Extremely improbable (less than 10^{-9} per hour)}
(an unlikely event in the operational life of the aircraft type)

- Multiple deaths
- Usually total loss of aircraft.

Table 4.1 shows how effects are categorised and the numerical values assigned to their maximum occurrence. A slight confusion may arise by the fact that the loss of an aircraft might sometimes be non-fatal, but this is regarded as an exception to the general spirit of the definition.

The intention is that catastrophic effects should virtually never occur in the total operating life of an aircraft type fleet. Less hazardous failures may be accepted with the general rule that the probability of occurrence of a failure is inversely proportioned to its severity.

To understand the consequences of the quoted probabilities it is worth relating them to aircraft utilisation. A single civil aircraft flying scheduled routes may be expected to be flown for about 3000 hr/yr for an operational life of 15–20 years. There may be 200 such aircraft in use, giving about 50 000 operational hours per aircraft making 10^7 hours per type.

Safety levels

There are four main aspects to the regulations controlling the level of safety (Fig. 4.1). All but the crashworthiness sections are aimed at avoiding accidents. Crashworthiness regulations provide a standard for surviving an accident.

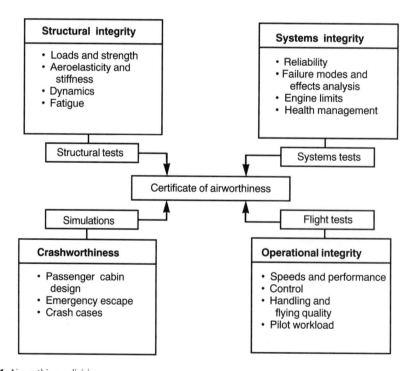

Fig. 4.1 Airworthiness divisions.

Structural integrity

This formed the first aspect of airworthiness (an excellent account of the historical development of structural airworthiness can be found in a paper by W. G. Heath in *RAeS Journal*, April 1980). The paper records letters written by the Wright Brothers as they anticipated the need for pre-flight structural analysis and tests in their original gliders and the first Wright-Flyer. The papers of Wilber and Orville contain these reports:

> *I am constructing my machine to sustain about five times my weight and am testing every piece . . . (circa 1900)* and about the Flyer aircraft:

> *We hung it on the wing tips some days ago and loaded the front set of trussing to more than six times its regular strain in the air . . . We also hung it by the tips and ran the engine-screw with a man on-board . . . The strength of the machine seems OK (November 1903).*

It is interesting to note the optimistic view that was taken. In these days, the rational analysis of the critical design case should have been based on the 'crashworthiness' of the vehicle!

The year 1928 saw the publication of the first set of British requirements (*Handbook of Strength Calculations*) although at this time they were not mandatory. This was a military document (*Air Publication 970*) which reflected the main area of aeronautical interest at the time. The early requirements considered the flight load cases as quasi-static conditions in which the flight manoeuvring accelerations were rationalised to inertia forces (load factors). This practice is still commonly used today for the symmetric load analysis. The original specific load factors (accelerations) varied according to the type of aircraft (aerobatic or normal category) but fell in the range 8.5 to 4.0. These were equivalent to modern ultimate load criteria, and considering the fact that in the early days they worked to a safety factor of 2.0, they seem low by modern standards for military aircraft. It was not until 1935 that the concept of 'proof load' cases was suggested. A proof to ultimate ratio of 0.75 was originally used and this compares favourably with the present civil aircraft value of 0.67. Eventually the ultimate factor was reduced to 1.5 and remains at this value today. With the passage of time no-one can give a rational explanation for the choice of this value. It is generally agreed that the factor is introduced to account for the variability in material properties, variability in manufacturing standards, deterioration in service and uncertainty (and inaccuracy) in stressing methods. It may be argued that all these aspects have improved over the last 40 years and therefore the factor should be reduced. Set against this possibility is the increased level of structural complexity, increased aircraft performance, greater utilisation and extra longevity of the aircraft.

The original load cases were effectively aircraft attitude dependent (low speed at high angle of attack, high speed at low angle of attack). This type of analysis led directly to the development of the flight and operational envelopes which are commonplace today. These methods assumed a rigid aircraft structure in which distortions due to load did not affect the flight action. In early braced-wing and fuselage structures this was acceptable but the development of the thin cantilever monoplane design gave rise to more flexible structures. This led to the development of dynamic response analysis and the introduction of flexural and torsional stiffness criteria. As early as 1930 four aeroelastic phenomena were recognised (aileron flutter, wing flexural/torsional flutter, control reversal and wing divergence). In fact, retrospective analysis of Langley's original aircraft suggests that aeroelastic flaws may have robbed him and the American scientific establishment of the prize for first powered flight. Since then many more instances of aeroelastic problems on new aircraft have arisen. The usual quick-fix for these effects is to increase wing torsional stiffness and mass-balance the control surfaces. Both these add weight to the aircraft and thereby reduce its effectiveness but the regulations insist on such measures. It is only in the past few years that more sophisticated methods of controlling aeroelastic effects have been allowed. These form part of 'active control technology' (ACT). However, the introduction of artificial coupling between structural and system aspects raises other airworthiness questions. These are considered in more detail in the 'system integrity' section described below.

The early flight envelope load cases assumed 'still-air' conditions. It soon became clear that air turbulence could give rise to critical flight cases. In 1942 the 'gust-encounter' was introduced into the regulations. These early calculations did not take account of the dynamic response of the aircraft or the variability in the gust profile. Only the single case of meeting a 25 ft/sec up and down gust at the aircraft design diving speed was considered. This was soon expanded into the 25, 50, 66 ft/sec gust series now specified and used to describe the aircraft gust envelope. To overcome the potential errors involved by ignoring the gust variability, the concept of the 'sharp-edge gust' and the application of 'gust alleviation factors' was introduced. The alleviation factor is a function of wing loading and lift-curve slope (with other parameters) but this evaluation still ignored aircraft structural flexibility. Eventually the requirements modelled more accurately the gust profile with either a linear or cosine function over the prescribed build-up distance (usually 100 ft). The requirement now is for the analysis of the dynamic amplification of the wing stress due to flexibility. This departure from the original 'rigid' aircraft analysis has brought with it the requirement to expand the number of stressing cases to cover the possibility that wing sections away from the bodyside may be critical (peak stress conditions could no longer be assumed to occur simultaneously.) In recent years, designing to meet the gust cases has become more complicated with the introduction of more sophisticated aerodynamic wing design, the introduction of devices that create local random turbulence conditions, and the consideration of gusts coming from any direction (not just vertically or horizontally). This has led to the development of dynamic response calculations being applied to the aircraft design in areas outside the gust environment (e.g. landing, impact loads control responsiveness). Interpretation of gust criterion can be found in airworthiness regulations (e.g. JAR 25-ACJ 25.305(d)).

The major omission in the development of the structural requirements has been the effect of the pilot on the aircraft. The regulations insist that the pilot is considered as taking no action that would alleviate the loading effect. This conservative criterion was based on the knowledge of the unpredictable behaviour and ability of pilots. As the pilot's workload is slowly being relieved by computer control and a more repeatable response can be assumed from the non-human system, it is becoming feasible to add the controller into the response equations (active-control). In fact it would now be erroneous to ignore these effects since they may represent significant input to the total aircraft system. Again, such developments link the structural airworthiness issues to system failure modes and system confidence levels.

Of all the developments in structural airworthiness, the necessary consideration of metal fatigue has made the greatest impact. The subject was largely ignored until the late 40s but now it represents the major element in structural air-worthiness. Initially fatigue was regarded as a phenomenon affecting only rotating machinery which accumulated millions of loading cycles in operation. The Comet crashes in the mid-50s highlighted the serious consequences of high stress/low cycle fatigue in pressurised fuselages. These accidents led to the forfeit of the commercial advantage held by De Havilland as a result of their initiation of the first jet airliner service mentioned in Chapter 1. Although much was learnt about fatigue analysis and the design strategies to avoid problems, De Havilland never regained the commercial initiative from their American rivals.

Strategies to combat fatigue failures in the early days involved the prediction of 'safe-life' stress levels. The difficulty of using this approach for aluminium alloy parts lay in the uncertainty of selecting safe stresses from the fatigue curves for the materials used. Furthermore, it was shown that the major aircraft assemblies (e.g. wing spars, landing gear) were subjected to a wide spectrum of load amplitudes and frequencies. The available data for the material was based on constant amplitude/cycle tests. Crude semi-empirical laws were developed to estimate fatigue damage per flight cycle and then to predict the aircraft fatigue life (to a chosen degree of probability). Nowadays direct stress (strain) measurements are used to estimate cumulative fatigue damage to critical parts of the structure.

In those areas where a safe-life could not be confidently predicted the concept of 'fail-safe' was applied. The inherent scatter of fatigue test results and the inability to predict the catastrophic failure of some early designs led to the adoption of redundant structures in which the various elements could be periodically inspected for fatigue damage. Once the concept of regular inspection was accepted as a justifiable structural design strategy the analytical interest centred on the rate of crack propagation. The requirements were set for a sub-critical crack length to be detected between scheduled service inspections. The development of methods for predicting the critical crack length and the rate of crack growth led to the introduction of 'fracture mechanics' and the concept of 'damage tolerance'. A combination of safe-life prediction to dictate stress intensity levels and fail-safe (plus inspection) is the current strategy for design to avoid catastrophic fatigue failure. The concepts of fail-safe and safe-life are often mis-understood (or misinterpreted). W. G. Heath suggested that a better description of the modern technique would be 'safe by inspection' or 'inspection dependent' safety.

In the early years of civil aviation structural airworthiness tests were completed ahead of the operational life to ensure safety. Nowadays it is not possible to design suitable tests to give adequate assurance from an airworthiness point of view. The specification of tests raises several awkward questions, including:

How do we predict the critical areas to be inspected?
How reliable are the inspections?
When do you terminate the life of a part?
How do you account for repaired structures?
How do you allow for aging of the structure?

To combat some of the fears, the airworthiness authorities now ask for a 'Structural Integrity Audit'. In this, the combined service experience of the fleet can be assembled and problem areas identified. As a consequence of this reporting, the degradation of the structure due to other causes (corrosion, accidental damage, manufacturing defects) can be recorded and used to indicate potential problem areas. Acceptable means of compliance for 'Damage Tolerance and Fatigue Evaluation of Structure' can be found in airworthiness regulations (e.g. JAC25-ACJ 25.571(a)).

System integrity

An aircraft may be regarded as a series of systems (or as one large system). A very useful description of system airworthiness may be found in a book by E. Lloyd and W. Tye.[1] Originally, the individual aircraft systems were 'self-contained' and purposely designed so that a failure in one would not influence the safe operation of the other systems and would not cause the aircraft to crash. This is still a design criterion but, following the introduction of the Autoland system in the early 70s and with the recent developments of automatic (computer-controlled) systems, more sophisticated safety assessments are necessary today. Modern systems contain many inter-connections between aircraft functions (e.g. automatic landing, auto-pilot, engine management, stall avoidance, auto-stabilisation and active load control). This means that a more systematic approach to the effects of individual component failure is demanded. For example, the relatively simple system controlling yaw-damping requires cross-coupling of electrical, hydraulic, mechanical, instrumentation and computer systems. Failure of a component in this system must not immobilise other, potentially more essential, flight systems.

The airworthiness requirements are purposely drafted in broad terms. The designers must conduct failure analysis for each system and also ensure safe operation of each aircraft function. Historical strategies for safe operation involving duplication or triplication of systems have now been superseded. Redundancy of these systems still features in the design but the analysis now requires the degree of risk to be estimated.

Assessment of the systems involves failure mode of effect analysis (FMEA). When making this assessment account must be taken of previous experience on similar systems, the variation in the performance of systems (the statistical distribution), awareness of the crew to the failure and their prescribed emergency actions, the probability (capability) of detecting the failure (warning systems), and aircraft inspection and maintenance procedures.

It must be remembered that although an engineering failure analysis can be conducted in a quantifiable way, any failure resulting from human error may be difficult to assess. This puts the emphasis on selection and training methods and management control. The regulations insist that the quantifiable assessment must be based on a pessimistic analysis based on the cockpit layout, speed of actions, simulation and operational awareness. Maintenance induced errors are equally difficult to quantify. The assessment must consider the likelihood of the maintenance staff 'modifying' the design concept and thereby increasing failure rate. Both flight crew and maintenance staff induced effects are only minimised by careful attention to detail design and good management control.

Operational integrity

For airworthiness purposes, performance data are needed for the complete range of airframe configuration and engine settings. This data must be provided for all the anticipated states for temperature, pressure, runway condition, runway gradient and wind effects. Also, the effect of engine failure on performance at critical flight phases must be determined. All these requirements represent a considerable amount of effort on the part of the manufacturer. Many flight tests must be conducted to evaluate the flight margins with adverse drag tolerances, adverse environmental conditions and minimum-level engine performance. The minimum tolerance for these margins is specified in the airworthiness requirements together with the definition of the acceptable flying techniques to prove the values. Examples of these conditions can be found in the specified take-off and landing techniques (see FAR/JAR-ACJ [Advisory Circular Joint]).

Apart from demonstrating the safe flying characteristics of the prototype aircraft, the flight tests also provide the information used for computing the aircraft flight manual data. Since most of the operational data are dependent on many variables the data in these manuals is condensed to derive simplified charts for safe flight (operation).

Many aspects of the aircraft design (structure, aerodynamics, powerplant and systems) are influenced by the operational limits imposed by the airworthiness regulations. Some examples are given below:

- minimum speed for rotation at take-off
- minimum speed to maintain lateral control in the event of an engine failure
- maximum speed to avoid aerodynamic buffet
- minimum speed to avoid stall (or alternatively limits on the angle of attack)
- minimum speed in the second segment climb phase
- maximum speed in the approach to landing phase (crosswind limits)
- minimum speed to operate thrust reverses
- brake system energy and torque limitations
- maximum speed for operation of high-lift devices
- speed and height limits imposed on engine operation in non-standard atmospheric conditions (hot and high)
- maximum speed for engine re-light after shut-down, including the effect of altitude

This list represents only a small selection of the limitations imposed by the airworthiness regulations. Speed limitations may also be imposed to protect the aircraft structural framework. These may be in the form of airspeed and Mach number boundaries. Typical constraints are shown in Fig. 4.2.

The maximum operating speed limit V_{MO} may not be deliberately exceeded in

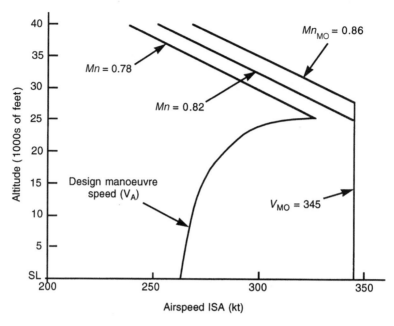

Fig. 4.2 Operating speed boundaries (source B Ae.).

any flight phase (climb, cruise or descent). Full application of controls should be only made at or near speed V_A.

The definition of speeds together with the stall boundaries (the limits of maximum lift coefficients, $C_{L_{max}}$, for the aircraft) and normal acceleration limits are combined to establish the flight and gust envelopes (see Fig. 4.3). These are used to produce the aircraft loading cases associated with weight, balance geometry and other configurational variations.

Worst case conditions are the basis for structural analysis. This forms part of the duties of the Structural Analysis Office. Such analysis provides confidence in the structural integrity of the aircraft to the hazardous (10^{-7}) and catastrophic (10^{-9}) levels of safety.

The airworthiness regulations specify the most likely loads (to the accepted level of risk) that the aircraft should resist. These are termed 'limit loads'. The aircraft must be capable of reacting these loads multiplied by a factor called the 'proof factor' without suffering substantial permanent deformation. Also, the structure must not fail below a load equal to the limit load multiplied by the 'ultimate factor'. For civil airliners the proof factor is fixed at 1.0 and the ultimate factor at 1.5. When stressing the aircraft the applied stresses resulting from the proof or ultimate design loads must be less than the allowable proof and failure stresses for the material and structural shape. The ratio of allowable stress (or load) to the applied stress (or load) is termed the 'reserve factor'. This must be greater than unity for the structure to be safe. The USA practice is to subtract 1.0 from the value of reserve factor and call it the 'margin of safety'.

Airworthiness requirements are set to match the expected ability of an average

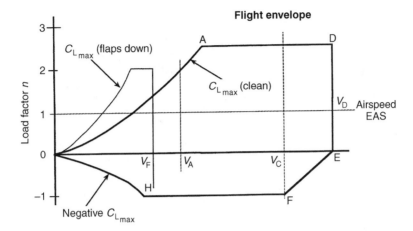

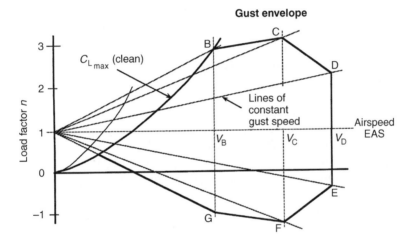

Fig. 4.3 Flight and gust envelopes (source FAR/JAR 25).

flight crew in relation to the handling and flying qualities of the aircraft. Such limitations assume the aircraft is under 'manual' control. As the flying qualities of the aircraft and control authority are transported to automatic (computer) systems some of the older regulations are inappropriate. However, in order to have confidence in the performance of such automatic flight control systems they are required to be designed to extremely high (10^{-9} to 10^{-11}) levels of safety as described earlier.

Some examples of limitations relating to flying and operational qualities are listed below:

- minimum times for the crew to recognise engine and other critical system failures
- minimum times for the selection of emergency braking/reverse thrust/spoilers for an aborted take-off

Table 4.2

Performance		Control and manoeuvrability	
101	General	143	General (definition)
103	Stalling speed	145	Longitudinal control
105	Take-off (TO)	147	Directional and lateral control
107	Take-off speeds	149	Minimum control speed
109	Accelerate-stop distances	161	Trim
111	Take-off path	173	Static longitudinal stability
113	Take-off distance and run	175	Demonstration of 173
115	Net take-off path	177	Static directional and lat. stability
119	Landing climb-all engines	201	Stall demonstration
121	Climb-one engine out	203	Stall characteristics
123	En-route flight paths	207	Stall warning
125	Landing	251	Vibration and buffeting
131	Abandoned TO (low friction)	253	High-speed characteristics
132	Abandoned take-off (slush)	261	Flight in rough air
133	Take-off from wet runways		
135	Landing (low friction)		

- similarly minimum times to select systems on landing
- maximum system operating times (e.g. flap deployment, undercarriage retraction, engine response)
- maximum rotation rate during take-off
- maximum landing approach angle and geometry of the flare manoeuvres
- factors to account for touchdown point and landing performance
- height limitations resulting from inadequate rate of climb performance
- height limitations due to cabin pressure differential constraint
- rate of descent limit to safeguard cabin repressurisation
- factor on minimum drag speed to reduce pilot (system) workload for maximum endurance operations
- limits on aircraft centre of gravity positions to match available control effectiveness

The above list is not exhaustive but is intended to provide a flavour of the type of limitations which result from the airworthiness regulations.

Many of the limitations in the above list may be minimised on aircraft with active control systems. For such aircraft the demonstration of control system reliability is a major element in the certification process as described in the previous section on system integrity.

The extent of the Operational (flight) requirements contained in FAR/JAR 25 is shown in Table 4.2.

Associated with each of these items is a data section on 'method of demonstrating compliance' (e.g. Section 2 – ACJ). The operational integrity of the aircraft is governed by these regulations.

Crashworthiness

This subject is dealt with in detail in Chapter 5, but a summary of the main points is given below for completeness in the discussion on airworthiness.

Even with the finest attention to airworthiness, accidents may occur. The crashworthiness aspects are aimed at improving the chance of surviving in the event of these accidents. It is difficult to set specific requirements in this area; the best that can be done is to make the most of what technology can offer. The main difficulty in this approach is the high emotive pressure, particularly from the press, that exists at the time of an accident.

Aircraft accidents vary enormously from the totally non-survivable mid-air collision, to the relatively inconsequential disputes with other aircraft and airport equipment during taxiing and docking. For the first, there is little the airworthiness regulations can do except provide for preventive measures (flashing beacons, better visibility, etc.) and improve operational management (air traffic control). Fortunately, the majority of accidents do present the possibility for survivability. Crash survival requires that the occupants both survive the impact and are then able to evacuate the aircraft before the post-crash hazards become intolerable.

The requirements therefore consider each of the following factors:

- crashworthiness of the structure
- adequate restraint of the occupants
- provide a non-injurious environment
- provide, where possible, reductions in the crash forces
- give protection against fire hazards
- ensure safe egress from the vehicle

The list above is presented in the relative order of priority for the designers. The general guidelines may be transformed into the following detail design requirements.

- Ensuring that the basic structure around the occupants provides a protective envelope which will be stiff enough to guarantee that the emergency exits still function.
- The seats and restraint straps are designed to ensure that occupant movement within the protective envelope is arrested before and during the impact.
- Elimination of dangerous features within the cabin and cockpit. This is often only a matter of careful detail design or passenger management (e.g. closed lockers on overhead luggage racks). Potentially serious hazards when identified can often be easily eliminated (e.g. by stowage in secure parts of the aircraft or by not carrying the cargo).
- Maximum crash forces (or accelerations) must be reduced to human tolerance levels. This can often be accomplished by the detail design of the surrounding structure, restraint system and seat design. Exact analysis is not possible. The designer has to show that providing the structural envelope, the seat and the restraint system remain intact, then the forces on the passengers are tolerable. Past experience has shown that this is acceptable.
- Designing the fuel system and fuel tanks to resist major damage in the crash, placing fuel as far away from the occupants as possible, and keeping fuel lines outside the occupied areas are all design requirements that will reduce the risk of fire. Unfortunately it is not always possible to follow these 'guidelines' due to aircraft layout. Research into the use of smoke-hoods and cabin water

spray systems are examples of the continuing development of such requirements.

- Avoiding or reducing the risk of disabling injuries and careful attention to the provision of escape routes for all crash situations is a natural design responsibility. Several airworthiness requirements are aimed at these issues (for example the recently introduced regulations concerning the location of emergency exits and the installation of floor track lighting).

Certificate of airworthiness

The conclusion from the airworthiness analysis of the aircraft is the issue of a Certificate of Airworthiness (C of A) for the design. This represents an expensive but unavoidable element in the total development of the aircraft project. The C of A is only valid for the design as tested and manufactured by the approved organisation. If modifications to the design are made or if the manufacturing methods or organisation are changed, the C of A has to be re-issued. This will only be done if the airworthiness authorities believe that the changes do not reduce the safety standard. Within this context it is inevitable that major changes will require extra flight, structural or system tests to be conducted. It has been shown in the past that even small alterations to the original design may lead to unsafe operation and therefore the Airworthiness Board is extremely cautious in the automatic extension to the original certificate. The cost of testing can be high and must be accounted for when considering the development of an existing aircraft configuration (e.g. re-engining or fuselage stretch). As such, the certification processes is a key element in the aircraft programme. The flight test programme forms only one element, albeit highly significant, in this process. Nearly all the specialist departments in the company are concerned with meeting the airworthiness requests in full. This involvement ranges from function tests to analysis of operational practice (failure analysis and reliability). To achieve a Certificate of Airworthiness the aircraft will be supported by numerous analytical reports and other documentation. These will cover all aspects of the design of the aircraft plus required operational practices (e.g. pilot training, maintenance requests, reporting procedures, etc.). When an aircraft is sold to an airline it will be accompanied by a C of A. This is specific to the aircraft and to the issuing authority (e.g. FAA, CAA). The manufacturer accepts responsibility for the safe operation of the aircraft and in so doing stipulates the procedures and practices required for maintenance and operation. Failure to conform to these will invalidate the C of A and result in the airworthiness authority withdrawing the licence to the operator. Although such practices seem to impose unnecessary overheads on the airline it is generally accepted that safety must be ensured.

Environmental issues

Even in the pioneering days, aviation had critics who commented that aircraft were noisy, smelly and an intrusion into their private lives. As aircraft have become more numerous, larger and more powerful the environmentalists have become more influential. The expansion of turbojet commercial services flying over (and into) already congested and polluted cities has highly politicised the environmental issues.

New York politicians in the early 60s under pressure from voters unilaterally imposed restrictions on the noise generated by aircraft operations and installed measuring equipment at and around the airport to monitor offending airlines and allow sanctions to be imposed. At about the same time other cities (notably Los Angeles) were concerned about the level of airborne pollution generated from engine emissions in climatic inversion-layer conditions. Although this fog was related principally to ground transport vehicles, the public visibility of aircraft smoke trails and the smell of unburnt kerosene near airports focused public attention which resulted in legislation on aircraft emissions.

Aircraft and engines are now regulated with regard to both noise and emission pollution. When designing aircraft it is necessary to understand the implications of these environmental controls on aircraft operations and the consequential effects on detail aircraft component design. Although noise and emissions are inter-related by common engine design parameters it is convenient to separate the two issues.

Each is described separately below.

Aircraft noise

Following several research studies, technical meetings, conferences and law suits, the International Civil Aviation Organisation (ICAO) eventually recommended national legislation that would match the existing US noise control rules. This was done by adding a formal appendix (Annex 16) to the 1944 Chicago Convention. The American legislation formed a separate section to the FAR (identified as Part 36). The established international and national regulations now impose maximum noise exposure at three critical operating conditions (see Fig. 4.4).

Three noise recording locations (under the approach, under the take-off paths and to the side of the runway) are used to enclose an airport box. The measurements are aggregated to form a single value for the operations at the airport. In the early form of the legislation this value was set at 108EPNdB (effective perceived noise scale).[2] At the time this requirement represented a severe challenge to the industry as it was 6–10 dB lower than the noise emitted by the then current jets. The regulations imposed stricter limits on smaller aircraft. Although smaller aircraft require less powerful engines which would individually produce less noise, the regulations took account of the larger number of movements from such aircraft due to the increased frequency of their associated shorter range operations. The current noise regulations are illustrated in Fig. 4.5.

To meet the regulations, existing aircraft (engines) were modified by the installation of 'hush kits'. Such modifications adversely affected the aircraft economics which hastened the introduction of replacement aircraft by quieter and more efficient designs.

Noise is generated when air pressure is rapidly varied. Such conditions are

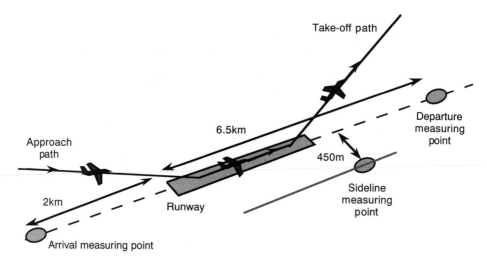

Fig. 4.4 Standardised noise measurement positions.

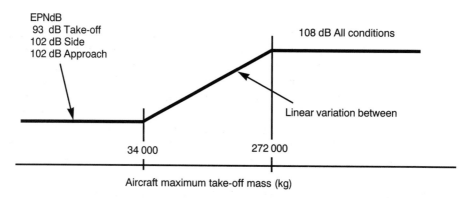

Fig. 4.5 Legal values (source FAR Part 36).

naturally created in and around engines due to the changes in pressure and temperature resulting from the generation of power. The aircraft airframe will also produce noise, particularly in those areas where airflow directions are intentionally varied (e.g. at control surfaces, high-lift devices, engine intakes), or when projections are immersed in the forward airstream (e.g. landing gear, airbrakes and lift dumpers), and where there are discontinuities of surface contours (e.g. over the windscreen, at the edges of the flaps when extended and at badly fitted access panels). In general, noise generated by the airframe is not significant when engines are at (or near) full throttle but can become important when the engines are throttled back in the approach phase prior to landing. Whereas in the early days of regulation the noise generated during the take-off and climb-out phases predominated, the development of quiet higher bypass engines and the careful control of aircraft departure flight paths has now focused attention to the approach phase.

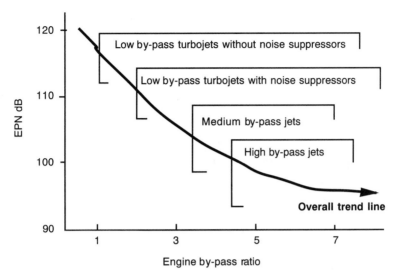

Fig. 4.6 Engine noise (source Rolls Royce data).

The long shallow (3°) approach path increases the noise exposure to the community under the fixed instrument controlled landing profile. With the engines at reduced power, airframe noise becomes a more significant component. In future, designers will need to give more attention to the reduction of aerodynamic noise from the aircraft structure in the landing configuration. For example, this may require the use of less complex flap designs to achieve the same lift. Alternatively, pilots may be required to make approaches at a steeper angle (e.g. 5–6°) to increase the height of the aircraft over the community near the start of the runway. Both these developments will have a significant impact on the design and operation of the aircraft.

As mentioned above, the most significant reduction in engine noise has resulted from the introduction of high bypass ratio engines with large diameter slower speed fan jet streams surrounding the hot exhaust, as shown in Fig. 4.6.

These developments have also altered the noise spectrum. Figure 4.7 shows that the main sources of noise from the pure jet and low bypass engine is from the hot exhaust gasses as they mix with the ambient airstream. Noise energy has been estimated as proportional to the seventh power of the exhaust velocity. It is essential that this velocity is made as slow as possible to reduce noise. Unfortunately the thrust generated by the engines is proportional to the square of this velocity so a technical compromise must be reached. More details of this compromise are given in Chapter 9.

The core exhaust velocity can be reduced by mixing slower speed air with the exhaust stream. Several different types of exhaust nozzle are available, one of which is shown in Fig. 4.8. In the high bypass engine the mixing is much easier to accomplish. This results in an average exhaust velocity which is much lower than in the above case. Where possible, the most effective method for noise reduction is to mix the bypass and core streams internally (within the cowling structure as this

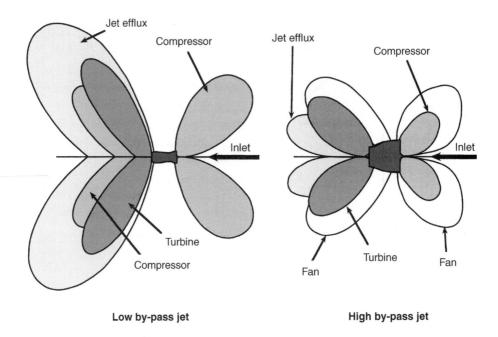

Low by-pass jet   High by-pass jet

Fig. 4.7 Noise spectrum (source Rolls Royce data).

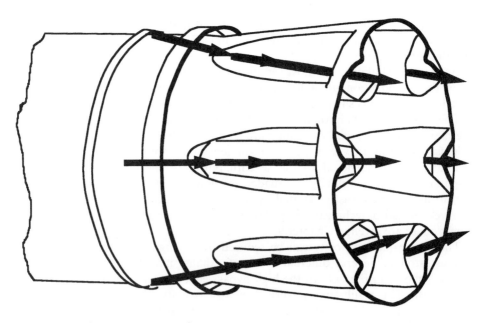

Fig. 4.8 Jet efflux (source Rolls Royce data).

can be sound insulated) and then to expel the resulting stream through a mixer nozzle. The lower velocities and the shorter distance over which the mixing is achieved give a substantial reduction in noise. This results in the exhaust noise being less dominant than that generated by the fan frequencies. These noise sources may be partly suppressed within the nacelle structure which will also incorporate sound absorbing materials.

To reduce engine noise intrusion into the fuselage/passenger cabin the engine must be separated from the fuselage side laterally (spanwise) or longitudinally (rear fuselage mounting) and the internal fuselage layout adjusted to position non-seating areas (e.g. galleys and toilets) at the anticipated noisy cross-sectional planes. Alternatively, the cabin structures must be sound insulated in these areas and allowance made for the increased weight of fuselage structure that would result.

The above descriptions relate to the control of noise at its source but the operation (flight profile) of the aircraft can have significant influence on the annoyance caused by the aircraft to the population over which it flies. In order to evaluate these effects there have been a number of noise models created. The noise 'footprint' from the aircraft and its flight path may be considered in several discrete segments.

The shape of the footprint shows contours of constant noise level. Figure 4.9 shows the contour for two different sized aircraft together with the airport measuring points (approach A, sideline S, departure D). Starting at the end of the runway prior to take-off the pilot will select full power from the engine which will be reacted by the brakes (to check the aircraft systems). This creates the small lateral and rearward lobes on the contour profile at the start of the runway. On release of the brakes the aircraft will accelerate along the runway creating relatively constant noise contours parallel to the runway. Due to the shielding of the aircraft offside engines, the noise measurements may be slightly variable. Sideways noise transmission is more readily attenuated due to natural ground absorption. When the aircraft lifts off the runway this attenuation is reduced and the noise contours swell laterally due to pure radial noise transmission. As the aircraft climbs the distance to the noise source from the ground increases and the noise is naturally attenuated in the atmosphere. This reduces the noise heard on

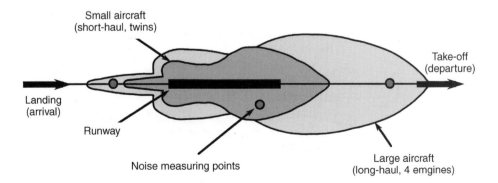

Fig. 4.9 Noise contours.

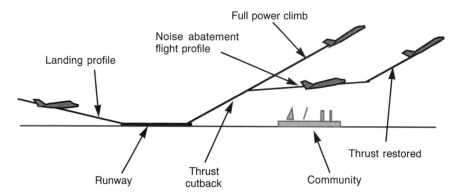

Fig. 4.10 Flight profiles.

the ground. The maximum noise occurs when the aircraft is at about 300 m (1000 ft). To reduce the climb-out noise disturbance some airports impose procedures which require the pilot to reduce thrust (power cutback) after the initial climb phase (as illustrated in Fig. 4.10). This has the effect of reducing the noise generation at source but reduces the aircraft climbing ability and therefore increases the duration of the exposure. For the legislators and operators a difficult compromise exists between 'full power' and 'cutback' flight profiles.

When the pilot restores full power the increased noise may generate a noise island downstream from the airport footprint (as shown in Fig. 4.11). To spread the annoyance, flight path directions may be varied. This avoids repeated exposure over the same population. The effect of cutback profiles on the footprint contour is shown in Fig. 4.11.

The noise contours generated by modern high by-pass engines do not show the same advantage for cutback as found on the older turbojets. This means that it is becoming less effective to use cutback flight profiles as a means of noise abatement except for the noisiest aircraft types.

The noise contours related to the approach path are much easier to define as

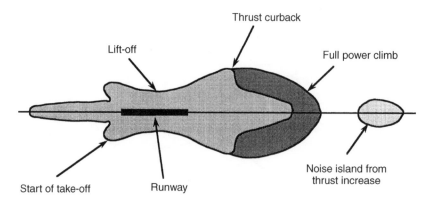

Fig. 4.11 Effect of cut-back.

all aircraft are flying down a fixed ILS (instrument landing system) beam and the engine is at a low thrust setting. Although the engine noise source is reduced at these settings the aircraft is in an aerodynamic dirty configuration with flaps and eventually landing gear fully deployed. The approach profile of 3° means that the aircraft flies low over the population for a greater distance. For this reason the approach noise contour is becoming increasingly significant as mentioned above.

Several attempts have been made to model the noise footprint and to simulate the noise produced at an airport. One of the most significant models has been developed by the FAA (the Integrated Noise Model, INM). Several countries have developed similar noise models to assist the planning and development of existing and new airports (e.g. in the UK the ANCON model and in Denmark the DANSIM model).

From this brief introduction to aircraft noise, it is clear that the design and operation of aircraft is now significantly influenced by noise regulations. In the past, airframe designers have been able to hide the noise generated from aerodynamic effects of the airframe behind engine generated noise. The adoption of high by-pass engines, which are intrinsically quieter than earlier engines, will force aircraft designers to consider noise suppression techniques for airframe, particularly in the approach configuration.

The introduction and adoption of Stage 3 noise regulations has resulted in reductions in perceived noise levels at airports over the past years but as more traffic is generated the airport will need to handle more and larger aircraft. This will eventually increase the noise in the airport community. These considerations have forced the regulators to consider introducing tougher noise regulations for future aircraft (Stage 3 plus). The nature of such regulations is not yet finalised but when decisions are made they will have a direct influence on future engine and airframe design.

Considerable improvements can be shown for new aircraft designs compared to older types. Airbus claim that the A320 is 30 dB quieter than the old 707s (this implies that you would need to have 16 A320s operating together to be considered by the human ear as noisy as one Boeing 707). The effect of noise decays rapidly with distance from the source. For this reason other transport systems present serious noise nuisance to the community:

93 dBA – high speed train, 100 m distant
87 dBA – express train, 100 m distant
82 dBA – bus in town, 8 m distant
80 dBA – A320 at take-off, 300 m distant
70 dBA – A320 at take-off, 700 m distant

Several textbooks and technical reports are available to allow a more detailed study of the effects of noise. An excellent introduction is given by MJT Smith.[2]

Emissions

Although aircraft systems may leak small quantities of various gases and liquids (e.g. hydraulic oil) during operation, the predominant source of emissions is from

Table 4.3 Flight profile for the LTO cycle

Flight mode	Duration	Thrust setting
Idling and taxi-out	19 min	7%
Take-off	42 sec	100%
Climb-out	2.2 min	85%
Approach to land	4.0 min	30%
Taxi-in and idle	7.0 min	7%

the engine exhaust. Burning jet fuel produces particulates (smoke) and various gases including carbon dioxide (CO_2), water vapour (H_2O), various oxides of nitrates (NO_x), carbon monoxide (CO), unburnt hydrocarbons (HC) and sulphur dioxide (S_2O). Both CO_2 and H_2O occur naturally in the atmosphere and are not regarded as harmful. All the other components are defined as pollutants. In 1981, ICAO set out recommendations to control the level of acceptable pollutants from aircraft emissions. To this end, engines are tested at various thrust settings (corresponding to those used in specific flight segments) and the levels of pollutants determined. These are then used to determine typical emission quantities.

For emission evaluation purposes, the aircraft flight profile is divided into that part occurring below 3000 ft (called the landing – take-off cycle, LTO) and that above (i.e. the climb and cruise segments). It is the LTO cycle that has attracted the legislation as this affects the pollution in and around the airport community. The cycle includes the approach, landing, taxi to stand, taxi from stand, take-off and the initial climb-out phase (from and to 3000 ft). Each aircraft type produces a particular emission profile in the LTO cycle (i.e. grams of specific pollutant per kilogram of fuel burnt). The calculation is based on average duration in each of the segments in the LTO cycle. The method assumes typical power settings of the engine in each of the manoeuvres. The specific data is shown in Table 4.3.

Emission data for all aircraft and engines is published by FAA and other regulatory bodies. Typical emission values (g/kg of fuel) for two different sized aircraft are shown in Table 4.4.

Table 4.4 Emission components[3]

Flight mode	Fuel burn (kg/sec)	HC	CO	NO_x	SO_2
(i) Fokker F28-100 with two RR Spey 555 engines (data from reference 3)					
Take-off	0.7225	5.14	1.1	19.0	0.54
Climb-out	0.5893	0.53	0.0	14.7	0.54
Approach	0.2197	8.20	20.0	5.8	0.54
Total taxi	0.1153	93.99	90.0	1.7	0.54
(ii) Boeing B747-400 with four PW 4056 engines (data from reference 3)					
Take-off	2.342	0.06	0.4	28.0	0.54
Climb-out	1.930	0.01	0.6	22.9	0.54
Approach	0.658	0.13	2.0	11.6	0.54
Total taxi	0.208	1.92	21.9	4.8	0.54

As the sulphur content of fuels from different oil wells is significantly different, the values quoted above for sulphur dioxide are average values. This approximation is acceptable as SO_2 only accounts for a small percentage of the total emissions from the engines (see below).

The above data shows that emission quantity and profile vary widely with regard to the type of aircraft (engine) and the mode of operation. Improvements made by the engine manufacturers from the old engines used on the F28 to the more modern power plants of the B747-400 can clearly be seen in the above data. By multiplying the fuel flow and duration with the emission production for the B747 the values (grams) shown in Table 4.5 are obtained.

These calculations illustrate the difference in emissions in airborne and ground manoeuvres. When the aircraft is flying with high power levels, the engines are operating efficiently. This means that most of the available energy is extracted from the fuel leaving low HC and CO emission but generating high quantities of NO_x. In the ground phase the engines are throttled back and operating at poor combustion conditions. This produces large amounts of HC and CO. Nearly half of the total amount of these pollutants is generated in the ground phase.

The calculation above shows that about 20 kg of emission products are exhausted by just one B747 aircraft in one LTO cycle. It has been estimated that millions of tons of pollution are produced by the total air transport industry in each year. During the LTO cycle, typically 834 kg of fuel will be burnt by the B747 and this will create 2.5 tonnes of carbon dioxide and about one ton of water vapour.

Emissions during cruise are difficult to estimate because, unlike the LTO cycle conditions, it is difficult to test engines in the rarefied upper atmosphere conditions. Coupled to this is the variability in operating conditions which affect cruise performance (stage length, aircraft weight, speed and altitude). The level of pollutant produced will be affected by all these operating conditions. Nevertheless, it is known that the engines will be operating at the design conditions and this will mean that the combustion process is most efficient. The legislators, recognising these facts and noting the intention of engine manufacturers to produce optimum design during cruise conditions, have chosen not to set requirements in this phase (i.e. above 3000 ft altitude). On the contrary, it is suspected that high flying aircraft (e.g. 70 000 ft altitude of supersonic aircraft) could cause upper atmosphere pollution problems associated with the ozone layer. Legislation may be proposed on future operations at high altitudes which, if fully imposed, may curtail future

Table 4.5 Total emission components (Boeing 747)[3]

Flight mode	HC	CO	NO_x	SO_2	Total (kg)
Take-off	5.9	39.3	2754	53	2.8
Climb-out	2.5	152.8	5834	137	6.1
Approach	20.5	318.8	1831	85	2.3
Total taxi	623.0	7106.1	1557	175	9.5
Total (kg)	0.651	7.614	11.98	0.45	20.7

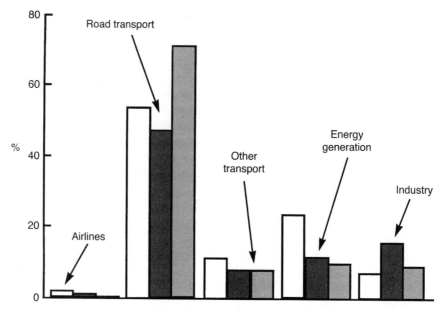

Fig. 4.12 Sources of pollution (source SAS data).

supersonic commercial aircraft operations and restrict high altitude cruise from subsonic scheduled services.

Air transport is a negligible contributor to global air pollution (see Fig. 4.12). Nevertheless, the industry continues to make advances in the reduction of emissions by improving energy efficiency. Taking road transport (mainly cars) as a datum (100%) source of energy per seat kilometre, rail uses less than 70% and air about 55%. Current modern aircraft require about half the fuel to fly a short range sector (500 nm) as older (1970) types. This large improvement over the past 20 years has come about due to a combination of engine, airframe and operating practices.

Several textbooks and reports are available which provide fuller explanations of aircraft emissions. A good place to start is a report by L. J. Archer.[3]

Ecology

Although the air transport industry has made considerable efforts over the past 20 years to reduce noise and emissions, the future is likely to be no less demanding on environmental issues. Legislation may be introduced that places an increasing burden on the use of fuel (e.g. carbon taxes) and imposes even tighter operating restrictions to reduce the annoyance due to noise and emissions. These will have considerable effects on the way that aircraft and engines are designed and operated. New considerations regarding the necessity to use renewable (recyclable) materials and the need to avoid wastage during operations (e.g. fuel spillage) may be forthcoming. The only certainty is that the industry must be, and be seen to be, responsive to all reasonable ecological pressures that arise.

One of the principal considerations to future operations may arise from the restriction of airport development due to strong ecological pressure groups protecting green field sites and changes to land-use. This may have considerable influence on aircraft design and operation as it may become increasingly desirable to reduce occupancy of the scarce runway area; for example aircraft may need to be designed for steeper approach flight paths and to land at slower speeds than existing practice.

Existing environmental issues and new considerations place a considerable burden on aircraft design teams. They must be politically and socially aware of these pressures and be prepared to respond to genuine issues. Past experience has shown that such demands on the design of new aircraft can be met by the emergence of new technologies and operating practices. It is to be hoped that this can be continued into the future.

References

1. Lloyd, E., Tye W, *Systematic safety*. CAA London, 1982.
2. Smith, M. J. T., *Aircraft Noise*. Cambridge University Press, England, 1989.
3. Archer, L. J., *Aircraft emissions and the environment*. Oxford Institute for Energy Studies (EV17), England, 1993.

5

Fuselage layout

Objectives

This chapter starts our detailed consideration of the main component parts of the aircraft. The fuselage layout is often examined first in the design process as the size and shape is dependent on the number of seats to be carried and this is related to the specified passenger load. Once the shape of the fuselage has been decided it forms part of the fixed aircraft configuration in later design studies.

After an initial consideration of the overall criteria for the fuselage design this chapter then describes the parameters that size the passenger cabin. This involves decisions on the fuselage cross-sectional geometry and the longitudinal layout of the seat rows. The distribution of seats between the different ticket classes is a major criterion in these decisions. Airworthiness regulations must also be taken into account so that passengers are protected in an emergency and can make a safe and quick evacuation of the fuselage.

The chapter also considers other factors that must be taken into account for the full design of the fuselage. These include the shaping of the front and rear sections to blend into the cabin geometry, providing accommodation for the flight crew, housing cargo and freight in standard containers and the provision of structural support for the wing and tail surfaces (and possibly rear engines). To provide a comfortable and safe environment for the occupants the fuselage design also requires several systems to be specified. These range from the flight deck avionics and displays to the passenger cabin conditioning systems. An introduction to these systems is given towards the end of the chapter. The description of fuselage layout and design concludes with an example of fuselage layout on a proposed new aircraft. This shows how the principles described in this chapter can be applied to a specific design philosophy.

After you have studied this chapter you will be able to set out the geometry of a fuselage to meet the aircraft specification and will appreciate the systems needed to support the overall design.

Passenger preferences

One of the principal purposes of air transport is to carry people comfortably and safely over the journey they want to make. The cabin interior more than any other feature on the aircraft will be carefully assessed by the customer. It is therefore essential that the starting point for a good layout must be the detail design of the passenger cabin. Although ticket price and journey time may be the main factors in choosing to make the flight, the passenger when boarding the aircraft is making secondary judgements on the comfort and standard of service offered by the airline. The airline therefore pays particular attention to the interior design of the cabin. The aircraft manufacturer will normally supply the new aircraft without cabin decoration. Each airline arranges to fit-out the interior of the passenger cabin to match their corporate identity and style.

The passenger will assess the environment aboard the aircraft in different ways.

Emotional

- Does it meet with expectations and is it aesthetically pleasing?
- Does it feel friendly, efficient and safe?

Physical

- Is it tidy?
- Is the air conditioning efficient?
- Is the cabin odour free?
- Are there non-smoking areas?
- Is it reasonably quiet?
- Is it vibration free?
- Is it claustrophobic?
- Is my personal space intruded?

Spatial

- How big is the cabin?
- What is the seating arrangement?
- How close are other people?
- Will I be disturbed by other passengers?
- Can I move from my seat without disturbing others?
- How convenient is the carry-on baggage facility?

To minimise boredom during the flight the cabin must allow:

- various types of entertainment
- comfortable conversation
- undisturbed reading and the possibility to write
- convenience when eating and drinking
- the possibility to sleep and relax

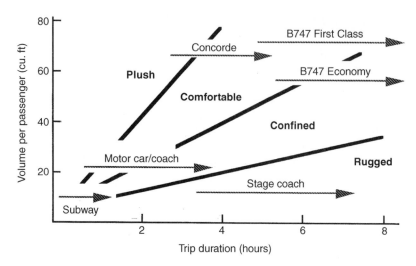

Fig. 5.1 Perceived passenger comfort with journey time.

All these requirements and expectations must be met within the technical constraints of the aircraft. A classical compromise must be struck between the provision of sufficient space and cabin service to make the passengers feel comfortable during the flight and the minimisation of the fuselage size to reduce structural weight and aerodynamic drag. Fortunately, the circular cross-section of the conventional fuselage matches the requirements of optimum enclosed volume, minimum structural weight and minimum wetted area (i.e. lowest drag).

The amount of space required by travellers to feel comfortable is directly related to journey time. Several years ago, some American researchers conducted tests to find out how passengers perceived comfort levels relative to journey time. Their results are diagrammatically presented in Fig. 5.1.

From our own travelling experiences we know that we can tolerate cramped conditions for a short time but for longer periods we need more space. It is this effect that makes travelling in first class compartments feel so much more comfortable than in the cheaper tourist section. Likewise, for a similar level of comfort it is possible to provide less space to those travellers lucky enough to fly supersonically and thereby get to their destination in half the normal flight time.

The provision of cabin services is also dependent on journey time (Fig. 5.2). For short trips a comfortable seat may be all that is required but for long journeys recreation, entertainment and social space are necessary. Unfortunately, most of these passenger facilities add weight to the aircraft. This means that aircraft on long distance flights require more service weight provision that those on short journeys. As competition for revenue increases the requirement for comfort enhancing features will grow and this will need to be reflected in new aircraft designs.

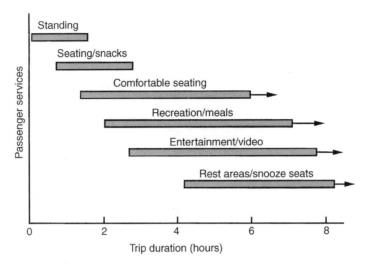

Fig. 5.2 Passenger service needs with journey time.

Passenger cabin layout

Two geometrical parameters specify the passenger cabin (diameter and length). Of these the cabin diameter is the most critical because once fixed it is impractical to change. On the other hand, part of the natural development programme of the aircraft will be to stretch the fuselage by adding extra length ('plugs') to increase capacity.

Cabin cross-section

The shape of the fuselage cross-section is dictated by the structural requirements for pressurisation. A circular shell reacts the internal pressure loads by hoop tension. This makes the circular section efficient and therefore lowest in structural weight. Any non-circular shape will impose bending stresses in the shell structure. This will add considerable weight to the fuselage structure. However, a fully circular section may not be the best shape to enclose the payload as it may give too much unusable volume above or below the cabin space. On some designs this problem has been overcome by the use of several interconnecting circular sections to form the cross-sectional layout. Figure 5.3 shows three non-circular fuselage shapes. The Boeing 747 (a) incorporates a smaller radius arc on the upper deck to provide adequate headroom to the top cabin. The upper and lower radii are blended with a short straight section near the upper floor position. In the proposal for a new large aircraft (b) the same principle is used but the blend is made with a circular arc section. A radical Airbus proposal (c) overcomes the double-deck problem by the use of a horizontal double-bubble. Structural and operational problems have led to most of the multi-arc sections being discarded. (The rectangular areas under the lower floors in Fig. 5.3 represent the cargo containers which must also be considered when deciding the best shape of fuselage section.)

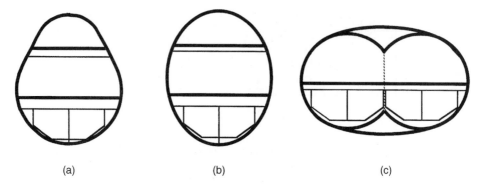

(a)   (b)   (c)

Fig. 5.3 Multi-arc fuselage sections.

To balance the hoop tension loads in the skin of multi-arc sections, the intersection points of the different radii are required to be connected to transverse members. In most cases this balancing structure is used for the floor support as shown in Fig. 5.3. Although the multi-arc sections may be regarded as feasible for the cylindrical cabin section care must be taken with the geometry as these need to be blended forward and aft into the nose and tail structures without causing drag raising profiles.

The determination of cabin cross-sectional shape and size is one of the first detail design studies to be undertaken on a new project. The overall size must be kept small to reduce aircraft weight and drag, yet the resulting shape must provide a comfortable and flexible cabin interior which will appeal to the customer airlines. The main decision to be taken is the choice of number of seats across the aircraft and the consequential aisle arrangement. For a specified number of passengers, the number of seats across will fix the number of rows in the cabin and thereby the fuselage length. The length-to-diameter parameter (sometimes referred to as the fuselage fineness ratio) is an influential factor in the design of the fuselage, as shown in Fig. 5.4. A low ratio (leading to a stubby fuselage shape) will result in a drag penalty but will offer potential for future stretch. A high fineness ratio gives a long thin tubular structure which may suffer from dynamic structural instability and will restrict future developments (stretch).

Selection of fuselage diameter is further complicated by the need to provide different ticket classes in the aircraft. The 'first-class' passenger will be given (and expect) more personal space, larger seat dimensions and better services than the cheaper 'business-class' passenger who in turn will expect more than is provided for the 'economy-class'. Three-class layouts are common in most scheduled long-haul services and two classes are provided on short to medium range flights. For commuter, leisure and charter operation single class layouts are commonly used. Charter operators often adopt layouts that are more cramped than scheduled economy-class services. In general, first- and business-class compartments will have fewer seats across the aircraft and more aisle width than the economy or charter layouts. In order to maximise the market for the aircraft the cabin designers need to consider the sectional layouts for each class plus the opportunity to meet charter configurations. Once the fuselage section is fixed the operational

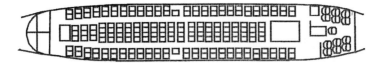

8 abreast layout (210 seats)

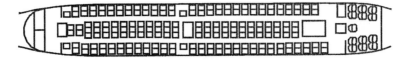

7 abreast layout (210 seats)

6 abreast layout (210 seats)

Fig. 5.4 Effect of cabin layout on fuselage fineness ratios.

layout options available within the cabin section are determined. Aircraft manufacturers are careful to take advice on the operational layout requirements from potential customers before selecting the fuselage cross-sectional shape as once it is fixed it is difficult to alter. The review of existing cross-sectional shapes shown in Fig. 5.5 may be used to suggest the first guess of the fuselage sectional layout.

Note, in Fig. 5.5 the larger aircraft use the space below the floor to hold freight containers. The size of these containers is standardised to allow use on different aircraft types. Some current layouts are shown in Fig. 5.6.

To load freight and baggage into the aircraft the goods are packaged onto pallets or put into containers. The major airlines use containers that are designed for their own aircraft and ground handling equipment. The most common arrangement puts the cargo in the lower holds (below cabin floor), as shown above but cargo can be loaded onto the main deck as an alternative to the seats as shown in Fig. 5.7.

Loading into the lower deck holds may require containers that are shaped to give clearance to the sides of the lower fuselage profile as shown in Fig. 5.6. The International Air Transport Association (IATA) has specified sizes for standard containers. Some of these specifications relate to the frequently used LD-designations shown in Table 5.1. Containers LD-1, LD-2, LD-3, LD-4 and LD-8 are the most common types. Data from this table can be used in the layout of the fuselage shape and to predict cargo capacity in the design of a new aircraft.

The number and type of containers to be carried may form part of the payload

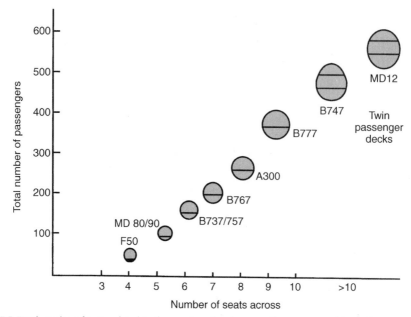

Fig. 5.5 Total number of seats related to the number of seats across the economy cabin section.

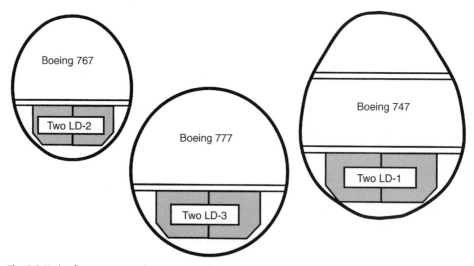

Fig. 5.6 Under-floor cargo container space requirements.

specification. The minimum number will correspond to the baggage allowance associated with the passengers. It is necessary to provide a volume of $0.125\,\text{m}^3$ per seat (this is equivalent to the normal 20 kg baggage allowance for the tourist class). The airlines will want to use the aircraft to carry cargo as well as passengers so careful consideration must be given to the provision of enough space to match the cargo requirements for potential airlines.

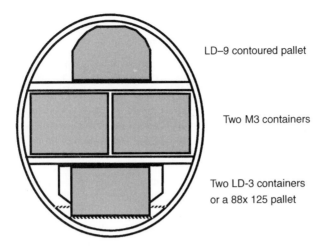

LD–9 contoured pallet

Two M3 containers

Two LD-3 containers
or a 88x 125 pallet

Fig. 5.7 Cargo version of new large aircraft (source MD 12).

Table 5.1 Standard sizes for freight containers (source Boeing)

Designation	Width	Height	Depth	Base	Maximum load (lb)	Notes
LD-1	92.0	64.0	60.4	61.5	3500	Type A
LD-2	61.5	64.0	60.4	47.0	2700	Type A
LD-3	79.0	64.0	60.4	61.5	3500	Type A
LD-4	96.0	64.0	60.4	–	5400	Rectangular
LD-5	125.0	64.0	60.4	–	7000	Rectangular
LD-6	160.0	64.0	60.4	125.0	7000	Type B
LD-7	125.0	64.0	80.0	–	13 300	Rect/Contoured
LD-8	125.0	64.0	60.4	96.0	5400	Type B
LD-9	125.0	64.0	80.0	–	13 300	Rect/Contoured
LD-10	125.0	64.0	60.4	–	7000	Contoured
LD-11	125.0	64.0	60.4	–	7000	Rectangular
LD-29	186.0	64.0	88.0	125.0	13 300	Type B

The size of passenger seats used in the aircraft is the choice of individual operators. The widths shown in Table 5.2 are representative of current practice.

Airworthiness regulations specify minimum aisle width (FAR 25.815 quotes: minimum aisle widths of 15 in (381 mm) and 20 in (508 mm) respectively below and above a reference height of 25 in (635 mm) above the floor) but most airlines will

Table 5.2 Typical seat widths

Class	Seat width (mm)	Seat width (in)
Charter	400–420	16–17
Economy	475–525	19–21
Business	575–625	23–25
First	625–700	25–28

(For reference a public service bus seat is approximately 425 mm [17 in] wide.)

fix aisle widths greater than this to ease congestion during cabin servicing. Figure 5.8 illustrates various layout options to suit a 229 inch (5817 mm) floor width (ref. Boeing 777).

Some airlines install a seat which is designed to be convertible from a double in first/business layout to a triple unit for economy. In these cases the aisle width will be dictated by the non-tourist configuration. The advantage of this type of seat lies in the flexibility it provides to change the ratio of seats in the various classes. Airlines require different proportions of business and economy seats to suit seasonal variations.

Apart from the first-class arrangement, the layouts in Fig. 5.8 do not show the cabin profile at the side of the outer seats. Providing adequate head-room for the passenger in the window seat places a further constraint on the shape of the section. For double-deck layouts in which the upper floor is positioned substantially above the centre of the circular section, or for those designs requiring large under-floor cargo holds, the fuselage profile will be inclined at the side of the window seat making it necessary to position the outer seats away from the edge

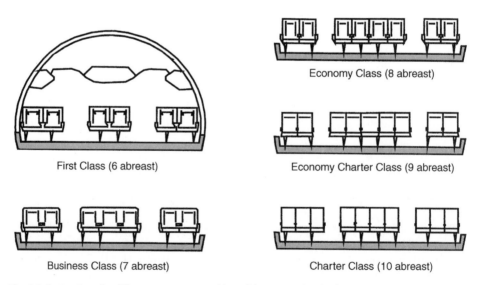

First Class (6 abreast)

Economy Class (8 abreast)

Business Class (7 abreast)

Economy Charter Class (9 abreast)

Charter Class (10 abreast)

Fig. 5.8 Seat options for different compartments (classes) (source Boeing data).

Conventional overhead lockers

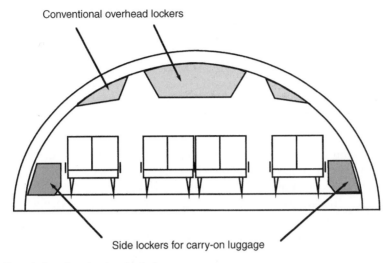

Side lockers for carry-on luggage

Fig. 5.9 Upperdeck seating showing side lockers.

of the floor, as shown in Fig. 5.9. On some designs the space generated by moving the seat away from the fuselage side allows extra floor mounted storage lockers to be positioned to the outside of the window seats.

Cabin length

Once the cabin cross-section has been decided, the number of seats across the cabin will be fixed. Dividing this number into the total number of seats in each class gives the average number of rows of seats to be installed. The required cabin length will be related to the leg-room provided for each class. For well designed seats this is related to the seat pitch (see Fig. 5.10).

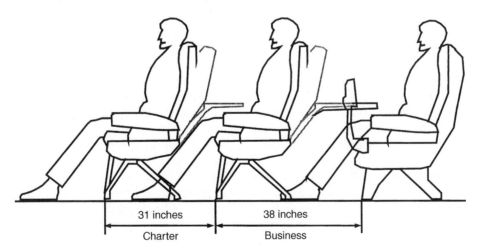

31 inches

38 inches

Charter

Business

Fig. 5.10 Seat pitch options.

Table 5.3 Typical seat pitch

Class	Seat pitch (mm)	Seat pitch (in)
Charter	700–775	28–31
Economy	775–850	31–34
Business	900–950	36–38
First	950–1050+	38–42+

The perception of comfort is directly linked to the seat pitch and the number of seats in a unit (a single seat requires less leg length than a double, etc.). It is obviously impractical to make the seat pitch variable with the actual seat unit as airlines prefer straight rows across the cabin to simplify servicing (passenger management, serving food, cleaning, etc.). The seat pitch is chosen by the operator within the ranges shown in Table 5.3 (for comparison bus/coach seats are pitched at about 725 mm).

The number of seats to be provided in each class is dependent on the type of operation and the demand for tickets in each class. The type of operation may vary due to seasonal demand for air travel. In the summer months there will be more demand for the cheaper seats from the holiday market, in the winter the total demand may fall but the business demand will not reduce proportionately. This means that the split of seats between the various classes varies throughout the year. Airlines may need to reconfigure the cabin layout to suit the demand pattern (e.g. winter and summer operations). Seats and internal partitions must be easily moved to avoid long change-over times. Rails running along the length of the cabin floor are used to hold the seat and partitions with sufficient security to meet the crash load conditions specified in the airworthiness regulations. All the various seating arrangements must be consistent with the lateral position of these rails, as shown in Fig. 5.8.

The total number of passengers that can be accommodated in the cabin will be limited by the aircraft maximum take-off weight limit. This is set by structural, aerodynamic and performance criteria and not related to the interior arrangement. Also, the cabin capacity may be limited by the type and number of emergency exits provided (e.g. see FAR 25.807 for full details). This may also restrict the maximum number of seats in the high-capacity charter role.

In order to fix the initial layout, an estimate of the proportion of passengers in each class has to be made. Evidence should be collected from aircraft of the same type and operated on similar routes as the proposed design. Typically for a three-class layout there will be 8% first, 13% business and 79% economy seats.

Different arrangements of seats will be possible within a fixed cabin length. It is important for the project design team to identify the special requirements for each airline to satisfy the overall layout. Figure 5.11 shows how a fixed cabin length can be arranged to give different seating for three-, two-, and single-class options. Note how the fixed cabin facilities (toilets and galleys) in this design have been cleverly positioned to act as partitions between the different compartments. More space is needed for the business/first-class accommodation and this will reduce the total number of seats that can be fitted into the fuselage. In turn this will

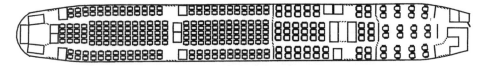

300 seats in three-class arrangement (three class at 60–38–32 pitch)

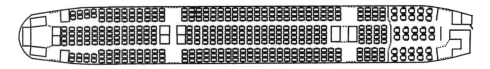

400 seats in two-class arrangement (two class at 38–32 pitch)

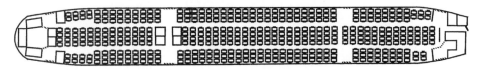

440 seats in a single class (single class at 32 pitch)

Fig. 5.11 Cabin layout options.

affect the revenue potential of the flight although the first- and business-class seats will attract a premium on the ticket price. The layout options will be carefully studied by the airlines to find a match for their market variations.

In the layouts in Fig. 5.11, the 300 and 400 seat aircraft will permit more fuel to be loaded (up to the design maximum aircraft take-off weight). This will allow the aircraft to fly further. This trade-off between seats and range is one of the main parameters that the airline will use to assess the suitability of the design to their route patterns. The balance between passengers and range forms the basis of the payload–range diagram for the aircraft as shown in Fig. 5.12.

Extra seats must be provided for the cabin attendants. The number of attendants is left to the discretion of the airline but must be sufficient to control passenger evacuation in an emergency and satisfy the licensing authority. The number of stewards is set by the airline to provide prompt service to the passengers. Typically, one attendant per 30–40 passengers is chosen for the economy class, one for 20–25 in business and one for 10–15 in first class. The attendants will be provided with 'flip-up' seats for use during the take-off and landing phases. These are generally positioned in the vicinity of the emergency exit and on other doors.

Service facilities (including galleys, toilets and wardrobes) must be provided in the cabin layout. These must be positioned to suit the proposed seating layouts. The galley and toilet units require built-in services for electricity, water and waste management. These facilities will need to be serviced during the aircraft turn-round

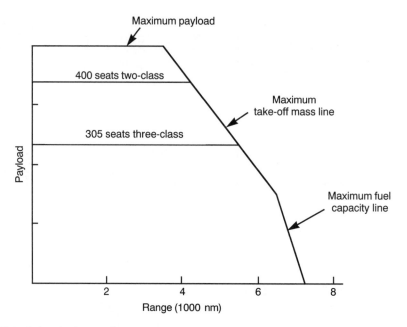

Fig. 5.12 Typical payload–range diagram.

phase. This will dictate the position of external access (doors and panels). It is not possible to quickly alter the position of these facilities although some units may be interchangeable. The provision of galley and toilet facilities is left to the operator. The number of each type is matched to the passenger capacity. For example, short-haul flights will require less galley service than long-haul flights. For an aircraft used for both long-haul and short-haul operations there will be more seats in the short-haul configuration and this will set the requirement for toilets and possibly galleys. The number of passengers for each facility is related to the ticket class with typically between 10 and 60 passengers for each galley and 15–40 passengers for each toilet (the lower numbers for first-class accommodation). The position of these units must not interfere with movement of passengers during loading or disembarkation. Typical sizes of the galley and toilet areas [30 × 36 in (762 × 914 mm) and 36 × 36 in (914 × 914 mm) respectively] are shown in Fig. 5.13.

Cargo space

Although the accommodation of passengers is the principal concern of the fuselage designer, it is also important to provide sufficient and convenient cargo space. In some configurations a mixture of freight and passengers is accommodated in the main cabin in separate sections. This is known as a Combi layout, shown in Fig. 5.14. Such a layout require a large freight door to access the cabin cargo area.

In most designs, freight will be housed under the floor of the passenger cabin,

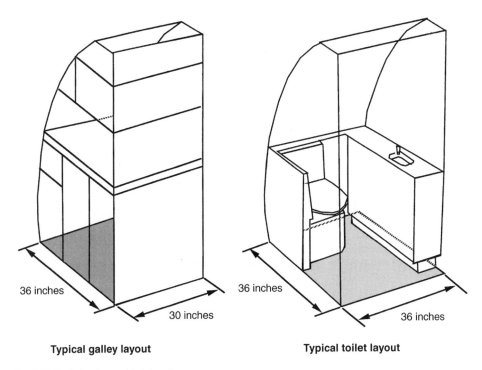

Typical galley layout

Typical toilet layout

36 inches

30 inches

36 inches

36 inches

Fig. 5.13 Typical galley and toilet units.

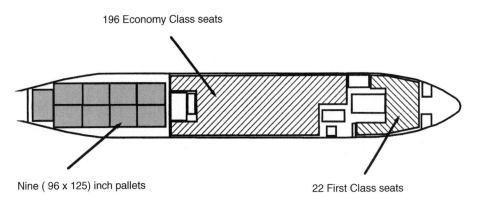

196 Economy Class seats

Nine (96 x 125) inch pallets

22 First Class seats

Fig. 5.14 'Combi' layout (source MD 12).

as shown in Fig. 5.6. The complete specification will include the disposition of cargo in the front and rear holds. A typical example is shown in Fig. 5.15. Large access doors will be needed to get the freight and luggage containers into the holds.

For short-haul types the aircraft turn-round time must be minimised to reduce block time and improve direct operating costs. The overall arrangement of the various panels and access doors around the fuselage must be considered in relation

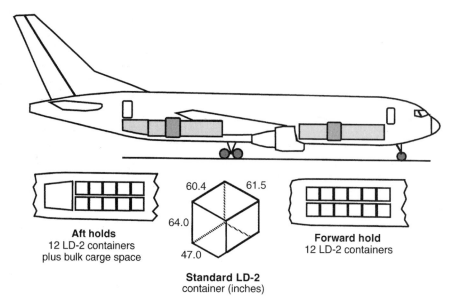

Aft holds
12 LD-2 containers
plus bulk carge space

60.4   61.5

64.0

47.0

Standard LD-2
container (inches)

Forward hold
12 LD-2 containers

Fig. 5.15 Freight layout (source Boeing 767).

to the airport services and the management of aircraft turn-round. The size and positioning of doors must suit airport ground equipment geometry. It is usual to provide passenger doors on the left (port) side of the aircraft which leaves access on the right side for aircraft servicing (catering, toilet cleaning, baggage handling, etc.).

Fuselage geometry

Once the internal details have been settled it is necessary to complete the external shape of the fuselage. The passenger cabin and associated under-floor cargo holds will be contained in a cylindrical shell. The external skin will need to be slightly larger in size than the interior space to allow for the fuselage structural framework, internal decoration and soundproofing panels. A thickness of 100–140 mm will be added around the internal profile to provide space for the fuselage structure, etc. The cylindrical section will be streamlined (faired) in front and behind the cabin. The addition of these shapes to the central section will complete the fuselage profile.

Front fuselage

At the front of the cabin the fuselage has to be streamlined to reduce drag. The front fuselage accommodates the forward-looking radar in the nose section, the flight deck with associated windscreen, and the nose undercarriage. The flight deck layout naturally revolves around the working environment for the flight crew.

Anthropometric data for flight crews has provided the basis for the arrangement of pilots' seats, instruments and controls. Apart from a comfortable seating arrangement, the flight deck must be suitable for the display of information to the pilots and to allow them to make the appropriate control responses to fly the aircraft safely. The development of electronic displays has transformed the traditional layout of the flight deck. Figure 5.16 shows a typical arrangement. The overall length of the flight deck varies according to aircraft type from about 110 inches (2.75 m) for smaller aircraft to 150 inches (3.75 m) for larger types. The larger space can accommodate an extra (third) flight crew member if required by the operators. The pilot's seat and controls must be adjustable to suit pilots between 5 ft 2 in (1.55 m) and 6 ft 3 in (1.88 m) in height and thereby to allow the pilot's eye to be positioned at the vision datum. The aircraft must be capable of being flown from either pilot seat position; therefore the windscreen and front geometry will be symmetrical about the aircraft longitudinal centreline. The pilot must have good visibility, in all flight and ground manoeuvres. He should be able to see below the horizon in the approach attitude and at least 10° below the horizon when climbing. In turning flight he should be able to see upwards at about 20° and sideways 110°. On the ground the pilot must be able to see the aircraft wing tips (albeit this may mean that he has to lean forward and sideways). The arrangement of instruments and controls is standardised to avoid pilot confusion in an emergency but with the introduction of synthetic displays some of the detailed arrangements are subject to airline preferences.

The nose wheel and leg will be stored (retracted) into a non-pressurised bay below the flight deck or on small aircraft partially between the pilot seat positions. Forward of the pressure bulkhead an equipment bay is provided. This will include the forward-looking weather radar system and storage for other avionics equipment. Access to this space will be necessary to service the systems. An external door or panel is usually provided and the most forward section is made from radar transparent material.

Modern 'glass' cockpit displays and side-stick controllers have transformed the layout of the flight deck from the traditional aircraft configuration. Airworthiness regulations specify the external vision requirements and thereby the window layout necessary for the fuselage profile. The front fuselage profile presents a classical design comprise between a smooth shape for low drag and the need to have flat sloping windows to give good visibility. Although the airworthiness requirements for the window area are the same for all aircraft, the designers have sufficient creativity to make the front fuselage profile distinctive between different manufacturers.

The layout of the flight deck and specified pilot window geometry is often the starting point of the overall fuselage layout.

Rear fuselage

The rear fuselage profile is chosen to provide a smooth, low drag shape which supports the tail surfaces and in some configurations the rear engine installation. The lower side of the profile must provide adequate clearance for aircraft to rotate

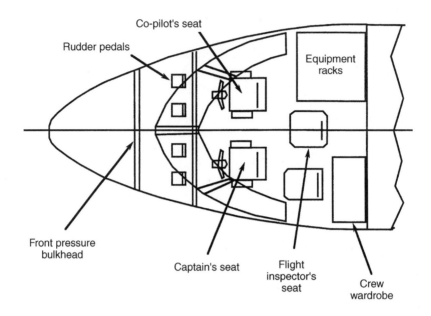

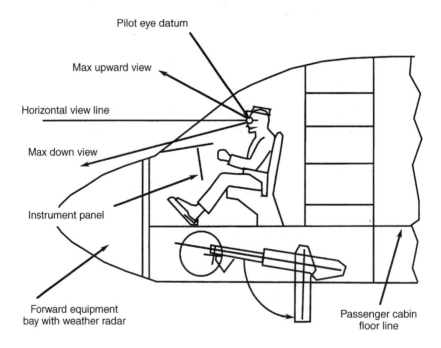

Fig. 5.16 Flight deck layout.

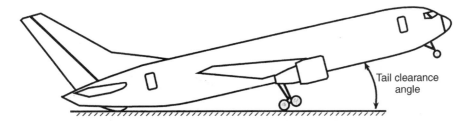

Fig. 5.17 Rear fuselage geometry.

on take-off as shown in Fig. 5.17. Initial flight tests will be used to demonstrate that maximum rotation angles are possible at the minimum unstick speed.

In most designs the cabin layout will be arranged to extend into the rear fuselage, terminating in a hemispherical pressure bulkhead. The reduced cabin dimension in this area means that the standard seat arrangement may need to be altered. Often services (e.g. galley, toilets and storage units) are positioned in the area.

In aircraft configurations which require access to the fuselage for freight loading the rear and front fuselage profiles and the associated structural framework will be compromised by the installation of large doors and loading ramps.

Airworthiness

For commercial operation most of the airworthiness regulations are aimed at preventing accidents. However, even with the best efforts of aeronautical engineers and operators the aircraft may crash. In such an event it is the responsibility of the fuselage designers to ensure that the passengers are given protection during and immediately after the accident and that they can quickly and safely evacuate the fuselage. The designers have often more than one option for the design of the aircraft. Crashworthiness and evacuation criteria must be considered when making decisions on fuselage layout (see also Chapter 4).

Aircraft accidents vary from non-survivable crashes (e.g. mid-air collisions) to small knocks against airport equipment on the airport stand. Apart from careful air traffic control and by making aircraft more visible there is little that the designers can do to protect the passengers in non-survivable crashes. Fortunately such accidents are rare and most crashes are survivable. In such accidents the deceleration forces experienced by the occupants are tolerable and the damaged fuselage structure and equipment does not cause injury to the occupants. Improving crashworthiness of the aircraft will increase the severity of the crash which will be survivable.

Crashworthiness

One of the main requirements for crashworthiness is to provide a protective envelope around the passengers. The stressed skin pressurised cylindrical fuselage usually contains enough strength to provide this feature. However, crash dynamics often exhibit low sink rates and either nose-up or tail-down impact with the ground. This causes substantial bending forces along the length of the fuselage. These frequently fracture the shell at sections with large shear inputs (e.g. wing mountings) breaking the fuselage into several sections. To offer the best protection to passengers it would be advisable to:

- avoid seating in the areas that are likely to fracture in a crash;
- avoid (if possible) seating passengers over those areas with stiff understructure (e.g. wing/undercarriage) as these areas will have less energy absorbing characteristics;
- reinforce the fuselage section below a high wing layout but ensure a crushable (energy absorbent) structure is maintained;
- increase space between seats but ensure that the passenger is securely held into the seat and that other items do not intrude into the seating area;
- use aft-facing seats or use upper torso support to restrain head, neck and body;
- avoid seating passengers in areas likely to be intruded by components in an accident (e.g. propeller/fan disc, engine nacelle undercarriage);
- use seat structures that are strong enough to meet the crash loading specified in airworthiness regulations but also have controlled collapsible features to reduce the peak acceleration felt by the occupant.

Obviously, some of these guidelines must be compromised in the light of the overall design of the aircraft and other safety considerations.

Apart from the crash considerations described above the interior of the aircraft must offer a safe environment to the passengers. This means that all loose equipment (including carry-on luggage) must be stored in lockable cupboards, that the interior trim is fire and smoke resistant and that essential safety equipment (e.g. oxygen, life rafts, evacuation chutes) is provided. If possible the aircraft structure and configuration must avoid the risk of fuel spillage in the crash especially if it could get into the cabin area or if it makes the escape routes dangerous.

Assuming that the occupants have survived the crash, the fuselage layout must also allow safe and rapid evacuation. The criteria are described below.

Evacuation

Emergency evacuation of the cabin plays an important part in deciding the fuselage layout. Before a certificate of airworthiness is granted to the type, the manufacturer will be expected to demonstrate to the airworthiness authorities that all occupants can vacate the aircraft in 90 seconds or less using the emergency equipment normally carried.

The airworthiness regulations state precisely the minimum number and type of emergency exits to be provided in the cabin structure. These must be positioned each side of the fuselage. Doors used for services (e.g. galley replenishment, toilet servicing, etc.) may be classed as emergency exits providing they can ensure unobstructed access during emergencies. The position of these doors for emergency evacuation may fix the location of the service units in the cabin. Exits not used for passenger loading or service (e.g. over-wing exits) will also require extra space to avoid congestion during evacuation. Cabin windows are often placed in emergency exit structures. The need to make more space available in these areas conflicts with equal seat spacing in the cabin.

(Note: in general, the windows are positioned to suit the fuselage structural frame geometry and not chosen to match the seat row geometry as the internal arrangement of seats may be changed to match different operations: e.g. seasonal variations in traffic.)

Doors and emergency exits are heavy and complicated structures. Designers will specify the number, position and sizes of these exits at the preliminary stage of the project. As mentioned earlier, the cabin layout options may be limited to match the number and type of emergency exits provided (e.g. see FAR 25.807 for full details). The minimum number of exits to be provided is related to the maximum number of seats.

The airworthiness regulations set out the precise requirements. In these regulations the type of exit is specified according to the size of the opening (inches):

Type I   24 wide × 48 high (610 × 1219 mm)   Type II   20 wide × 44 high (508 × 1118 mm)
Type III   20 wide × 36 high (508 × 914 mm)   Type IV   19 wide × 26 high (483 × 660 mm)
Type A   42 wide × 72 high (1067 × 1829 mm)

(The Type A exit is equivalent to a passenger or service loading door.)

Table 5.4 (from airworthiness regulations) specifies the minimum number and type of exits to be provided on each side of the fuselage. For capacities greater than 179, a pair (one each side of the aircraft) of additional Type A exits will allow an extra 110 seats, a pair of Type I exits will allow an extra 45 seats. Above 300 seat capacity all exits must be of Type A; for example, a new large aircraft with 600 seats will require six Type A exits on each side of the fuselage.

As mentioned above, to give uncongested access to these doors in an emergency, the internal cabin layout near these exits must provide more space than in other parts of the cabin.

Table 5.4 Requirements for emergency exits (source FAR/JAR)

SEATS	EMERGENCY EXITS			
Less than	Type I	Type II	Type III	Type IV
10	–	–	–	1
20	–	–	1	–
40	–	1	1	–
80	1	–	1	–
110	1	–	2	–
140	2	–	1	–
180	2	–	2	–

Systems

Human life cannot be naturally sustained at the normal cruising altitude for jet transport aircraft. Environmental systems will be essential to make the passengers safe and comfortable. Passengers need protection from outside air which is at a temperature of minus 50°C or less, a pressure less than 30% of the sea level value, and with insufficient oxygen to sustain life. It would be possible to pressurise the fuselage to an equivalent sea level condition but this would impose high hoop tension loads in the fuselage skin (requiring a thicker and heavier shell structure). To reduce these loads a compromise is necessary. The cabin environment is conditioned to a higher equivalent altitude than sea level. This reduces these stresses. The equivalent altitude is kept low enough to make the passengers feel comfortable. For most commercial flights the cabin height is set at an equivalent altitude of about 8000 ft (as shown in Fig. 5.18). The structure will be designed to a specified differential pressure loading and the pumping system will have a maximum pressurisation/depressurisation rate.

The passengers will require an air conditioning system that will provide comfortable room temperature and freshness in the cabin. When the aircraft is in the cold environment of the high altitude the conditioning system will bleed hot air from the engine compressors to use as the heat source. On the other hand, when the aircraft is on the ground in hot climates, the cabin conditioning system will need to extract heat to cool the cabin environment.

The cabin services put a high demand on the aircraft electrical system. The provision of lighting, illuminated signs and entertainment throughout the cabin requires careful positioning of electrical looms and internal panelling.

Carry-on baggage requires either overhead storage lockers or other cupboards. Passengers require easy access to these areas during the flight but they must be securely locked during take-off and landing and in emergency conditions.

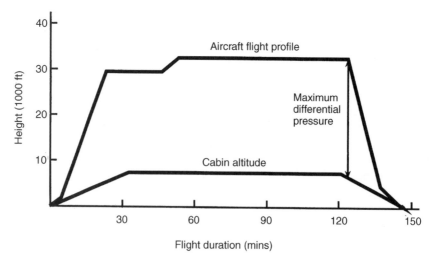

Fig. 5.18 Cabin pressurisation requirements.

The type/aesthetic of internal decoration for the cabin is an airline option. The operators purchase the aircraft with the bare internal fuselage. In most cases specialist companies design and fit-out the cabin to the aesthetic choice of the airline. The cost of furnishings is expensive (6–9% of the bare aircraft price) and the weight of components adds significantly to the aircraft empty weight. The internal trim consists of plastic and composite panels and structures which must look nice, be hard-wearing and rot proof, but above all, be fire proof and smoke resistant. Passenger seats are also supplied by specialist companies. They are designed to meet airworthiness crash resistance loads and other safety criteria. They also provide stowage for safety equipment, folding table, entertainment facilities (audio and video) and passenger service controls.

Fuselage layout exercise

Assume that you are working in the advanced project office and your manager has asked you to size the fuselage of a new 300 seat aircraft. The marketing strategy for the new aircraft is to offer more comfort to the passenger than is given by current designs. The aircraft is designed to travel on medium to long stage routes having a maximum range of 4000 nm with full payload.

Answer

From Fig. 5.5 a conventional 300 seat aircraft would have nine seats across in the economy configuration. A typical layout for nine abreast with twin aisles is shown in Fig. 5.8 (a 2–5–2 arrangement of seats). From the same diagram a more comfortable seating arrangement would offer eight abreast (2–4–2) in the business class. This avoids the undesirable middle seat in the five seat unit for the conventional economy configuration.

The nine abreast seating in the 229 in (5817 mm) width layout shown in Fig. 5.8 provides only a 19.5 in (495 mm) aisle width. This is regarded as somewhat tight for a comfortable cabin; therefore the minimum aisle width will be increased to 20.5 in (521 mm). This increases the overall internal width of the cabin to 231 in (5867 mm). The fuselage structural framework must enclose the cabin. Allowing for the necessary shell structure and the internal decorative panelling a cabin wall thickness of four inches is required. This gives an external diameter of 239 in (6.07 m). From published data of other aircraft it is possible to compare fuselage diameters, as shown in Table 5.5.

Our chosen cross-sectional layout is shown in Fig. 5.19. The business/first seats are each 29 in (737 mm) wide and the aisles are 28.5 in (724 mm). The economy seats are 23.75 in (603 mm) wide and the aisles are 20.5 in (521 mm).

For a 300 seat aircraft about 20% of the seats will be in the non-economy class. As we are providing business-class space for the economy passengers we will need to upgrade the business section to six abreast seating (2–2–2) and to extend the facilities/space beyond current standards for the first-class customers.

Table 5.5 Comparison of aircraft fuselage diameters

	(m)	(in)
A300/330/340	5.64	222
B767	5.03	198
DC10	6.02	237
L1011	6.06	238
*IL96	6.08	239
*B777	6.20	244

* (These two aircraft have potentially a much larger passenger capacity than the example aircraft.)

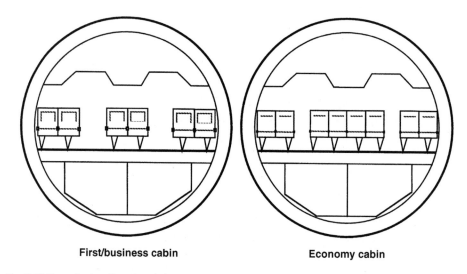

First/business cabin   Economy cabin

Fig. 5.19 Example aircraft section choice.

We can now determine the number of rows of seats required for each of the eight and six abreast sections of the cabin:

Economy-class seats (eight abreast) = 80% of 300 = 240 seats
(eight seats across the cabin gives exactly 30 rows)
first/business-class seats (six abreast) = 20% of 300 = 60 seats
(six seats across gives exactly 10 rows)

For this type of aircraft the normal seat pitch for economy class is 32 in (813 mm).

This will be increased to 36 in (914 mm) to match the project design philosophy.

Correspondingly, the business-class passengers would currently expect 38 in (965 mm) pitch. This will be increased to 40 in (1016 mm) pitch.

To provide a first-class compartment upgraded from the current standard, two rows of business-class seats will be extended to 60 in (1524 mm) seat pitch. This will

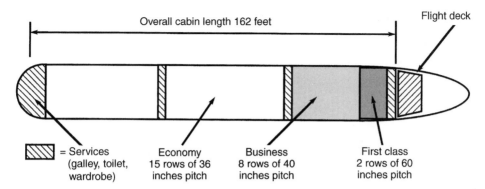

Fig. 5.20 Example aircraft cabin layout.

allow a reclining (sleeper) seat to be provided for these passengers. Hence the total length of the cabin seating is:

economy = 30 seat rows @ 36 in pitch = 1080 (27.43 m)
business = 8 seat rows @ 40 in pitch   =   320 (8.13 m)
first = 2 seat rows @ 60 in pitch   =   120 (3.05 m)
                                    total = 1520 (126.7 ft = 38.6 m)

The cabin length must include the service areas (catering, toilets and wardrobes) and some 'lost' space must be expected around entrances and emergency exits. From the layouts shown in Fig. 5.11, an estimate of 35 ft (10.5 m) is made for this passenger capacity. This makes the overall cabin length approximately 162 ft (49 m). We can assume that some of the extra cabin length is accounted in the non-cylindrical parts of the fuselage shape (i.e. forward and behind the main cabin section). We will assume that the total length is divided as shown in Fig. 5.20.

Forward of the cabin the front fuselage will house the flight deck, equipment bay, etc. This will account for approximately 20 ft extra length. The fuselage shape behind the cabin will be longer than the front section as it supports the tail surfaces. These need to be at sufficient distance from the wing to provide adequate stability and control. The fuselage shape forward and aft of the passenger cabin will add about 50 ft (15.24 m) to the fuselage length. The aircraft side view will therefore appear as shown in Fig. 5.21.

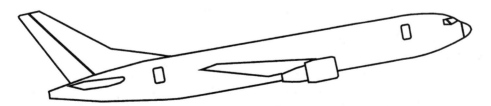

Fig. 5.21 Example aircraft fuselage profile.

Table 5.6 Comparison of aircraft fuselage lengths

	(m)	(ft)	Number of passengers (3 class)
A300	53.3	175	228
A330/340	62.5	205	295
B767	53.7	176	248
DC10	52.0	170	231
L1011	54.2	178	304
IL96	60.5	198	335
B777	62.8	206	310

All these aircraft have much shorter fuselage length than the example aircraft due mainly to the current (more confined) seating arrangements.

The overall length of the example aircraft is therefore expected to be:

$$(162 + 20 + 50) = 232\,\text{ft}\ (70.7\,\text{m})$$

From published data the fuselage lengths of some existing aircraft can be compared, as shown in Table 5.6.

The analysis above shows that it will be necessary to add about 8 m to the length of a conventional fuselage to give a better level of comfort in the cabin. This extension will have the effect of increasing the fuselage structure mass and wetted area. Both of these reduce the range the aircraft can fly within the maximum take-off weight. In the following chapters methods are presented that will enable you to quantify the mass and drag of the aircraft and to conduct performance calculations to determine this loss of range if you want to take this study project further.

6

Wing and tail layout

Objectives

This chapter describes how the wing and tail surfaces are defined by aero-dynamic, stability/control, layout and structural requirements. The size of the wing (area) will usually be dictated by aircraft performance requirements (e.g. field length) but the shape of the planform and other geometry may be influenced by wing layout factors.

In the early design stages choices need to be made on the position of the wing relative to the fuselage (e.g. high, mid or low position) and then on the overall envelope. This will include selection of aspect ratio, taper ratio, sweepback angle, thickness ratio and section profile, and dihedral. Each of these decisions is explained so that you could set out the shape of your preferred planform. Civil aircraft normally have flaps and high-lift devices installed in the wing geometry. Although the aerodynamic (lift and drag) characteristics of flaps are described in Chapter 8, a brief introduction is included in the section on flap design so that these can be added to your wing layout.

This chapter provides the baseline shape of the wing which will be refined by more detailed analysis later in the design process. The parametric design techniques by which this is done are explained in Chapter 13.

The shape of the tail surfaces (horizontal and vertical) are not only influenced by the same parameters as wing design but also need to provide safe handling qualities to the aircraft. To ensure such safety involves detailed stability and control calculations but these require a knowledge of the geometry of the tail surfaces. To start the process it is necessary to make a sensible choice of tail shape. This chapter describes how this may be done at the initial project design phase.

After reading this chapter, and with a knowledge of the chosen wing area, you should be able to size the wing and tail surfaces which together with the fuselage layout will allow you to make an initial drawing of the aircraft.

Wing design

In the conceptual design phase of the project an estimate of wing area and a crude guess at the wing shape are made. As the design progresses to the more detailed design phases it is necessary to carefully consider the wing geometry to obtain the shape which is optimum for the missions envisaged. Although the wing cannot be considered in isolation from the other components on the aircraft (e.g. engine installation, empennage layout, noise) it is essential to concentrate on the main wing parameters in the early stages of the layout and to consider the overall factors that are dictated by the aircraft configuration/specification later in the design process.

The main wing design parameters can be identified under four headings:

- performance requirements
- flying qualities
- structural framework
- internal volume

Performance requirements

The performance requirements will have to be considered in the conceptual phase. The size of the wing necessary to meet the mandatory (airworthiness) requirements and the design (specification) requirements will need to be evaluated. Certain climb rates and operating speeds will be laid down in the airworthiness requirements. The design requirements will specify the field length and cruise speeds.

Flying qualities

Ensuring that the aircraft flight handling qualities are acceptable will affect the choice of wing geometry (e.g. wing planform will dictate the stall and post-stall behaviour of the aircraft at low speed). The design requirements will specify the field length and cruise speeds. At high speed, aeroelasticity and aerodynamic buffet will be criteria to be considered. Vehicle ride will be affected by the gust responsiveness of the wing but this may be alleviated by automatic flight control systems coupled to wing surface controls. For control and stability, dutch roll and lateral response will be important parameters. Again these may be beneficially influenced by the aircraft automatic flight control systems. The influence of wing planform on flying quality is difficult to predict and often results in 'fixes' to correct inherent deficiencies.

Structural framework

The main criteria for structural considerations are safety and minimum weight. The wing structural framework must support all the non-wing components (e.g. engine, undercarriage), house all the flying controls and the high-lift devices (e.g. ailerons, flaps, airbrakes, etc.). Within all these requirements, the wing must be easy to manufacture and simple to maintain throughout its full service life.

Internal volume

The internal volume of the wing should be sufficient to hold the required fuel, (in separate tanks), the landing gear (if retracted into the wing profile) and the high-lift devices and other controls.

Wing layout

One of the first considerations on the conventional monoplane arrangement with which we are concerned is the wing placement relative to the fuselage. The choice lies between high-, mid- or low-wing positions. In each case there are structural and aerodynamic considerations along with effects on engine ground clearance, undercarriage layout, fuselage structure and cabin layout.

The various considerations that influence the choice of wing position are given below:

- Aerodynamic
- Structural
- Wing – fuselage attachment
- Effect on cabin
- Ground clearance of wing mounted engines
- Servicing of wing mounted engines
- Undercarriage configurations and installation
- Safety in the event of the aircraft striking the ground or ditching
- Effect on passengers.

The majority of turbofan powered civil transports have low wings. The advantages of this wing position are:

- Structurally it is simple, the main wing structural items can pass beneath the cabin floor thus making the wing-fuselage attachment comparatively easy.
- Undercarriage can be short and simple. The wing mounted engines and wing flaps are more easily reached from the ground for servicing.
- The wing, one of the strongest parts of the aircraft, shields the fuselage in the event of the aircraft striking the ground or water. In the event of ditching at sea, the wing also will act as a float thus keeping the fuselage out of the water.

The disadvantages of the low-wing layout are listed below.

- Aerodynamically, the wing upper surface is distorted thus affecting the ability of the wing to generate lift. Also, a suitable wing-fuselage fairing is required to minimise the interference drag.
- It is more difficult to install the engines such that there is sufficient ground clearance to avoid debris being ingested from the runway.
- In certain seats in the cabin the downward view is obscured by the wing.

The high-wing position relative to the low-wing is much better aerodynamically as the upper surface is undisturbed, although the wing–fuselage fairing is probably more difficult to manufacture. Structurally it is just as simple because the structure can pass above the cabin ceiling. The problems arise with the height of the engines and flaps above the ground for servicing, and the design and installation of the undercarriage. Most high-wing turbofan-powered aircraft have the undercarriage mounted on the fuselage. One advantage of the high-wing is that the aircraft centre of gravity is lower than the wing plane hence augmenting the natural stability of the aircraft.

The mid-wing position is aerodynamically easier to fair into the fuselage but otherwise is no better than the low-wing. Structurally it is much worse than either of the other two; the wing structure has to be terminated at the fuselage size and a joint made to a reinforced fuselage frame. With regard to the other considerations such as undercarriage layout and engine installation and servicing it lies between the high- and the low-wing.

For a particular design each of the wing positions must be considered on its merits. Except in the case of a military transport where the fuselage floor needs to be reasonably close to the ground to allow vehicles to be loaded using a ramp, the current view of the aircraft manufacturers for civil passenger aircraft favours the low-wing position.

Once the wing placement relative to the fuselage has been decided the next consideration is the geometry of the wing envelope. Values for the following geometrical characteristics must be selected.

1. Aspect ratio (span)
2. Taper ratio
3. Wing section profile
4. Wing/body setting
5. Thickness ratio
6. Sweepback angle
7. Dihedral angle

Each of these characteristics will now be described.

Aspect ratio

When the wing area is known, the selection of aspect ratio automatically sets the span loading:

$$\text{aspect ratio} = \text{span}/\text{mean chord} = \text{span}^2/\text{gross area}$$

As such it is influential in the generation of drag due to lift. This affects the overall climbing ability of the aircraft and its cruise efficiency. For example, climb performance with an engine failure will have to be guaranteed to meet airworthiness requirements. A wing planform with a low aspect ratio will have significantly more difficulty in meeting this requirement than one with a larger span. Obviously, a delicate balance has to be made between high aspect ratio designs leading to high-wing weight and long span, and low aspect ratio designs with lower wing weight but increased drag.

Considering the cruise condition, as aspect ratio is increased for a given wing area the overall drag of the aircraft will be reduced and the cruise fuel efficiency increased. This will lead to a reduction in the requirement fuel load to fly the range but this mass reduction will be offset by the increase in wing weight resulting from the larger span. Ultimately the choice of optimum aspect ratio will be determined by a trade-off study between aircraft wing mass and fuel mass leading to a total variation on aircraft all-up weight and the effect on operating costs. Long-range aircraft will benefit more from the fuel savings and therefore they may be expected to have a larger value for the aspect ratio. However, the combination of cruise efficiency and engine-out climb performance requirements (which are influenced by the number of engines in the aircraft layout) may also affect the preferred choice of aspect ratio. An educated guess based on current practice and the effect of any new technologies to be incorporated into the design can be made at the preliminary design stage. Typical aspect ratio values lie in the range 7–11 for commercial airliners.

When the aspect ratio is very low (less than 3) the wing shape will transform from the conventional trapezoidal wing into a delta planform and the airflow characteristics will change. Tip vortex flow will dominate and the wing aerodynamics will alter due to the high angle of attack and vortex flow conditions generated to achieve the required lift.

Taper ratio

The selection of taper ratio (i.e. tip chord divided by aircraft centreline root chord) involves several aerodynamic considerations. A constant chord rectangular planform may be easier and cheaper to manufacture than a tapered wing but it is aerodynamically and structurally less efficient. Aerodynamically an elliptical planform is regarded as ideal because it reduces the tip vortex effect. The geometry of some straight tapered layouts can be shown to approximate to the elliptical shape and obviously reduce the constructional complexity of a curved planform. Taper ratio in the range 0.4–0.5 can be shown to be only 2–3% less aero-dynamically efficient than the elliptical shape of equal area. The spanwise position of the centre of lift moves inboard as taper is increased. This reduces the root bending moment which consequently reduces the wing structural weight. The combined effects of increased aerodynamic efficiency and low weight tend to lead to a lower value for taper ratio (0.2–0.4) with a structural lower limit arising due to the need to provide sufficient wing tip stiffness to react forces from the aileron hinges.

The main drawback from low values of taper ratio comes from the fact that the tip sections are operating at a lower Reynolds number than the root sections. This will mean at high angles of attack the tip sections will stall before the root. A stalled tip will cause the aircraft to roll at the same time as losing lift (resulting in a nose-down pitch). These are classic symptoms for entry into a spin. The stalled tip flow reduces the aileron effectiveness making it more difficult for the pilot to lift the dropping wing to restrict roll.

There are a number of 'fixes' to counter the above stall/spin problems:

- progressively twist the wing section profile so that the tip sections are flying at a lower incidence than the root (called washout);

- progressively change the section profiles such that the tip section has a different thickness and camber to the inboard sections to delay the wing incidence at which the tips stall;
- introduce stall trip devices to the wing root sections (e.g. increased nose profile or stall strips).

All the above methods have the disadvantage of reducing aerodynamic efficiency of the aircraft in cruise and climb (i.e. at angles of attack lower than stall). Operational methods have been used on some aircraft to avoid the aerodynamic penalties. These include 'stick pushers' which are operated automatically at an angle of attack slightly lower than stall, and stall warning devices that are triggered by the detection of the stalled flow over the wing. These devices may be specified as 'advisory' in which case the pilot has the authority to disregard the warning, or may be classed as 'primary' in which case the stick pusher is actuated.

As in the case of aspect ratio, the adverse aerodynamic effects associated with increasing taper are counteracted by a reduction in wing mass due to the movement of the centre of pressure to the inboard sections and therefore a trade-off study is necessary to confirm or otherwise the choice of taper ratio. At this stage in the design process an initial choice can only be based on existing aircraft and the effect of any new technologies. It is in the selection of aspect and taper ratios that parametric trade-off studies become important (see Chapter 13).

In many wing planforms a double (or multiple) taper is used as shown in Fig. 6.1.

Such planforms may be chosen for essentially non-aerodynamic reasons (to reduce manufacturing complexity, to increase root thickness to provide storage depth for the main landing gear or to reduce wing/fuselage interference).

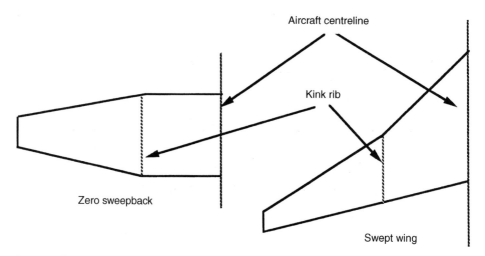

Fig. 6.1 Multi-tapered wing planforms.

Wing section profile

The selection of wing section profile is one of the most important decisions of the aircraft design process. It is extremely difficult to correct a poor choice as the whole of the wing geometry will be based on the chosen wing section co-ordinates. The aerodynamic characteristics of the section will be designed to provide an acceptable compromise between the operational requirements of the aircraft (e.g. high lift/drag ratio in cruise, good climb performance particularly in emergency flight conditions, good low speed lift, smooth high speed flight, delay of critical Mach number, good propulsion integration, etc.). The most critical operational aspects will take priority.

Traditionally, wing section data has been presented in a series of curves which show the principal aerodynamic characteristics (lift coefficient, C_L, drag coefficient, C_D, and aerodynamic moment coefficient, C_M) as shown in Fig. 6.2.

From a design point of view the following wing section parameters are of interest:

- maximum lift coefficient
- lift/curve slope
- incidence for zero lift

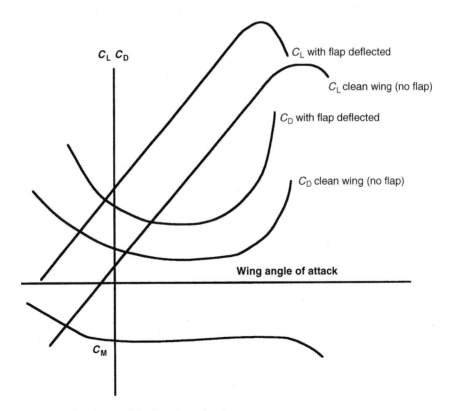

Fig. 6.2 Wing section characteristics (two-dimensional).

- drag/incidence ratio
- minimum drag incidence
- lift coefficient at minimum drag (maximum L/D)
- aerodynamic moment coefficient/incidence

Correction of the above sectional data for three-dimensional flow (e.g. accounting for aspect ratio, Mach number, Reynolds number and tip shape) is necessary. The methods used for such corrections are somewhat empirical and may lead to inaccurate estimates. It is therefore essential to conduct more detailed analysis and wind tunnel tests at an early stage in the design.

Modern computational fluid dynamic (CFD) methods now allow the full three-dimensional flow regime to be studied. It is possible to suitably modify wing section shape along the wing span to suit the critical conditions (e.g. engine pylon intersection). These methods are now used to specify the full wing surface geometry and to generate the aerodynamic parameters at various operating speeds. One of the benefits from this type of analysis has been the design of the wing profiles which delay the onset of critical Mach number. Such 'sections' are sometimes referred to as 'supercritical'. Compared with older aerofoil types, the advanced wing sections give the following options:

- at the same cruise Mach number enable the wing to have a higher thickness ratio or less sweep without detriment to the cruise drag; or
- at the same thickness ratio delay the drag rise to a higher Mach number.

Wing/body setting

One of the wing parameters which is dependent on sectional geometry is the wing incidence at cruise. It is important to ensure that the fuselage is at its minimum drag attitude at the mid-cruise condition. This condition dictates the wing/body setting angle. One of the characteristics of modern section geometry is the reflex curvature on the under surface towards the trailing edge. This shaping may not suit the flap profiles in this region and may require the section to be modified (i.e. flattening of the section at the trailing edge). The influence of flap geometry on wing layout is dealt with later.

Thickness ratio

The thickness (or thickness/chord ratio) of the wing section will often be decided by the total CFD wing design. Thickness is normally variable along the span to suit the local flow conditions. For structural (minimum weight) and volumetric criteria the thickness should be as deep as possible. Wing bending moment and shear force gradually increase from the tip to the root; therefore wing thickness is frequently chosen to be smaller at the tip and progressively increased along the span to the fuselage shear connections at the root. For subsonic aircraft an average thickness of 10% is typical. For supersonic aircraft wave drag is important. This can be shown to be proportional to the square of thickness ratio, therefore much thinner sections are specified (5–8% for such wings).

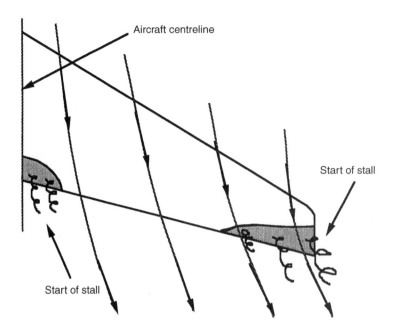

Fig. 6.3 Spanwise flow over swept wing planforms.

Sweepback angle

Sweepback is mainly used to reduce drag from local flow velocities at or close to supersonic speed. Sweeping the wing planform back allows thicker wing sections to be used and delays the onset of the critical Mach number. An equivalent effect can be shown for forward swept wings but these are intrinsically structurally unstable and therefore need a heavier structure to avoid undesirable aeroelastic effects. The main aerodynamic penalty from sweepback is the generation of spanwise flow over the wing planform as shown in Fig. 6.3.

The spanwise drift of the flow reduces lift, increases boundary layer thickness, increases drag, reduces aileron effectiveness and increases risk of tip stall. There are several 'fixes' which can be applied to reduce these effects. These include aerodynamic or physical fences to impose a straightening of the flow (engine nacelle supports (pylons) are sometimes used in this way). Creating a small vortex by kinking the wing planform can also be beneficial (e.g. multi-tapered planform shapes) as shown in Fig. 6.4.

Flap effectiveness is reduced by the swept trailing edge which reduces the maximum lift coefficient from the deflected flap. It is therefore common to make the trailing edge sweep much less than the leading edge.

Sweep has been used on some aircraft layouts to balance the wing centre of pressure with the aircraft centre of gravity, or to place the wing carry through structure at a more convenient section in the fuselage as shown in Fig. 6.5.

Adding sweep to the wing planform presents a major structural complication. The main wing spars, which are usually arranged to follow constant percentage

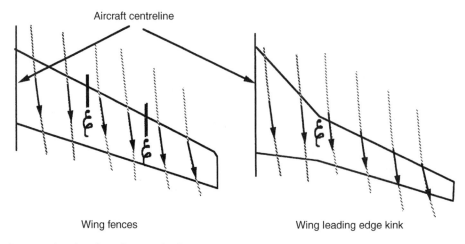

Wing fences   Wing leading edge kink

Fig. 6.4 Reduced tendency for spanwise flow.

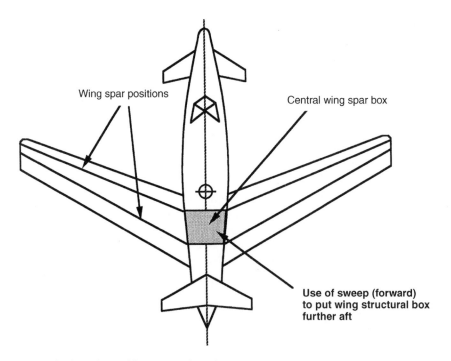

Fig. 6.5 Sweepback may be used for non-aerodynamic reasons.

chord positions and therefore run spanwise, will introduce a kink load at the fuselage side. This will necessitate a heavy bodyside rib to redistribute the loading. Also as outboard sections of the wing have a centre of pressure aft of the spar reaction points a swept wing is automatically loaded with an extra torque load (acting nose downwards for swept back wings and dangerously upwards for swept forward wings). This extra torque will require increased skin thickness to provide torsional stability to the wing section. The combination of the extra kink loads and the torque makes swept wings heavier than straight ones. This presents another 'optimisation' opportunity (sweepback reduces aircraft drag and therefore fuel required but increases aircraft structure weight).

Sweep angle is only matched to one flight condition and this may not be acceptable for aircraft with complex operational patterns. Due to the reduced lift potential for swept wings it would be an advantage if the sweep angle could be reduced in the take-off and landing phases (to increase flap effectiveness). To satisfy such design criteria it is possible to pivot the wing at the body sides and adjust the sweep angle to suit the flight regime (variable-sweep, swing-wing). Apart from the weight and complexity of this feature, the trailing edge (flap) and leading edge geometry is compromised by the overlapping of the fuselage bodyside structure. This feature has so far been used only on military aircraft because of the complex operational requirements. Up to now no civil aircraft has gone into airline service with a variable sweep wing but the early US supersonic civil aircraft proposals did use this layout to match the supersonic and subsonic cruise requirements.

Sweepback introduces two aerodynamic effects to the wing. Its primary purpose is to delay the drag divergence Mach number, but at the expense of a decrease in the maximum lift coefficient achievable by the wing. The choice of sweepback angle is connected to the type of wing section and particularly with the section thickness/chord ratio. For the same thickness/chord ratio, sweepback (Λ) will increase drag divergence Mach number (Mn) as follows:

$$(Mn_{sweep})/(Mn_{zero\ sweep}) = 1/\cos \Lambda$$

where Λ is taken as the spanwise quarter-chord sweep line. The result of sweepback on $C_{L_{max}}$ is as follows:

$$(C_{L_{max}})_{sweep}/(C_{L_{max}})_{zero\ sweep} = \cos \Lambda$$

Note: the general definition of drag divergence Mach number is somewhat ambiguous. Boeing define it as the Mach number at which the aircraft drag coefficient reaches a value 0.007 above that in the mid-subsonic flow region. Douglas defined it as the Mach number at which the shape of the aircraft drag coefficient versus Mach number reaches 0.10 from the subsonic flow region. Whichever method is used it is clear that the drag divergence Mach number will depend on the section profile, thickness/chord ratio and sweepback angle.

The drag divergence Mach number will be close to the normal operating Mach number as quoted in the aircraft specification. A statistical method based on the analysis of some modern aircraft is described below. This will enable a wing geometry to be selected at the preliminary design stage.

Figure 6.6 shows the relationship of Mach number to wing quarter chord sweepback. The line shows a (cosine)$^{-1}$ function.

Figure 6.7 shows the thickness chord at the wing root. This appears to be largely independent of Mach number with values between 14% and 15.3%.

From the two figures a choice of sweepback angle and root thickness can be

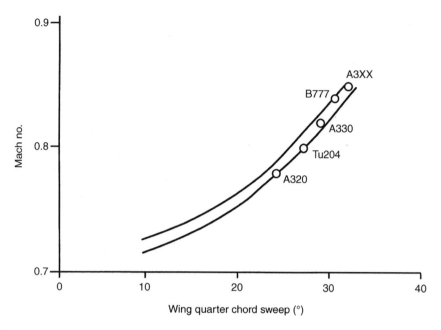

Fig. 6.6 Wing sweepback angle vs cruise mach number.

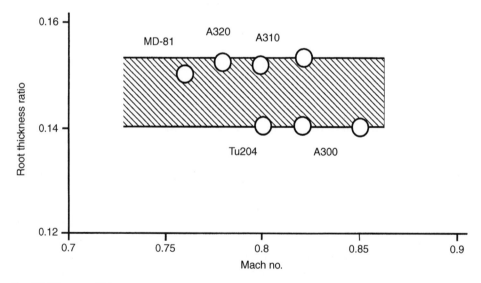

Fig. 6.7 Wing root thickness.

made. The formula below will enable the thickness of the outer wing to be calculated.

$$Mn = 0.877 - (1.387 \cdot T) + (0.431 \cdot \Lambda^2 \cdot 10^{-4}) + (0.1195 - 0.18 \, C_{L_{des}})$$

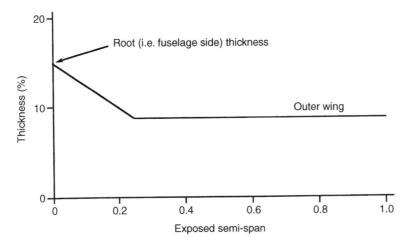

Fig. 6.8 Wing thickness spanwise distribution.

where: Λ = sweepback angle of the quarter chord in degrees
T = thickness ratio
$C_{L_{des}}$ = design lift coefficient found from Fig. 8.3

A typical spanwise thickness distribution is shown in Fig. 6.8. This can be used as typical for initial project studies. For the purposes of drag estimation an average thickness based on the formula below can be used.

$$(\text{average thickness ratio}) = \frac{(3 \times \text{outer wing value}) + \text{root wing value}}{4}$$

Dihedral angle

Dihedral is the angle the wing plane makes with a horizontal plane as viewed from the front as shown in Fig. 6.9.

The wing geometry is given dihedral to increase lateral stability in yaw. It would be easier to build a wing without dihedral but this can lead to aircraft which are difficult to fly in cross-wind conditions. The placement of the wing on the fuselage also affects lateral stability (high-wings are more stable than low-wings). For a conventional unswept trapezoidal wing the dihedral angle would typically be 0–1° for high-wings, 2–4° for mid wings and 3–6° for low wings. For low wings the dihedral angle may be higher to provide sufficient ground clearance for wing mounted engines. Wing sweepback naturally increases the aircraft yaw stability. For swept wing aircraft dihedral angle is therefore less. To reduce the required size of control surface on some aircraft the de-stabling influence of negative dihedral (anhedral) may be used to increase roll sensitivity (e.g. high-wing large transport aircraft show this effect).

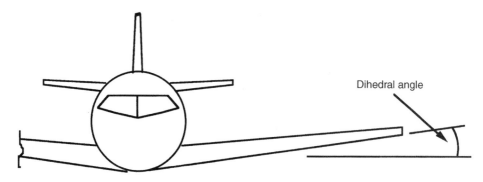

Fig. 6.9 Wing dihedral.

Flap type and geometry

There are a variety of trailing edge and leading edges devices available to augment the lifting capability of the wing. These are discussed briefly in Chapter 8 and an indication given of their effect on the aerodynamics of the wing. Conventional mechanical flaps have only been considered as to date these are the only types to have been used on civil jet transport aircraft. A survey of the landing $C_{L_{max}}$ of the aircraft in the aircraft data file (Data A) is shown in Fig. 6.10. An indication of the effect of sweepback is shown allowing for the difference in sweep between the wing and the flap hinge line.

The chordal length of the flap (as a percentage of the local wing chord) will

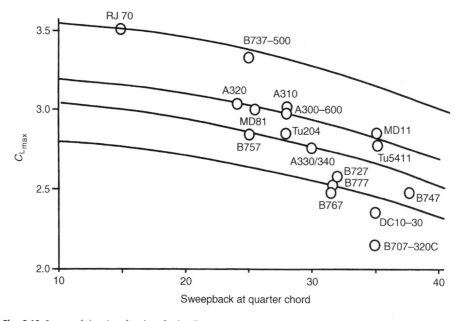

Fig. 6.10 Survey of the aircraft values for landing $C_{L_{max}}$.

influence the achievable maximum lift. However, at high values, the effectiveness of increasing chord is reduced giving practical upper limits for chord lengths as:

25%	for split flaps
30%	for plain flaps
30–40%	for slotted or Fowler flaps

Similarly the angle of deflection of the flap loses effectiveness at larger angles. Typical maximum deflection angle for various types of flap are:

split	55–60°
plain	40–50°
fowler	30–40°
leading edge	15–20°

The lift curve slope for a section with flap remains approximately unchanged from the unflapped section but the angle of incidence at stall is reduced. This has the beneficial effect of triggering stall at the inboard flapped sections and away from the outboard section thus helping to prevent tip stall.

Structural design

The aerodynamic advantages of the flap do not come cheap. The flap structure and operating system complicate the wing trailing edge structure and introduce mechanical and electrical systems into the relatively thin sections at the trailing edge. Introducing flaps onto the wing will increase aircraft structure and system masses and add extra cost to the aircraft manufacture. However, in general the improvements to the aircraft performance from the introduction of flaps are worth these penalties.

Wing planform studies

Careful selection of each of the above geometric parameters will need to be made during the project design phase. The wing layout must satisfy a number of competing demands from aerodynamic, stability, control, internal space, structures, manufacture, maintenance and overall costs. Once the aircraft has passed into the detail design phase only minor changes to the wing parameters are usually possible. The more time and care taken in establishing the most efficient wing geometry at the project design stage the better. Chapter 13 provides details of how such studies can be undertaken.

Tail layout

The main criterion for the design of the tail surfaces is the provision of adequate stability and control for the aircraft. There may be other requirements that have to be met by the tail (e.g. fuel tankage, structural support, etc.) but these are regarded as secondary to the main stability and control criteria. Technically this resolves into the provision of adequate moments about the aircraft centre of gravity to counteract de-stabilising forces from the aircraft geometry. The degree of stability and control to be provided will depend on the aircraft operational requirements. A large transport aircraft will need to be stable under various loading configurations and have enough control response to be able to gently position the aircraft, fly in emergency (engine-out) conditions and react to side winds. A light executive aircraft will need quick response from the controls which may dictate artificial stability from other aircraft systems in conjunction with the pilot input.

The aircraft stability and control requirements are usually considered in three flight regimes. Roll response (motion about the x axis) is conventionally provided by ailerons but for some aircraft layouts this is not feasible so differential motion of the horizontal tailplane is used. The pitch response, often termed longitudinal stability (motion about the y axis), dictates the size of horizontal stabiliser(s) (conventional tail, or canard, or both!). The yaw and sideslip response, termed lateral stability (motion about the z axis), dictates the vertical stabiliser(s). Each of the three types of motion are not independent since motion about one axis causes an effect on the others but for simplification in the initial project stage and for conventional layouts it is acceptable to consider them both de-coupled. The aircraft layout will have a considerable effect on stability and control aspects and careful thought must be given if unusual arrangements are proposed. The following list identifies some of the layout considerations for each flight action.

Pitch

The following points have to be taken into account when sizing the tailplane.

- The tailplane has to cope with the required centre of gravity (c.g.) travel in the en-route flight regime. A typical c.g. range would be from a forward c.g. of 10% of the mean aerodynamic chord to a rear c.g. of 35% of the mean aerodynamic chord. It must have enough power to provide the necessary trim and control.
- The forward c.g. position with a typical tricycle undercarriage and with take-off flap will give the highest load on the nosewheel. The tailplane needs to be sized to provide enough force to lift the nose at the required rotation speed.
- The tailplane must be large enough to be able to trim the aircraft on the approach with full landing flap at the worst c.g. position and at the same time provide enough power to control and flare the aircraft at touchdown. During the flare manoeuvre ground effect can have a significant influence.
- The wing section profile, wing planform and position relative to the aircraft centre of gravity will affect aircraft stability due to the pitching moments created by the lift, drag and aerodynamic pressure distribution. More wing camber, thicker section and flapped wings have intrinsically larger aerodynamic moment which will demand a larger tail to balance the aircraft.
- Engine installation has an influence due to the vertical offset of the thrust line from the aircraft centre of gravity and the pitching inertia contribution from the engine mass position.

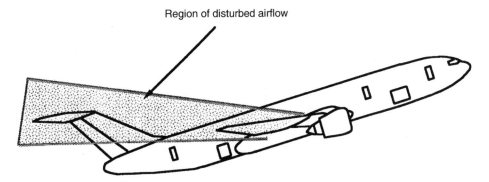

Region of disturbed airflow

Fig. 6.11 Avoidance of aircraft 'deep stall' condition.

- The position of the horizontal tail relative to the wing and fuselage (and rear engines) influences the effectiveness of the tail to produce the balancing force. A high (tee) tail positioned away from these interferences is the most effective. The tail should be positioned relative to the jet efflux so that effects from throttle changes are avoided or kept to a minimum.

The position of the horizontal tail relative to the wing in side view will also determine the aircraft's ability to enter a deep stall. A deep stall occurs at high angles of attack when the wake of the stalled wing blankets the horizontal tail making it ineffective and thus virtually impossible to recover from the stall.

Figure 6.11 shows the effect on the aircraft of the deep stall. Notice that the tailplane lies in a region where the airflow has little net velocity in the longitudinal direction. This degrades the effectiveness of the tailplane to produce a lift force to take the aircraft out of the stall attitude. This produces a stable flight condition with little forward speed but with a steady vertical descent. Without the ability to recover from this condition the aircraft will eventually crash. This unsafe situation can be avoided by positioning the tailplane outside (usually below) the area of the stalled wing wake. Wind tunnel tests would be used to verify the safe effectiveness of tail position in the wing stall attitude.

Modern airliners are equipped with systems that warn the pilot if the aircraft is approaching a stalled attitude. The pilot, unless conducting a highly controlled flight test, will normally respond to such warnings by pushing the aircraft nose down to reduce wing angle of attack and thereby increase aircraft speed.

Yaw

The primary control for yaw is the fin with its rudder. The following points need to be taken into account when sizing the fin.

- The fin size must be such as to cope with the required c.g. travel in the en-route flight regimes.
- In the event of an engine failure particularly for engines mounted on the wing,

the fin must be capable of generating a sufficient side force to balance the resulting de-stabilising moment.
- The cross-wind requirement in the landing configuration can often size the fin.

The engine failure case above, in the take-off condition is usually the critical sizing criterion for the fin, for aircraft with wing-mounted engines.

Roll

The primary controls here are usually wing mounted ailerons or spoilers or a combination of both. An alternative would be to use differential controls on the horizontal stabilisers. These have mainly been used on fighter aircraft where the inertia in roll is so much lower than on the larger civil transports. The roll controls have to be sized to produce rates of roll and acceleration in roll to meet the appropriate requirements. At the initial project stage on a civil transport using wing mounted controls, a study of current aircraft will enable an initial assessment of the size of these controls to be made.

As noted earlier the motion of the aircraft around each of the three axes has been considered to be independent. However, there is a considerable dependence between them, particularly between yaw and roll. Hence a full stability and control analysis will be necessary at a later stage in the design process. At the initial project stage insufficient is known about the aircraft to carry out such an analysis. An estimate based on volume coefficients (V) is sufficiently accurate at this stage.

$$V_{\text{tail}} = (SHT \times LHT)/(S \times c)$$

$$V_{\text{fin}} = (SVT \times LVT)/S \times b$$

where: S, b and c are the gross wing area, span and mean aerodynamic chord of the wing;

SHT, SVT are the areas of the tailplane and elevator, and the fin and rudder;

LHT, LVT are the tail arms measured from the aircraft centre of gravity to the quarter mean aerodynamic chord positions on the horizontal and vertical surfaces.

The tail arm geometry is shown in Fig. 6.12.

Typical values for the volume coefficient vary widely between different aircraft types (V_{tail} from 0.5 to 1.2, V_{fin} from 0.04 to 0.12). For initial project design purposes it is necessary to evaluate the volume coefficients for aircraft with similar layout, operation and weight to the proposed design and then use this value for estimation of tail areas. Care must be taken to allow for the introduction of new technologies (e.g. relaxed stability) as this may influence tail sizing.

There are many different possibilities for the layout of the tail surfaces, these include:

- conventional low tailplane with central fin
- mid-position tailplane

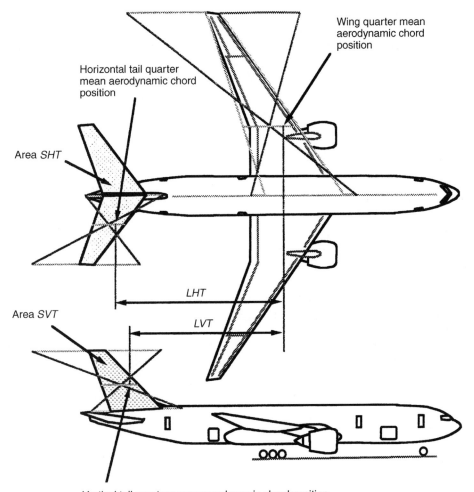

Wing quarter mean
aerodynamic chord
position

Horizontal tail quarter
mean aerodynamic chord
position

Area *SHT*

LHT

Area *SVT*

LVT

Vertical tail quarter mean aerodynamic chord position

Fig. 6.12 Tail arm geometry.

- tee-tail
- twin booms
- twin fins
- butterfly tail
- canard
- dorsal and vertical fins

All the unconventional configurations have advantages in some applications but care must be exercised when specifying such unusual configurations.

Airworthiness requirements and flight testing will dictate handling qualities which must be met by the tail design (e.g. mass balancing of the controls, control force harmonising).

The correct sizing of the tail surface seems to be more of an art than a science due to the often conflicting requirements of the various control demands and the inter-connections of the various flight modes. It is interesting to look at existing aircraft and notice the 'fixes' that have been found to be necessary to correct faults that were not predicted by the original design team.

7

Aircraft mass and balance

Objectives

Aircraft mass estimation and the associated 'balancing' of the layout are two of the main decision areas in the project phase of the design. The initial crude 'guesstimates' at the aircraft take-off weight will be progressively refined as more details of the aircraft configuration and systems are settled.

This chapter is intended to show how the mass estimation process is progressed and how details of the configuration and systems are related to the total aircraft mass. After an initial explanation of the factors which affect mass, a simplified method to determine the first estimate of aircraft take-off mass is presented together with an example calculation. A more detailed mass estimation method is then described together with formulae and expressions to predict each of the main components of the aircraft. After this it is possible to list the items in a standardised mass statement for the aircraft. The form of this list is described and the various sub-groupings shown.

With a detailed list of component masses it is possible to determine the aircraft centre of gravity (c.g.). Methods by which the component masses are positioned are given. Some of the aircraft masses are dependent on the type of operation (e.g. fuel and payload); this must be analysed to determine the furthest forward and aft limits to the position of the aircraft centre of gravity. The methods by which the aircraft c.g. range is predicted are shown and the influence of aircraft configuration described.

The way that the aircraft mass is affected by the stage-distance and passenger-load leads to the generation of the standard 'payload-range' diagram. The influence of aircraft weight and other geometry restrictions on the diagram is described. Finally a method is presented which balances the wing and fuselage group masses and sets the relative position of the wing to the fuselage geometry.

After studying this chapter it will be possible to estimate the overall mass of the aircraft, determine the mass of component parts, balance the aircraft and determine the centre of gravity limits that result from the loading procedures. All these aspects are essential to the specification of the baseline configuration.

The following chapters amplify the inter-relationship between weight and other specialist types of analysis. A knowledge of mass estimation and aircraft balance is necessary to fully appreciate this inter-relationship and the compromises that are necessary to produce a successful overall design. These aspects are covered in detail in this chapter.

Introduction

Aircraft mass (commonly called weight) more than any other design parameter, is seen to influence the design of the aircraft in the project stages. From the earliest days of flight (and for some time before) it was appreciated that the aircraft had to overcome the effects of gravity. With limited power from the engine it was soon realised that the most significant improvements in aircraft performance could be achieved by a reduction of aircraft mass. This understanding of the direct relationship between aircraft performance and mass made it important for the pioneers to produce aircraft as light as economically feasible. These same influences are felt by present-day designers but they are now aware of many more factors which influence aircraft mass. Performance aspects are still highly significant but they are now compromised by environmental restrictions (e.g. noise regulations), airworthiness regulations (e.g. climb rate after engine failure on take-off), and operational aspects (e.g. extended twin-engined operation, ETOPS). The designers also have a wider choice of technologies available and this requires careful judgement with respect to aircraft mass reductions and the technical risks involved. Two of the most crucial decisions in this respect are the choice of new materials (e.g. composites) and the relaxation of inherent stability (e.g. automatic computer stabilisation). Both technology advances offer substantial mass savings but confidence in their adoption requires the assessment and acceptance of the level of technical and financial uncertainty associated with their development. During the project design stages it is possible to assess the advantages and drawbacks for the introduction of such new technologies in theoretical 'paper' designs. This requires the study of the trade-offs between improved aircraft efficiency and the associated increase in commercial risk.

Aircraft weight is a common factor which links all the separate design activities (aerodynamics, structures, propulsion, layout, airworthiness, environmental, economic and operational aspects). To this end, at each stage of the design, a check is made on the expected total mass of the completed aircraft. As described in Chapter 2, a separate design organisation (weights department) is employed to assess and control weight. In the early stages of the design, estimates have to be made from historical statistical data of all the component parts of the aircraft. As parts are manufactured and the aircraft prototype reaches completion it is possible to check the accuracy of the estimates by weighing each component and where necessary instigate weight reduction programmes.

Overweight aircraft will suffer reductions in range, reduced climbing ability, reduced manoeuvrability and increased take-off and landing distances. If the operating requirements do not permit these reductions in performance (e.g. specified

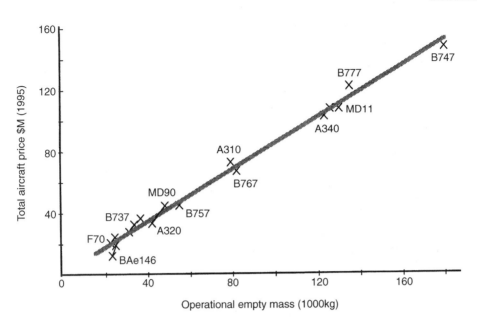

Fig. 7.1 Aircraft price relative to aircraft operational mass (source Avmark data).

maximum runway length) the increase in aircraft empty weight must be offset by a reduction in useful load to ensure that the maximum take-off weight of the aircraft is not exceeded. This reduction in useful load will make the aircraft less competitive and affect the commercial success of the project.

For aircraft with conventional layouts and structural form there is a direct relationship between operational empty weight and basic aircraft price (Fig. 7.1). Hence increases in aircraft weight will not only be reflected in increased operational costs but will add to the purchase price of the aircraft.

Extra weight will affect the fuel used to fly the specified range. The cost of this extra fuel and the increase in aircraft price shown above combine to adversely affect the direct operating costs for the aircraft.

Estimating the value of weight saving

In aircraft project studies it is possible to show the effects of increased aircraft structure weight on the aircraft design. If the aircraft specification (e.g. range, take-off performance, etc.) is to be maintained then increase in structure weight due to inefficient design will lead to greater fuel usage, larger engines, stronger landing gear, larger wing and tail areas. These increases will lead in turn to a demand for a heavier structure. This vicious circle effect is known as 'weight growth'. Project studies show that for every kilogram of unnecessary structure mass on the aircraft, the maximum take-off mass of the aircraft will increase by about three kilograms. To illustrate the influence of weight growth two aircraft were designed to fly the same range but with different numbers of passengers

Table 7.1 Influence of aircraft size on mass parameters

Aircraft size: number of seats	300	600
Increase in structure mass (kg)	1000	1000
Increase in operational empty mass (kg)	1879	1756
Increase in fuel requirement (kg)	1255	1431
Increase in aircraft maximum TO mass (kg)	3034	3188
Weight growth factor	3.03	3.19
Increase in aircraft price ($M)	0.87	1.19

(300 and 600). A structure mass penalty of 1000 kg was then added and the aircraft redesigned. The changes in the main design parameters are shown in Table 7.1. These figures provide a strong incentive to produce low weight designs.

The weight growth phenomenon can be utilised in weight saving programmes. If a new manufacturing process or material can be introduced that will save aircraft structural mass, the cost of the new technology can be offset by the full weight saving on the aircraft. For such reasons the manufacturers introduce advances in technology as soon as they are shown to be commercially viable.

In the above study, the cost of weight increase (and correspondingly the value of a weight reduction program) is assessed by the increase in aircraft price and cost of the extra fuel used. Dividing the increase in aircraft price by the weight increase shows a basic value of weight saving of about $500 per pound (lb). This compares well to published data from the Society of Allied Weight Engineers.[1] The cost penalty resulting from the increased fuel used can be aggregated over the operational life of the aircraft (assuming a typical utilisation). This shows a similar increase in operating cost and therefore a further $500 per pound value of saving weight. The study was done using 1995 price values; therefore these costs may be expected to rise with inflation over the life of the aircraft. It is clear from these studies that $1000 per pound is a conservative estimate for the value of weight saving on civil aircraft in the 300–600 seat size, providing that the weight savings can be identified during the initial design stage. If weight growth is identified later in the design process when the aircraft maximum take-off weight has been frozen, the payload or fuel load must be proportionately reduced. Reducing the number of passengers or cargo load directly affects revenue potential. Reducing fuel load restricts range and therefore the operational potential. Over the total life of the aircraft such penalties can be shown to be expensive. In such circumstances the value of weight saving can be shown to be much higher than the value assessed during the project design phase.

The initial mass estimate

The importance of aircraft weight with regard to performance, design, economics and regulatory aspects is shown in Fig. 7.2.

In the initial stages of the project all that can be done is to estimate the overall

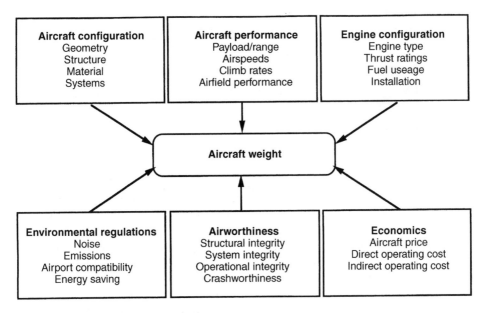

Fig. 7.2 Influences on aircraft weight.

aircraft take-off mass. This is then used to make initial performance estimations. A crude estimate can be made by considering the three components below:

$$M_{TO} = M_{UL} + M_E + M_F$$

where: M_{TO} = take-off mass
        M_{UL} = useful load
        M_E = aircraft empty mass
        M_F = fuel mass

The three unknowns on the right hand side of the take-off mass equation are considered separately below.

Useful load

Useful load (M_{UL}) can be determined from the original aircraft specification (passengers + cargo). Some operational items and the number of crew are linked to the number of passengers or payload capacity and to operators' practices and will affect the operational empty weight of the aircraft.

Aircraft empty mass

Aircraft empty mass (M_E) will vary for different types of aircraft and for different operational profiles. All that can be done in the initial stage of the design is to roughly size the aircraft by considering the expected ratio of empty mass to maximum take-off mass (M_E/M_{TO}). It is necessary to use data from existing and

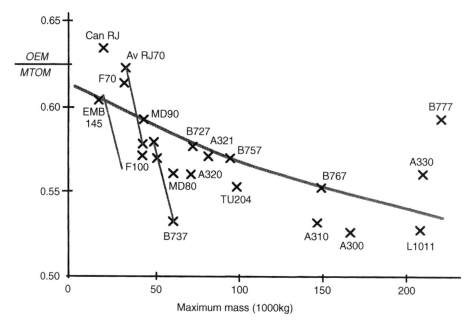

Fig. 7.3 Aircraft empty mass to take-off mass fraction.

previous designs to provide a guide to this ratio. Figure 7.3 shows the weight fraction for a wide range of current aircraft sizes. Some of the aircraft may have higher values than expected because they include structural provision for later stretched versions of the type.

It is wise to select aircraft at sizes (number of passengers and range) close to the proposed aircraft. Select aircraft from the aircraft data file (Data A) as a starting point. Plot the data and select an appropriate value for the empty mass ratio. Values in the range 52–62% are representative with the lower values associated with long-haul aircraft.

Fuel mass

Fuel mass (M_F) will be dependent on the mission profile for the aircraft. In the initial design phase it is difficult to accurately predict the fuel used in each of the non-cruise flight segments (take-off, climb, cruise, descent and landing) so an estimate has to be made based on an extended cruise phase. Cruise fuel can be determined from a modified range equation:

$$M_F = \text{fuel flow} \times \text{flight time}$$

$$= SFC \times \text{thrust} \times \text{time}$$

where SFC is the engine specific fuel consumption at a representative cruise condition. The above equation can be rewritten in terms of 'fuel fraction' (M_F/M_{TO}):

$$M_F/M_{TO} = SFC \times (\text{thrust}/M_{TO}) \times \text{time}$$

As the aircraft is in equilibrium in the cruise condition (i.e. not manoeuvring or accelerating):

$$(\text{thrust} = \text{drag}) \text{ and } (\text{lift} = \text{aircraft weight} = M_{TO} \cdot g)$$

where g is the gravitational acceleration $= 9.81 \text{ m/s}^2$.

Substituting the lift (L) and drag (D) conditions in the fuel fraction equation gives:

$$M_F/M_{TO} = \frac{SFC \times (D/L)}{g} \times \text{time}$$

$$= SFC \times g \times (D/L) \times \text{time}$$

A fuller explanation of SFC is given in Chapter 9. Typical values can be found in the engine data file (Data B). Typical low bypass ratio engines have SFC values 1.0–0.7 lb/hr/lb, and high bypass ratio engines have values down to 0.5. Values will also depend on cruise speeds.

From existing data select a value for an engine type similar to the one to be used in the project. The lift/drag (L/D) ratio varies with the aircraft aerodynamic design and the cruise lift coefficient. Transport aircraft are designed to be most efficient in the cruise phase so will have a reasonably high L/D ratio. Typical values are between 14 and 19 with the higher numbers related to larger long-range aircraft with large span. It is possible to determine the approximate L/D ratio for existing aircraft from the aircraft data file (Data A) using the modified range equation shown in Chapter 11 (p. 282).

Flight time is estimated by dividing the stage distance by the cruise speed. The specified design range is increased to allow for non-cruise flight segments (take-off, climb, descent and landing), the fuel reserves (diversion and hold), and other contingencies (e.g. wind). The total range to be allowed for in the fuel weight estimation is called the equivalent still air range (ESAR). The determination of $ESAR$ from the specified aircraft range is complicated because different operators assume their own conditions. The expression below can be used for medium- to long-haul flights (>2000 nm). For shorter stages the coefficient will be less and the reserves smaller.

$$ESAR = 568 + 1.063 \times \text{Specified range}$$

The main equation for aircraft take-off mass $(M_{TO} = M_{UL} + M_E + M_F)$ can be written as:

$$M_{TO} = M_{UL}/\{1 - (M_E/M_{TO}) - (M_F/M_{TO})\}$$

This form of the equation is useful as it only requires a knowledge of the mass ratios for aircraft empty weight and fuel load. These ratios can be assumed to be similar to aircraft of the same type and size if the absolute mass is difficult to determine.

Example

To illustrate the use of the initial estimation method described above, consider the design of a medium size (180 seats), medium range (2000 nm) aircraft.

The useful load is estimated as:

$$\text{passenger mass } (180 \times 75\,\text{kg}) \quad 13\,500$$

$$\text{baggage } (20\,\text{kg each}) \qquad\qquad 3600$$

$$\text{total } (M_{\text{UL}}) = 17\,100\,\text{kg}$$

From a review of similar aircraft the operational empty weight fraction is assumed to be 51%.

The 'fuel fraction' is estimated from the formula below:

$$(M_{\text{F}}/M_{\text{TO}}) = SFC \cdot (D/L) \cdot (ESAR/V)$$

From the engine data file the SFC for a medium bypass engine is 0.7 lb/hr/lb. This is in weight (force) units; therefore the 'g' has been accounted for in the value. Assumed cruise speed is 500 kt (257.7 m/s). Cruise L/D ratio is assumed to be 18 (current technology value). From the expression shown earlier and some allowance for the type of operation the equivalent still air range appropriate to the design range of 2000 nm is assumed to be 2750 nm, therefore:

$$(M_{\text{F}}/M_{\text{TO}}) = 0.7 \cdot (1/18) \cdot (2750/500) = 21.4\%$$

We can now use the M_{TO} equation to estimate the aircraft take-off mass:

$$M_{\text{TO}} = 17\,100/(1 - 0.51 - 0.214) = 61\,957\,\text{kg (approx. } 62\,000)$$

The estimation of M_{TO} is very sensitive to the prediction of empty and fuel mass ratios. If both of these are changed by only 1% (i.e. 2% in the total) the resulting estimation of M_{TO} would vary between 60 000 and 64 000 (i.e. ±about 2000 kg). Because of this sensitivity the initial estimation must be checked against more detailed methods as soon as the aircraft layout is defined in more detail.

Detailed mass estimation

As more details of the aircraft become known it is possible to use more accurate methods to predict component masses. Ultimately, when the detail drawings of all the components are available, more accurate estimates can be made by calculating the volume of each part and multiplying by the density of the material. In the project design stages we are unlikely to know the size of individual aircraft components to this level of detail but it is possible to use prediction methods that progressively become more accurate as the aircraft geometry is developed. Most aircraft design textbooks[3] contain such methods. A set of equations using some of the aircraft geometrical data is presented below.

At the preliminary stages of mass prediction, the masses for each main component of the aircraft are summed to give the maximum total mass [often incorrectly termed the 'take-off mass' ($MTOM$)]. Operationally the aircraft will have various take-off masses depending on the stage length and payload.

A more detailed estimation of aircraft take-off mass can be determined by summing the main component masses as shown below:

$$MTOM = M_W + M_T + M_B + M_N + M_{UC} + M_{SC} + M_{PROP}$$
$$+ M_{FE} + M_{OP} + M_{CR} + M_{PAY} + M_F$$

where:

M_W = mass of wing including control surfaces and flaps
M_T = mass of tail surfaces (tailplane/elevator + fin/rudder)
M_B = mass of fuselage (body), including wing attachment structure
M_N = mass of engine nacelles (not including engine or propulsion system)
M_{UC} = mass of undercarriage (nose + main units)
M_{SC} = mass of surface controls
M_{PROP} = mass of propulsion system (engines + all systems)
M_{FE} = mass of fixed equipment (electric, hydraulic, etc.)
M_{OP} = mass of operational items (residual fuels and oil, safety equipment, etc.)
M_{CR} = mass of crew
M_{PAY} = mass of payload (passengers, baggage, freight, etc.)
M_F = mass of fuel (including reserves)

Component mass estimation

If possible, several different methods should be used to estimate each of the component masses in the equation above. This will give confidence to the prediction and guard against inappropriate use of formulae. When it is impossible to determine the absolute mass of a component (perhaps because insufficient details are available) it is acceptable to use a standardised mass ratio, i.e.

mass of component/$MTOM$

Such ratios can be determined from existing aircraft of a similar type to the one under investigation. This procedure is often used in the early stages of project design.

The component mass estimating methods presented below are derived from statistical data of existing aircraft. In general, the aircraft on which the data is based will be of conventional layout with a semi-monocoque aluminium alloy structural framework. For designs not of this layout or manufactured from other materials the estimates will need to be suitably adjusted. Detail mass formulae will vary for different types of aircraft (e.g. business jets, large transport aircraft, etc.). The formulae quoted in the sections below are taken from detailed mass statements of existing civil transport aircraft.

Each of the component mass groups in the mass list above will now be considered separately.

Wing group (M_W)

The wing group is assumed to comprise all the structural items on the wing including surface controls (ailerons, flaps, lift dumpers, etc.) but not the systems within the wing (e.g. flight control systems, fuel tankage and system, anti-icing, etc.).

Previous designs show that the ratio $M_W/MTOM$ will be in the range 9–14% with the range 10–12% covering most medium- to long-range aircraft. As the design unfolds and more detailed geometric data are known it is possible to use formulae which include specific wing parameters. The expression shown below has been developed for conventional aircraft wing geometry made from aluminium alloy. The estimate may be reduced by up to 20% to account for the introduction of lighter composite materials for control surfaces and for lightly stressed parts of the structure.

$$M_W = 0.021\,265\,(MTOM \times NULT)^{0.4843} \times SREF^{0.7819} \times ARW^{0.993}$$

$$\times (1 + TRW)^{0.4} \times (1 - R/MTOM)^{0.4}/(WSWEEP \times TCW^{0.4})$$

where: $MTOM$ = aircraft maximum take-off mass (kg)
$NULT$ = aircraft ultimate load factor (typically 3.75)
$SREF$ = gross wing area (m^2)
ARW = wing aspect ratio
TRW = wing taper ratio
TCW = wing average thickness/chord ratio
$WSWEEP$ = wing quarter-chord sweepback angle (deg)

R is the effect of inertia relief on the wing root bending moment given by

$$R = \{M_W + M_F + [(2 \times M_{eng} \times B_{IE})/0.4B] + [(2 \times M_{eng} \times B_{OE})/0.4B]\}$$

where: M_W = wing mass estimated or iterated with this method (kg)
M_F = mission fuel ($= MTOM - M_{PAY} - M_{OE}$) (kg)
M_{PAY} = mass of payload (kg)
M_{OE} = operational empty mass of aircraft (kg)
M_{eng} = individual engine plus nacelle mass (kg)
B_{IE} = distance between inboard engines (m)
B_{OE} = distance between outboard engines (if appropriate) (m)
B = aircraft wing span (m)

A more detailed wing mass estimation method is described in the *AIAA Journal of Aircraft*.[2] The method takes into account the concentrated loads from wing-mounted engines (if present) and is applicable to multi-tapered swept wing geometry. It accounts for different materials used in the structural framework. The method is long and complex and should therefore be used when the aircraft configuration is frozen (i.e. at the end of the preliminary project phase).

Wing flaps, etc. For initial stages of the design process when details of the flaps, etc. are unknown the following typical values may be assumed. Trailing edge flap mass ranges from about 70–20 kg/m^2 with the lower values related to simple flap mechanisms with small chord length. Leading edge devices are about 30 kg/m^2.

Spoilers and lift dumpers (air brakes) are about $15\,kg/m^2$. A reduction of up to 20% can be assumed for the use of composite materials for control surfaces.

As more geometrical data is known the mass of the flap and other wing structures can be evaluated using the modified Torenbeek method[3] as shown below:

$$M_{FLAP} = 2.706 \times K_{FLAP} \times S_{FLAP} \times (B_{FLAP} \times S_{FLAP})^{0.1875}$$

$$\times [2.0 \times V_{AIAS}^2 \times 10^{-4} \times \sin(D_{FLAND})/TCWF]^{0.75}$$

where: B_{FLAP} = flap span (m)
S_{FLAP} = flap area (sq. m)
D_{FLAND} = flap deflection in landing (deg)
V_{AIAS} = aircraft approach speed (m/s)
$TCWF$ = flap thickness/chord ratio
K_{FLAP} is a coefficient to account for flap complexity:

= 1.0 for single slotted
= 1.15 for double slotted
= 1.15 for single slotted with Fowler action
= 1.30 for double slotted with Fowler action

Multiply the value of K_{FLAP} above by 1.25 if they are the extending type.

Other factors. For aircraft with rear fuselage mounted engines the wing mass (M_W) will be increased due to the lack of inertia relief on the wing structure (cf. R on p. 134). For aircraft with a fuselage-mounted main undercarriage, the wing mass (M_W) should be reduced by 5% to account for simplified (less stressed) inner structure.

Tail group (M_T)
Although tail mass is not a large fraction of $MTOM$, it is necessary to evaluate it accurately because it is positioned well aft of the aircraft centre of gravity. The tail mass will therefore affect the overall balance of the aircraft. Typical values of the mass ratio $M_T/MTOM$ range from 1.5 to 3.0% with 2% being a good initial guess.

More accurately the tail mass may be considered as the sum of the horizontal surfaces (tailplane, elevator, stabiliser) and the vertical surfaces (fin, rudder):

$$M_T = M_H + M_V$$

where: M_H = mass of horizontal tail surfaces = $S_H k_H$
M_V = mass of vertical tail surfaces = $S_V k_V$

S_H and S_V are total tail areas (i.e. tailplane plus elevator, fin plus rudder). k_H and k_V are determined from statistical data of existing aircraft. Values range between 22 and $32\,kg/m^2$ with $k_H = 25$ and $k_V = 28$ being typical values. For tail configurations with higher than normal span and for large areas, the higher values should be substituted. Add 10% to the above figures for fully variable tee-tails. When more details are known an analysis similar to the main wing mass estimation can be used.

The above analysis is based on conventional aluminium structures. If modern composite materials are used the mass may be reduced by 20%.

Body (fuselage) mass (M_B)

The mass of the body is obviously dependent on the size of the fuselage, the aircraft layout (e.g. engine and undercarriage positions) and operational aspects (large cargo or freight doors). Typical values for the ratio ($M_B/MTOM$) range from 7 to 12% with the higher figure associated with smaller executive-style aircraft.

There are many different methods of more accurately estimating body mass but the formula recommended for civil aircraft (50–300 seats) is the Howe[4] formula below:

$$M_B = 0.039 \, (L_F \times 2 \times D_F \times V_D^{0.5})^{1.5}$$

where: L_F = fuselage overall length
D_F = fuselage diameter (or equivalent diameter)
V_D = aircraft maximum speed (design driving speed)

It is recommended that the above mass be amended as follows:

increased 8% for pressurised cabin
increased 4% for fuselage mounted engines
increased 7% for fuselage mounted main undercarriage
increased 10% for large cargo door (etc.) discontinuity
reduced 4% if the fuselage is free from structural discontinuity

All the available methods are based on historical data so care must be taken when applying them to novel or very large (e.g. twin-deck) designs.

Nacelle mass (M_N)

Mass items to be attributed to the nacelle group are difficult to specify since some of the structure may be regarded as part of the wing, body, propulsion, or undercarriage groups. Care must be taken when assigning detailed estimates to avoid double accounting of mass items with other component mass groups.

Typical values for the mass ratio ($M_N/MTOM$) are 1.2 to 2.2%.

The nacelle mass is proportional to the size of the engine and the type of cowling specified (short, three-quarter, or full). In the initial design stages an estimate of the installed (required) thrust will have been made but the engine configuration may not be known. Details of the nacelle mass for some existing aircraft are shown in Fig. 7.4.

The 'best fit' quadratic function shows a 98% correlation with the data. To simplify the calculation this function can be approximated to the linear function shown below:

$$M_N = 6.8 \times T \qquad\qquad \text{(for } T < 600 \, \text{kN)}$$
$$= 2760 + (2.2 \times T) \qquad \text{(for } T > 600 \, \text{kN)}$$

where: T = total installed take-off static sea-level, SL thrust (kN)

The values calculated above should be slightly reduced if a short length bypass duct is used and increased for full length ducting. A further reduction of 10% can

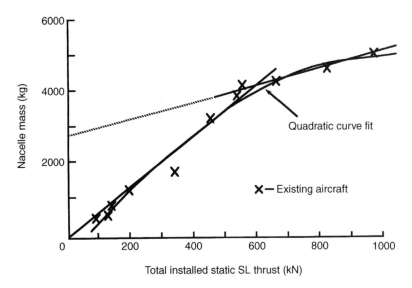

Fig. 7.4 Nacelle mass estimation.

be made if thrust reversers are not included but increased by 10% if extra noise suppression material (hushing) is necessary.

Landing gear mass (M_{UC})

The mass of the undercarriage will depend on the specified maximum landing mass and the rough field landing capability of the aircraft. In the initial design stages decisions will not have been taken on the landing mass specification. To avoid the need for fuel dumping many aircraft are now designed for a landing weight close to the maximum take-off weight. When no other information is available use the aircraft $MTOM$ value as used in the expression below. As most airlines are operated from good quality paved runways the main variation lies in the degree of complication, and the degree of compactness necessary (i.e. retraction mechanism). Figure 7.5 shows undercarriage masses for some current aircraft.

This landing gear mass data shows a 99% correlation by the function below:

$$M_{UC} = 0.0445 \, MTOM \text{ (i.e. 4.45%)}$$

The data above does not take into account recent improvements in undercarriage materials and wheel brake design. For a new aircraft design the landing gear is expected to weigh less than given by the above expression (assume 4.35%$MTOM$). (Of course this saving will be associated with a slight cost increase for the landing gear.)

Aircraft configuration will affect the design of the undercarriage and therefore its weight. Multiply the estimated weight by 1.08 if the aircraft has a high-wing and the main undercarriage is wing mounted.

Note, care must be taken when comparing landing gear mass from different aircraft as some manufacturers install gear that will be suitable for future larger

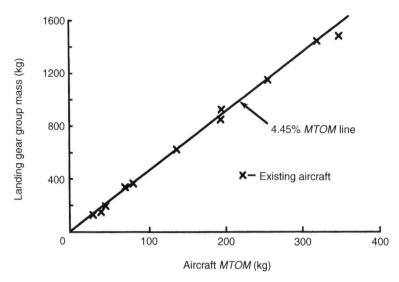

Fig. 7.5 Landing gear mass estimation.

stretched designs. This avoids some of the undercarriage design and certification costs for the development aircraft.

For aircraft balance assume the main units are 43.5% each and the nose unit is 13% of the total undercarriage mass.

Surface controls (M_{SC})

The surface controls include all the movable surfaces on the wing that have not been included in the flap mass calculation. The group includes all the internal wing controls and the controls for external leading edge devices (e.g. slats/slots) and lift dumpers/air brakes. It accounts for between 1 and 2% of the $MTOM$. Figure 7.6 shows values for some current aircraft. Although there is some scatter in the data the function below gives a 90% correlation:

$$M_{SC} = 0.4\,MTOM^{0.684}$$

The expression above assumes conventional aircraft configurations. If more complex or extensive leading edge flap system or lift dumpers are specified the value should be increased. For simple control systems (e.g. non-auto-pilot) and less complex structures (e.g. no leading edge devices) the value may be reduced by up to 25%.

Σ Structure mass ratio (M_{STR})

The sum of the above masses (wing, tail, body, undercarriage and surface controls) is called the aircraft structural mass. The ratio ($M_{STR}/MTOM$) will usually lie in the range 30 to 35% for conventional layouts. As a general guide in the early (less detailed) project phase it is acceptable to assume the structure mass to be about one third aircraft take-off mass (i.e. 0.33 $MTOM$).

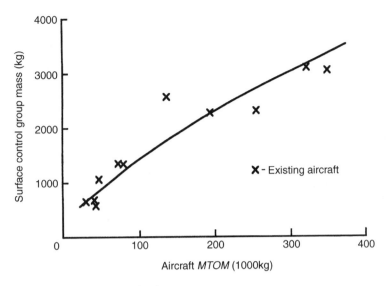

Fig. 7.6 Surface controls group mass estimation.

Σ *Total propulsion group mass* (M_{PROP})

In the early parts of the design process details of all the components in this mass group are unknown. It is therefore necessary to base estimates on the bare engines mass. The mass of the bare engines may be assumed to be proportional to the static take-off thrust/weight ratio for similar types of engine. For current large turbofan engines the specific mass is a function of bypass ratio as shown below. The relationship in Fig. 7.7 has been derived from a range of current and projected engine data.

The mean line is represented by:

$$\text{specific mass (kg/kN)} = 8.7 + (1.14 \times BPR)$$

The ratio of propulsion group mass to bare engine mass for a similar type of aircraft should be determined. The increase in mass over the bare engine value is difficult to assess in general terms as it is associated with the operational specification of the engine (e.g. thrust reversers, system complexity, power take-offs, etc.).

Figure 7.8 shows values from some current aircraft.

The following functions can be assumed from the propulsion group data:

$$M_{PROP} = 1.43 \, M_e \qquad (\text{for } M_e < 10\,000 \text{ kg})$$
$$M_{PROP} = 1.16 \, M_e + 2700 \qquad (\text{for } M_e > 10\,000 \text{ kg})$$

where: M_e = engine bare mass as calculated from the required engine thrust

The values from the above expressions assume a conventional installation with

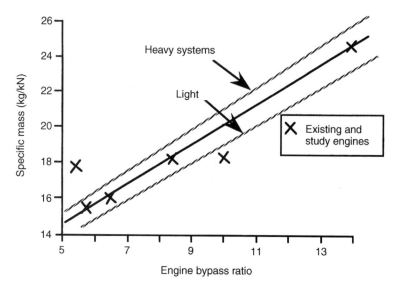

Fig. 7.7 Engine specific mass estimation.

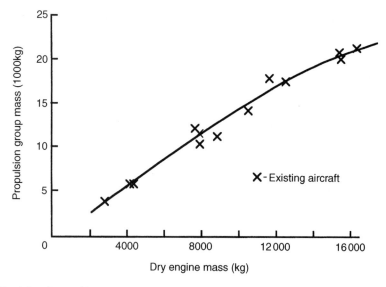

Fig. 7.8 Total aircraft propulsion group mass estimation.

thrust reversers and normal engine accessories. A more comprehensive analysis should be attempted as soon as the appropriate aircraft/engine details are known.

Σ *Total fixed equipment mass* (M_{FE})
The fixed equipment mass group is highly dependent on the type of aircraft under consideration. Items in this mass group are very variable in type (e.g. flight control systems, furnishings, etc. [refer to the mass statement on pp. 142–144]). The items

to be included depend on operational practices and aircraft specification. Various textbooks and aircraft manuals list the actual masses for components in this mass group for a variety of different aircraft. Without details on the types of system to be specified the value for the ratio $(M_{FE}/MTOM)$ can be related to the type of operation:

short-haul transport	14%
medium-haul transport	11%
long-range transport	8%

The values above are based on historical data. Advances in computer and control systems technology will change the size and mass of systems considerably. Individual components have been 'miniaturised' but to offset this advantage more extensive systems will be installed; for example increased electrical demand requires a larger auxiliary power unit than on older aircraft. You are advised to consider this mass group in detail as soon as the aircraft system information and specification is known so that a check can be made on the value assumed from the ratio above.

Σ Operational items mass (M_{OP})

Some of the operational items are specific to the type of aircraft and other operational practices and must be assessed individually. The following general data may be used (airline standard):

- Crew provisions including maps and reference manuals. Allow 10 kg total although with the introduction of on-line computer stored data this could be reduced (e.g. 2 kg) or even ignored as insignificant in the initial mass estimation phase.
- Passenger cabin supplies (kg)
    0.45 per passenger – commuter service
    2.27 per passenger – snacks only service
    6.25 per passenger – medium-range service
    8.62 per passenger – long-range service
    add 2.27 extra per first-class seat
- Water and toilet provision (kg)
    0.68 per passenger – short-range
    1.36 per passenger – medium-range
    2.95 per passenger – long-range
- Safety equipment (kg)
    0.9 per passenger – short over-land journeys
    3.4 per passenger – over-water and extended flights
- Residual fuel and oil
    $0.151 V^{2/3}$ – turbofan engines (V = fuel capacity in litres)

Crew mass (M_{CREW})

The number of crew will be dependent on the type of operation and the aircraft specification (number of passengers and airline practice). Minimum numbers are set by the airworthiness authorities. For a conventional aircraft a minimum of two flight crew will be necessary but this must be increased for long-haul (duration) flights to allow the pilots rest periods. The number of cabin staff will be a function of the seating layout and the emergency evacuation process. For normal scheduled services allow one attendant for 30–40 economy-class seats, one for 20–25 business-class seats, and one for each 10–15 first-class passengers. Allow 93 kg for each flight crew member and 63 kg for each of the cabin staff (presumably they are usually much slimmer!).

Σ *Mass of payload* (M_{PAY})

The payload (i.e. passengers, baggage, freight) will be set by the aircraft specification (payload/range). The number of passengers to be carried in each class and the extra cargo will form part of the design brief. The range over which this load is to be flown is also likely to be specified. This range will need to be related to a fuel load, taking into account reserve fuel and contingencies. Some items may be particular to the type of aircraft and its operation. The following data may be used for general (civil transport) assessments:

passengers (including carry on baggage) – 75 kg each
extra baggage (tourist) – 20 kg (other classes) – 30 kg each passenger

(Note: Baggage density is 192 kg/m³. This dictates the space required in the cargo holds for passenger luggage but extra volume will be required for the specified freight load.)

Mass of fuel (M_F)

The mass of fuel to be carried is related to the design payload/range specification for the aircraft. Reserve and contingency allowances will be added to the fuel needed to fly the operation, to allow for any flight diversions and hold and to provide for adverse weather or for other extra fuel use. The total (maximum) fuel will be limited by the size of the available fuel tanks. The density of fuel used in aircraft engines varies according to the type. Specific gravity will be between 0.77 and 0.82 kg/litre, with an average value of 0.80 typical. With an assumed density and a calculated amount of fuel required a check must be made on the volume required and this compared to the available space in the wings (etc.) to hold fuel.

ΣΣ *Aircraft maximum take-off mass* (*MTOM*)

The maximum take-off mass for the aircraft will be equal to the sum of the aircraft empty mass, operational items mass, payload and the fuel mass. The *MTOM* will assume a combination of payload and fuel load appropriate to the design specification. The aircraft will be designed to fly safely at this value of *MTOM* and the Certificate of Airworthiness will permit operation of the aircraft up to but not beyond this limit. The *MTOM* is therefore one of the most significant parameters of the design.

Aircraft mass statement

It has become normal practice in aircraft design to list the various components of aircraft mass in a standard format. The components are grouped in convenient sub-sections as shown below.

1. Wing (including control surfaces)
2. Tail (horizontal and vertical including controls)
3. Body (or fuselage)
4. Nacelles
5. Landing gear (main and nose units)
6. Surface controls

$$\Sigma \text{ TOTAL STRUCTURE MASS} = M_{STR}$$

7. Engine(s) (dry weight)
8. Accessory gearbox and drives
9. Induction system
10. Exhaust system
11. Oil system and cooler
12. Fuel system
13. Engine controls
14. Starting system
15. Thrust reversers

$$\Sigma \text{ TOTAL PROPULSION MASS} = M_{PROP}$$

16. Auxiliary power unit
17. Flight control systems (sometimes included in M_{STR})
18. Instruments and navigation equipment
19. Hydraulic systems
20. Electrical systems
21. Avionics systems
22. Furnishing
23. Air conditioning and anti-icing
24. Oxygen system
25. Miscellaneous (e.g. fire protection and safety systems)

$$\Sigma \text{ TOTAL FIXED EQUIPMENT} = M_{FE}$$

All of the mass sections in the mass statement are totalled to give:

$$\Sigma\Sigma \text{ (BASIC) EMPTY MASS (WEIGHT)} = M_E$$
$$\text{(i.e. } M_E = M_{STR} + M_{PROP} + M_{FE})$$

The following items may be added to the aircraft empty weight (mass) to give various definitions for operational conditions.

1. Crew provisions (manuals, maps)
2. Passenger cabin supplies (passenger seats are sometimes included)
3. Water and toilet chemicals
4. Safety equipment (e.g. life jackets)
5. Residual fuel
6. Residual oil
7. Water/methanol (if appropriate)
8. Cargo handling equipment (including containers)
9. Oxygen

$$\Sigma \text{ OPERATIONAL ITEMS} = M_{OP}$$

Added to this are

1. Flight crew
2. Cabin staff

$$\Sigma \text{ CREW MASS} = M_{CREW}$$

Crew mass is sometimes included in the operational items.
   The operational items (including crew) are added to the empty mass to give:

$$\Sigma\Sigma \text{ OPERATIONAL EMPTY MASS (WEIGHT)} = M_{OE}$$

$$\text{(i.e. } M_{OE} = M_{OP}\text{(including crew)} + M_{E}\text{)}$$

Now the payload must be added on:

1. Passengers and baggage
2. Revenue freight

$$\Sigma \text{ PAYLOAD} = M_{PAY}$$

The payload is added to the operational empty mass to give the maximum zero fuel
mass.

$$\Sigma\Sigma \text{ MAXIMUM ZERO FUEL MASS (WEIGHT)} = M_{ZF}$$

$$\text{(i.e. } M_{ZF} = M_{OE} + M_{PAY}\text{)}$$

Finally the fuel is included:

3. Usable fuel $(= M_F)$.

   The fuel load is added to the zero fuel mass $(M_F + M_{ZF})$ to give:

$$\Sigma\Sigma \text{ MAXIMUM TAKE-OFF MASS (WEIGHT)} = MTOM$$

Note the take-off weight of the aircraft will not always be at the maximum
weight. Weight at the start of take-off will vary depending on the payload and fuel
required to fly the range. These are operational considerations relating to the
service to be flown.
   The component mass list allows various mass items to be grouped together and
this has led to a variety of definitions for different conditions of the aircraft (e.g.
basic empty weight, operational weight, zero fuel weight). Some commonly used
terms are listed in the reference section at the end of this chapter. Because of the
lack of standardisation of terms, care must be taken when comparing quoted
weight or mass figures from different sources.

Limiting speeds

For some of the components in the mass list the estimation of structural mass
requires a knowledge of aircraft speeds. The speed limits may be set by operational
requirements. Typical limits are shown in Fig. 7.9.
   The maximum speed limit is defined as the design diving speed $(Mn_D$ or $V_D)$.
(Although civil aircraft are unlikely to be performing high speed diving
manoeuvres the maximum speed is still defined by the old terminology.) When
conducting project studies it is unlikely that the maximum speed of the aircraft will

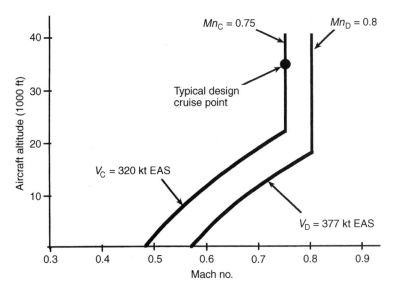

Fig. 7.9 Typical aircraft speed limits.

be specified; however the design cruise speed, Mn_C or V_C (i.e. maximum cruise speed) will be known. Airworthiness requirements for civil aircraft set a minimum margin of Mach 0.05 between the design cruise speed and the maximum (diving) speed. Data from current aircraft designs can be used to show current practice.

Figure 7.10 can be used together with a knowledge of the cruise speed to estimate the design diving speed (V_D) used in the detailed mass estimation formula.

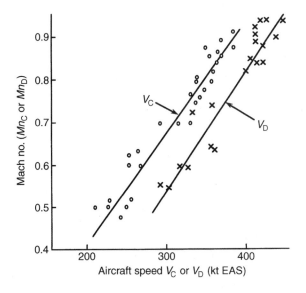

Fig. 7.10 Aircraft speed relationships.

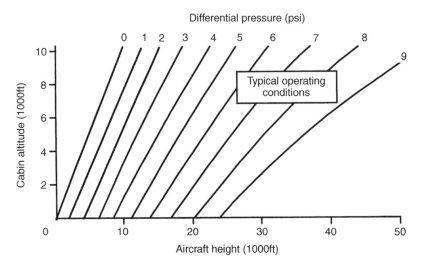

Fig. 7.11 Cabin pressure requirements.

Airworthiness regulations will specify the maximum accelerations the aircraft has to withstand (for flight manoeuvre and gust envelopes, see Chapter 4). These requirements will dictate the loading on the aircraft structure and thereby the size and weight of the structural members. In the early design stages a maximum value of $2.5g$ can be assumed (where g = gravitational acceleration). This value must be multiplied by an ultimate factor of 1.5 to get the ultimate design factor.

Cabin pressure

The fuselage skin will be designed to resist the loads arising from the maximum cabin pressure differential. This will be dependent on the specified cabin altitude (i.e. cabin pressurisation) and the maximum altitude at which the aircraft will be flown. The relationship for standard atmospheric conditions is given in Data D. The effect is shown in Fig. 7.11.

Cabin altitude is often set at a maximum height of 8000 ft. The maximum aircraft altitude will be determined from aircraft performance and operational limitations. The mass estimations quoted earlier for the fuselage assumed a conventional cabin differential pressure. For aircraft flying at unusually high altitudes or with a lower cabin pressure than normal the estimated mass should be increased.

Aircraft centre of gravity

Once all the component masses are known it is a relatively easy task to determine the aircraft centre of gravity position (Fig. 7.12). It will be necessary to know this

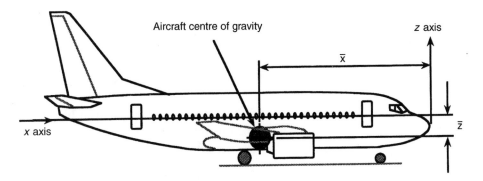

Fig. 7.12 Aircraft centre of gravity.

position at an early stage of the project design to enable the wing to be correctly positioned along the fuselage on the general arrangement drawing of the aircraft.

The standard list of component masses shown in the mass statement can be extended to determine the centre of gravity and pitching moment of inertia of the aircraft. Extra columns are required that specify the distances of the component centre of gravity positions from a given set of aircraft axes. The position of the aircraft axes can be chosen arbitrarily. Often they are chosen to coincide with principal features of the aircraft (in the diagram above the x axis is chosen as the body datum and the z axis is chosen at the aircraft nose. Note this may add confusion if the aircraft fuselage is stretched. To avoid this difficulty, on some designs a remote origin ahead of the aircraft nose is used for the reference datum. The mass list can be extended as shown in Table 7.2. The aircraft centre of gravity position can be found from:

$$\bar{x} = \frac{\Sigma Mx}{\Sigma M}, \qquad \bar{z} = \frac{\Sigma Mz}{\Sigma M}$$

The centre of gravity of the aircraft along the y axis is not usually evaluated since most aircraft are symmetrical about the xoz plane (i.e. the vertical aircraft centre plane).

Evaluation of the aircraft moments of inertia using the data in Table 7.2 is shown at the end of this chapter.

Table 7.2

Item	M	x	Mx	Mx^2	z	Mz	Mz^2
Wing . . .							
Tail . . .							
Body . . .							
etc.							
Totals	ΣM		ΣMx	ΣMx^2		ΣMz	ΣMz^2

Position of mass components

When the position of component mass is unknown the following rules can be used.

Wing: (unswept)   38–42% from leading edge at 40% semi-span from aircraft centreline

(swept)   70% of distance between front and rear spar, behind front spar, and at 35% semi-span from aircraft centreline

Fuselage: distance from fuselage nose   (% fuselage length)
wing-mounted–propfan engines   (38–40)
wing-mounted–turbofans   (42–45)
rear fuselage-mounted engines   (47)
engines buried in fuselage   (variable)

Tailplane: 42% chord at 38% semi-span from aircraft centreline

Fin: (T-tail)   42% chord at 55% above root chord
For other types of tail layout you will have to guess.

Nacelles: 40% nacelle length from nose

Surface controls: at trailing edge at mean aerodynamic chord position

Landing gear: at aircraft centre of gravity of the aircraft if detailed positions of the legs have not been decided, or at the wheel centres if the geometry is known

Fuel tanks (full)   for a prismoid tank (Fig. 7.13) with a length L between parallel end faces of areas $S_1 S_2$, the centre of gravity can be calculated to be at a distance (from plane S_1) by the formula below:

$$\frac{L}{4}\left(\frac{S_1 + 3S_2 + 2\sqrt{S_1 S_2}}{S_1 + S_2 + \sqrt{S_1 S_2}}\right)$$

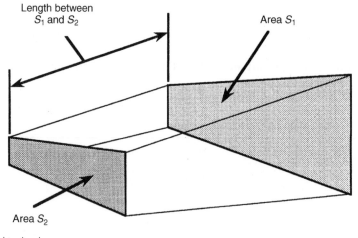

Fig. 7.13 Fuel tank volume.

Aircraft balance diagrams

The previous chapter has dealt with aircraft in only one loading condition (usually at maximum take-off mass). The aircraft will be flown at weights less than the maximum due to variations in the stage distance flown and the number of passengers carried. For full confidence in the design it is necessary to consider all the variations that are possible in the load and payload distributions to establish the most forward and rearward centre of gravity positions. These positions will affect the stability and control characteristics of the aircraft and will dictate the critical loads on the tail surfaces and landing gear. Not all aircraft have a wide variation in payload and in these cases the centre of gravity limits are easily determined in the tabular manner described earlier. For other aircraft, the centre of gravity positions are best described using a graph called the balance diagram.

The balance diagram

A typical balance diagram is shown in Fig. 7.14. The following notes describe the construction of the diagram for a conventional single or twin-aisle airliner.

1. The position of the aircraft centre of gravity in the operational empty condition (i.e. without payload and fuel) can be found by summing the influences of all component masses using the tabulation method described earlier (Table 7.2). It is usual to allow 1–3% variations about the nominal position to account for variations of operational items (passenger movement, undercarriage retraction, etc.) (see points A and A′).
2. The loading of passengers is assumed to take place progressively (seat by seat). Line A to B assumes the window seats from front to back are occupied first. Line A′B′ considers window seats from back to front to be chosen first.
3. The passenger loading is considered to continue with the seats next to the aisles (BC from front to back, B′C′ from back to front).
4. The remaining seats are then filled in the same way (CD and C′D′).
5. For the most forward centre of gravity position, the cargo in the front bay is added to the most forward position of the passenger loops (point E). However, this case may be considered as too critical as the aircraft load master (i.e. the ground engineer responsible for loading the aircraft) may decide to load the rear cargo if the forward passenger location is considered as critical.
6. The rear cargo is then added (point F).
7. The fuel is then added (point G).
8. Cargo is loaded similarly for the most rearward aircraft c.g. position, (I) (J).
9. At the most critical point (K) on the rear loading position the fuel can be added to indicate the most aft centre of gravity position (L).

The positioning of the cargo in the various holds can be used to bring the aircraft centre of gravity into an acceptable central position. This task is the responsibility

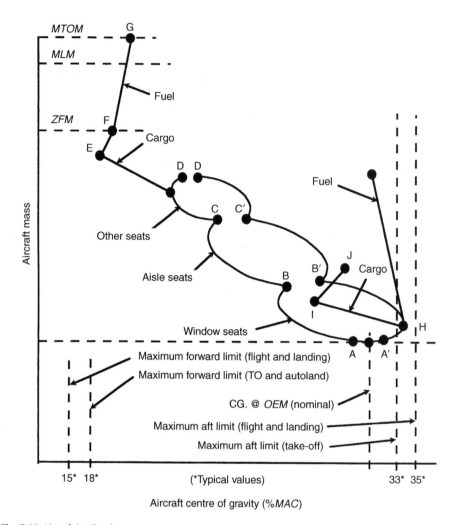

Fig. 7.14 Aircraft loading loops.

of the airline load master in consultation with the aircraft captain or other flight crew.

In some aircraft (e.g. Concorde) the change in aircraft centre of gravity and centre of lift positions in flight is stabilised by transferring fuel from forward to rear tanks during flight (as ballast) but this technique is not recommended on conventional commercial aircraft designs.

Aircraft layout can have considerable influence on the shape of the aircraft balance diagrams. Configurations with engines either forward of the wing or on the rear fuselage cause the loading loops to be inclined as shown diagrammatically in Fig. 7.15. To provide most flexibility to the operation of the aircraft the centre of gravity range (maximum rear to maximum forward positions) must be kept as small as possible. For large civil transport aircraft the maximum aft position

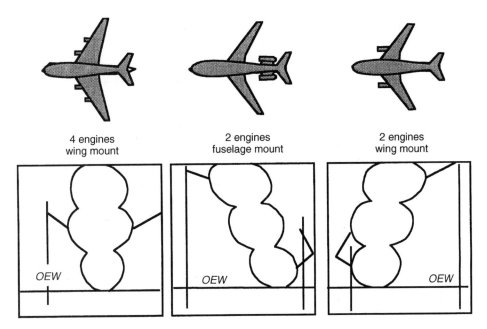

4 engines
wing mount

2 engines
fuselage mount

2 engines
wing mount

OEW

OEW

OEW

Fig. 7.15 Effect of aircraft configuration on the loading loops.

should be limited to about 35 to 40% aft of the leading edge of the wing mean aerodynamic chord (*MAC*) and the maximum forward position of 8% *MAC*. A typical configuration would be 11%–31% (e.g. A300). In general, the greater the centre of gravity range and the larger the payload the bigger and heavier will be the required tail surfaces.

Payload–range diagram

The loading options for the aircraft are often shown in the form of a diagram giving various proportions of fuel mass and payload in terms of range. This is shown in Fig. 7.16.

A description of the diagram (for a typical airliner) is given below. The numbers on the diagram refer to the notes below.

1. Start the payload–range diagram by drawing the vertical axis for mass and the horizontal axis for range.
2. Draw the horizontal axis (range) at the aircraft operational empty mass (*OEM*).
3. For the aircraft to fly, fuel must be added. Note the usable fuel mass line starts above the *OEM* position due to the mandatory fuel reserves (hold, diversion and contingency) that must be carried. As fuel is added, the aircraft is capable of flying further but note that the relationship is not linear.

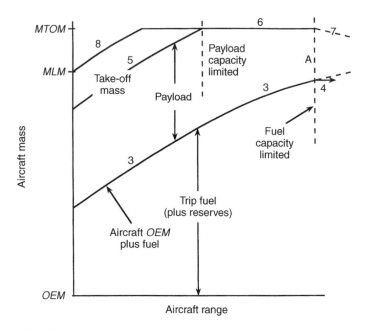

Fig. 7.16 Aircraft payload–range limitations.

4. The aircraft will have a limit (line A) at the maximum available volume for fuel. This is determined by the physical capacity of the fuel tanks available within the aircraft structure. This limitation is sometimes overcome by the addition of long-range external wing- (or fuselage-) mounted tanks.
5. The maximum payload that can be carried may be set by either fuselage space limits or by structural strength (*MZF*). A line can be constructed above the 'aircraft empty plus fuel' line, displaced at the maximum allowable payload. This represents the take-off mass for the aircraft to fly a particular range.
6. The aircraft will be designed to a maximum value of take-off mass (*MTOM*). This will limit the upward growth of line (5). This represents the maximum range for the aircraft carrying full payload. The aircraft can only fly further by reducing the payload but keeping the aircraft at *MTOM*.
7. Without extra fuel tanks, the payload of the aircraft has to be further reduced, below the *MTOM* limit, to allow flight beyond the 'maximum fuel tankage' range (A).
8. The mass–range diagram is also restricted by the allowable maximum landing mass (*MLM*). This will limit the maximum additional fuel (plus reserves) that can be carried above that needed to fly the range.

Figure 7.16, although representing all the constraints on aircraft mass is too complicated to illustrate the operational limits of the aircraft. These are best described by constructing a graph of the aircraft payload that can be carried over a specified range as in Fig. 7.17.

The fuel required is assumed to be implicit in the diagram. Often the part of

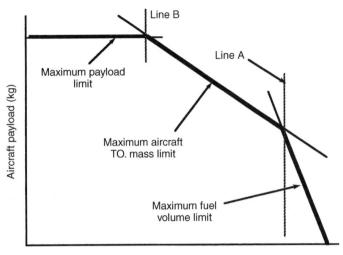

Line B

Line A

Maximum payload
limit

Maximum aircraft
TO. mass limit

Maximum fuel
volume limit

Aircraft payload (kg)

Aircraft range (excluding reserves) (nm)

Fig. 7.17 Aircraft payload–range diagram.

the diagram to the right of line A is not drawn by the manufacturer unless extended range requirements (e.g. ferry) are demanded from an operator.

The portion of Fig. 7.16 bounded by lines 3, 4, 5, 6 and 7 is used for the construction of the conventional payload–range graph shown in Fig. 7.17.

The payload–range diagram can be used to illustrate different versions of the same aircraft model by superimposing (a) lower numbers of passengers, (b) lower maximum weight limits, and (c) different maximum fuel capacities as shown in Fig. 7.18.

Aircraft layout and balance

During the initial project work the position of the wing along the fuselage will need to be decided. The chosen location will affect the position of the aircraft centre of gravity. This centre of gravity (c.g.) position will influence the payload loading limits for the aircraft. If the c.g. position of the OEM is too far aft the subsequent loading will be unacceptable for nosewheel undercarriage loading and if too far forward there will be insufficient tail power to balance and control the aircraft. There are several empirical methods that can be used to locate the correct 'matched' position of wing and body but they all use iterative techniques.

The simplest method is to consider the aircraft operational empty mass in two separate groups.

1. *The wing and associated masses* (M_w)
   which includes wing, fuel system, main undercarriage (even if attached to fuselage) and wing-mounted engines and systems.

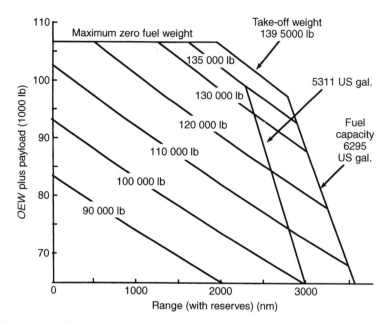

Fig. 7.18 Actual aircraft payload–range diagram (source Boeing data).

2. *The body-associated components* (M_f)

which includes fuselage, equipment, furnishings and operational items, airframe services, crew, tail, nose undercarriage and fuselage-mounted engines and systems.

Note, if the position of wing-mounted engines is linked to the fuselage layout (e.g. by the position of the fan or turbine noise planes), then the engine mass would be transferred to the body group.

If a sensible guess can be made for the position of the *OEM* centre of gravity position as a percentage of wing mean aerodynamic chord, then moments may be taken about the wing leading edge to give:

$$x_f = x_{OEM} + (x_{OEM} - x_w) \frac{M_w}{M_f}$$

where: x_f = the position of the fuselage group c.g. (%*MAC*)

x_{OEM} = the chosen position of aircraft *OEM* c.g. (%*MAC*)

x_w = the determined position of the wing group c.g. (%*MAC*)

To ensure that none of the mass components have been omitted, make sure that

$$M_w + M_f = M_{OEM}$$

An overlay of the wing and fuselage layouts is therefore possible with the centre of gravity position of the body group set at x_f behind the leading edge of the wing mean chord.

To select a suitable value for the aircraft *OEM* centre of gravity position it is necessary to follow previous experience on aircraft of similar layout and type. For the configurations studied in the balance diagrams shown earlier (Fig. 7.15) the following values are appropriate:

A (conventional wing-mounted engines): $x_{OEM} = 25\% MAC$
B (fuselage-mounted engines): $x_{OEM} = 35\% MAC$
C (wing forward-mounted engines): $x_{OEM} = 20\% MAC$.

For aircraft in which the c.g. range is to be restricted use $x_{OEM} = 25\% MAC$.

References

1. SAWE, Paper No. 2228, May 1994.
2. Wing mass formula for subsonic aircraft, S. V. Udin and W. J. Anderson. *AIAA Journal of Aircraft*, Vol. 29, No. 4, 1991.
3. Synthesis of subsonic airplane design, E. Torenbeek. Delft University Press, Netherlands, 1981.
4. Aircraft weight prediction – Part 1, D. Howe. CofA Design Note DES126/1, Cranfield University, UK, 1970.

Reference data

Definitions of wing geometry

In the estimation of mass and for other calculations different definitions for the wing geometry are quoted. To understand these the following notes are given.

Mean aerodynamic chord (MAC)

$$\bar{c} = \frac{2}{S} \int_0^{b/2} c^2 \, dy$$

which for straight tapered wings is equal to:

$$\frac{2}{3} c_r \frac{1 + \lambda + \lambda^2}{1 + \lambda}$$

where: $\lambda = c_t/c_r$

c_r is the root (centreline chord) and c_T is the tip chord; the position of the MAC can be determined graphically as shown in Fig. 7.19.

Standard mean chord (SMC)

$$SMC = \frac{S}{b}$$

where: $S = $ gross wing area
$b = $ wing span

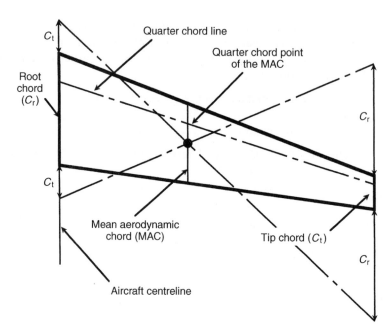

Fig. 7.19 Graphical estimation of *MAC* position.

Aerodynamic centre

This is defined as the point in the *xoz* plane of the wing about which the aerodynamic pitching moment coefficient of the wing is sensibly constant up to the maximum lift in sub-critical flow. For moderate sweep and taper you can assume it to act at the quarter chord position aft of the leading edge on the *MAC*. To make analysis easier all wing aerodynamic forces are assumed to act at this point.

Definitions of aircraft mass

The main sub-groups of mass (e.g. structure, payload, etc.) may be grouped in several different ways. There appears to be no internationally preferred definition and you may find any of the definitions below quoted in reports and in aircraft data sheets.

MEM	manufactured empty mass = the weight as produced at the factory (i.e. the sum of [structure + propulsion + fixed equipment]).
DEM	delivered empty mass = *MEM* plus mass of standard removable items (i.e. those attributable to particular customer requirements).
EM (Dry)	empty mass dry = *DEM*.
BEM	basic empty mass = *MEM* plus removable items (sometimes referred to as basic mass).

OEM	operational empty mass = mass of aircraft that is operational but without payload and fuel.
APS	aircraft prepared for service = OEM, sometimes referred to as basic operational mass BOM.
ZFM	zero fuel mass = OEM plus payload.
TOM	take-off mass = mass of aircraft at start of take-off run, within the structural limit $MTOM$ (maximum TOM).
Ramp mass	TOM plus fuel used for engine run-up and taxiing out to the runway prior to take-off.
LM	landing mass = the mass of the aircraft at moment of landing with structural limit MLM (maximum LM).
Gross mass	the aeroplane weight at any point in the flight but sometimes confused with AUM below.
AUM	all-up mass = gross mass (within a maximum structural limit of $MAUM = MTOM$).
Operating mass	OEM plus fuel = zero payload mass.
Useful load	(payload + fuel) = disposable load.
Payload	all 'commercial' load carried. This may be limited by volume capacity in the fuselage and cargo holds or by overall structural loading.

Moments of inertia

In order to analyse the control characteristics of the aircraft it is necessary to determine the moments of inertia (I_x, I_y, I_z) about the three axes passing through the aircraft centre of gravity. The conventional parallel axis theorem and summation rules from basic mechanics can be used:

$$Mx^2 = \Sigma Mx^2 - \bar{x}^2 \Sigma M$$
$$My^2 = \Sigma My^2$$
$$Mz^2 = \Sigma Mz^2 - \bar{z}^2 \Sigma M$$

and

$$I_x = My^2 + Mz^2 + \Sigma \Delta I_x$$
$$I_y = Mz^2 + My^2 + \Sigma \Delta I_y$$
$$I_z = Mx^2 + My^2 + \Sigma \Delta I_z$$

where, ΔI_x, etc. are the moments of inertia of large items of mass about their own centre of gravity (for most components these terms are often small compared to the total aircraft inertia value and can be ignored).

To determine the principal moments of inertia it is necessary to estimate the product of inertia term:

$$M_{xz} = \Sigma M_{xz} - \bar{x}\,\bar{z}\,\Sigma M$$

and the principal angle (θ) between the x and z axes

$$\tan 2\theta = \frac{2M_{xz}}{I_z - I_x}$$

The principal moments of inertia are:

$$I'_x = I_x \cos^2 \theta + I_z \sin^2 \theta + M_{xz} \sin^2 \theta$$

$$I'_y = I_y \qquad \text{(since } xoz \text{ is plane of symmetry)}$$

$$I'_z = I_z \cos^2 \theta + I_x \sin^2 \theta + M_{xz} \sin^2 \theta$$

For many aircraft designs the principal angle θ is small, making it possible to assume that: $I'_x = I_x$, $I'_z = I_z$.

By convention, the moments of inertia evaluated above may be referred to by letters:

A = roll moment of inertia $= I'_x$
B = pitch moment of inertia $= I'_y$
C = yaw moment of inertia $= I'_z$.

This notation is used in aircraft stability and control equations and for aircraft load analysis.

8

Lift and drag estimates

Objectives

A knowledge of the lift and drag characteristics of the aircraft is fundamental to determination of its flying qualities. This chapter describes how lift and drag can be derived from a known aircraft geometry and conversely, how the selection of aircraft layout influences the generation of lift and drag. As the fuselage shape is predetermined by the payload requirements, the nacelle geometry by the engine size, and the tail surfaces by stability considerations, the wing parameters are the principal variables to be considered in respect to lift and drag aspects. To this end, guidance is given on the selection of wing section profile and planform geometry to provide specified performance requirements. A drag estimation method covering all the aircraft components (fuselage, nacelle, wing and tail surfaces) is also provided. Such methods are suitable for use in the conceptual design phase and provide an introduction to more detailed estimation procedures for later design stages.

Introduction

The ability to estimate the lift and drag capability of the aircraft in various configurations (e.g. aircraft mass and flap variations) and in different flight conditions (e.g. cruise, climb, take-off, landing) is fundamental to the process of aircraft design. All components (e.g. wing, fuselage, tail, engine) will need to be considered in the total estimation process but the wing is the most significant. Fuselage shape and size is mainly determined by the requirements associated with the payload (e.g. passenger seating arrangements, cargo container sizes). Nacelle shape is associated with engine geometry and flow requirements into and away from the core engine. Tail sizing is mostly a function of stability and control requirements. On the other hand the wing size and shape has considerable impact on the lift and drag of the aircraft. Careful selection of the wing geometrical features will be a central issue in the design of the aircraft to meet lift requirements and to reduce overall drag. As many of the issues in this process have been discussed in Chapter 6, only those aspects directly related to the estimation of lift and drag are discussed here.

Lift and drag forces on the aircraft are determined as the vertical (normal to the flight direction) and horizontal (along the flight direction) components of the overall pressure field around the aircraft. As such it is not surprising that lift and drag are not independent parameters. As we shall see in the estimation process drag will be assessed partly as a function of shape and surface condition (profile drag component) and partly as a function of lift (lift induced drag component). The generation of sufficient lift for a flight condition may require changes in wing profile (e.g. flap deployment). Such changes will have considerable impact on drag evaluation.

The overall pressure distribution around the aircraft will also contribute a pitching moment about the point which the lift and drag components are assumed to act. This pitching moment will vary for different aircraft configurations (e.g. flap and undercarriage deployment) and must be considered when the balance of forces on the aircraft is assessed as it will be one of the major contributors to the requirement for tail size and efficiency.

The sections below describe initially the inter-relationship of wing shape with the generation of lift and drag and subsequently a method of estimating lift and drag for the whole aircraft from a knowledge of the component geometry.

Wing geometry selection

Assuming that the gross wing area required for the aircraft has been determined from initial estimates (see Chapter 11), there are two geometrical parameters to be determined; namely, the aerofoil section profile and the wing planform shape. Three operational cases need to be considered when making these selections; the wing must be efficient in the en-route (cruise) configuration; it must have sufficient lift generation capability in the take-off and landing configurations (i.e. with flaps deployed), and it must have the lowest possible drag in the critical engine-out certification climb segments after take-off.

Aerofoil section profile

The detailed specification of aerofoil profile shape requires the definition of wing chord length, camber shape, maximum thickness and the leading edge, LE, radius as shown in Fig. 8.1. To generalise the geometry, the lengths are non-dimensionalised by dividing each parameter by the chord length to create a ratio (e.g. thickness/chord ratio).

The sectional shape will vary along the wing span to provide smooth chordwise flow conditions over the different parts of the wing and to guard against outboard (tip) stalling of the wing. Aerofoil section data available at the conceptual design stage will be related to two-dimensional flows over a constant section shape. This data will need to be modified to take account of three-dimensional flow conditions and to integrate the changing section shapes along the wing span.

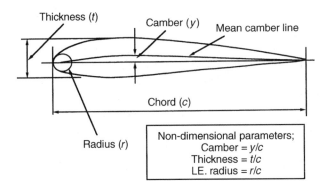

Fig. 8.1 Section geometry definition.

The wing section shape will determine the maximum practicable lift coefficient C_L, the stall characteristics, the lift curve shape (C_L versus wing incidence, α), the profile drag, the critical Mach number and the shape and extent of compressibility drag rise. All these criteria are important as they dictate overall aircraft operational and design parameters. For example, the maximum lift capability of the clean wing will dictate the type and size of high-lift devices necessary to meet the field performance requirements (e.g. approach speed); the lift curve shape will determine the wing incidence in the cruise phase and thereby the fuselage to wing setting angle to minimise the aircraft overall drag; and the drag rise curve will dictate the maximum cruise speed with engines at the cruise setting.

In general, sections with a high thickness/chord ratio and high camber give high maximum lift coefficient but they lead to higher drag and a lower critical Mach number (lower cruise speed). All modern turbofan aircraft have section shapes known as 'super-critical'. These are designed to have positive pressure gradients up to about the half chord position. This helps to delay the formation of local shock waves on the upper surface. Such profiles have high values for critical Mach number (about 0.06 Mach number increase over a conventional section of similar thickness/chord ratio) but lose out slightly on the generation of maximum lift.

A more detailed account of sectional geometry changes can be found in standard textbooks on aerodynamics and wing section design.

Wing planform geometry

The planform of a wing, for a given gross area, is defined by the geometric parameters, aspect ratio, taper ratio and sweepback. These are shown in Fig. 8.2(a). Many wing planforms incorporate a cranked trailing edge (see Fig. 8.2(b)). This improves the efficiency of the flap and moves the centre of lift of the wing closer to the centre of the aircraft creating a lighter wing structure. For initial design studies it is sometimes necessary to avoid such complications in the planform geometry and to approximate the total area to an equivalent straight tapered shape as shown in Fig. 8.2(b). The choice of the wing thickness and planform parameters is discussed in Chapter 6.

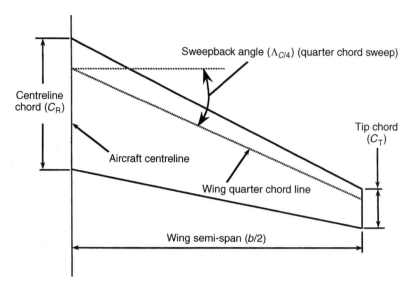

Fig. 8.2(a) Wing planform geometry (straight taper).

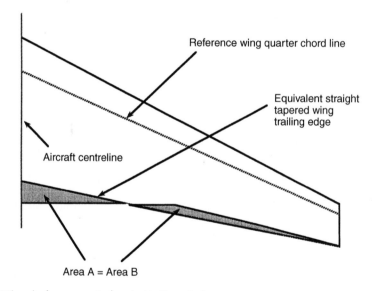

Fig. 8.2(b) Wing planform geometry (cranked trailing edge).

Lift estimation

Design lift coefficient

On a modern civil jet aircraft the design lift coefficient will be that appropriate to the cruising condition at the design cruise Mach number. A simple formula

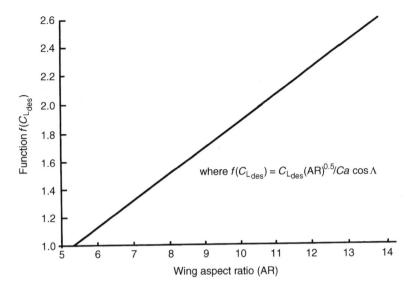

where $f(C_{L_{des}}) = C_{L_{des}} (AR)^{0.5}/Ca \cos \Lambda$

Fig. 8.3 Derivation of design lift coefficient.

Table 8.1

M_{des}	$C_{L_{des}}$	Ca
0.65–0.70	0.14–0.60	1.10
0.70–0.75	0.30–0.55	1.07
0.75–0.85	0.30–0.45	1.05
0.85–0.95	0.20–0.30	1.02

defining design lift coefficient as a function of basic wing geometry is given in Fig. 8.3 (see also ref. 1).

The factor Ca in the above relationship reflects the influence of wing camber on the design lift coefficient. For super-critical sections where the deviation of the aerofoil meanline is zero (near zero) until at least 70% of the wing chord, Ca may be taken as 1.0. For conventional sections where camber is distributed along the aerofoil chord, values of Ca will vary from 1.02 and 1.05 for low camber sections to 1.05 and 1.15 for significantly cambered sections. Some guidance on likely values for Ca for conventional sections is given in Table 8.1.

Maximum lift coefficient

The choice of section profile will dictate the section data used in the assessment of the section maximum lift coefficient. Due regard to the effect of Reynolds number must be given in the process. On a three-dimensional wing with taper, the section maximum lift coefficient will vary along the span due to the effect of reducing Reynolds number. The spanwise variation of lift coefficient will depend

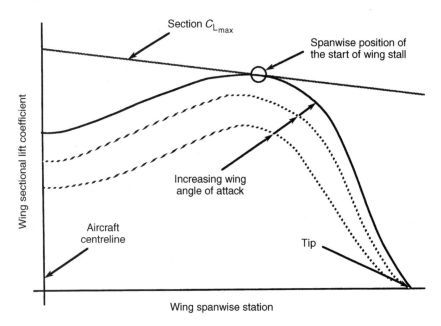

Fig. 8.4 Spanwise lift distribution.

mainly upon the spanwise loading as shown in Fig. 8.4. The angle of attack at which the two curves coincide will indicate the spanwise position of the start of the wing stall.

The maximum lift coefficient achieved by the wing will always be less than the section maximum value as shown in Fig. 8.4.

The effect of sweepback must also be taken into account in the estimation of lift from the wing. A typical relationship is shown below:

$$(C_{L_{max}})_{3D} = 0.9 \, (C_{L_{max}})_{2D} \times (\cos \Lambda)$$

Modern turbofan aircraft have a clean wing maximum lift coefficient of about 1.5 when taking into account the above effects. Without other consideration such a value would result in a very large wing area to meet the field and climb requirements. To increase the maximum wing lift coefficient, high-lift devices are used.

Effect of high-lift devices

High-lift devices generally fall into two categories: trailing edge (TE) devices and leading edge (LE) devices. Trailing edge devices decrease the zero lift angle and thus increase the lift for a given angle of attack as shown in Fig. 8.5.

Several flap types are available. These range from the simple split flaps to complex triple-slotted flaps. A selection of trailing flaps is shown in Fig. 8.6.

Trailing edge flaps are of two types; those that do not increase the chord of the wing and those that employ chord extension. The latter type are termed Fowler flaps. These are more effective as they produce extra lift by increasing the overall

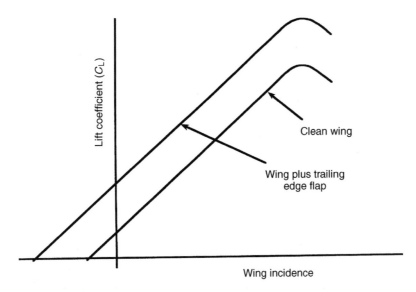

Fig. 8.5 Effect of trailing edge flaps on lift curve.

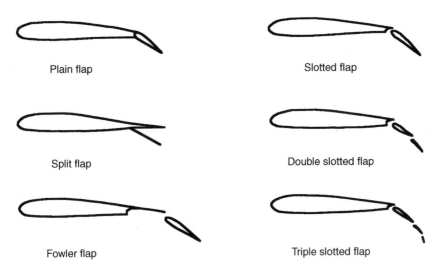

Fig. 8.6 Types of trailing edge flaps.

wing area as well as adding extra profile camber from the flap deflection. The maximum lift increment for a simple flap typically occurs at a deflection angle of 50°. Above this angle the flow separates and extra lift is lost. To increase the lift increment further, double- and triple-slotted flaps are often employed. Each section is deflected at a greater angle to increase lift whilst preventing separation of the flow. However, such double- and triple-slotted flaps increase mechanical complexity and wing mass.

More recently, some civil transport aircraft have coupled aileron deflection to

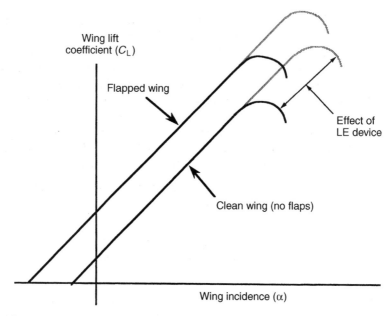

Fig. 8.7 Effect of leading edge devices on lift curve.

further increase the maximum lift coefficient and improve control during approach. An example of this type of design is the Airbus A300–600 aircraft with a single-slotted Fowler flap which at maximum deflection is linked to an aileron droop of nine degrees.

Leading edge devices increase the maximum lift on the wing by preventing wing leading edge stall increasing the maximum angle of attack as shown in Fig. 8.7.

As with trailing edge flaps, there are several different types of leading edge devices available as shown in Fig. 8.8.

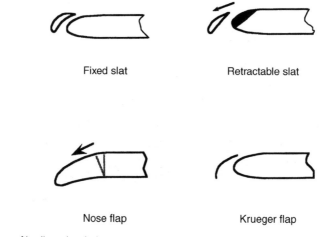

Fig. 8.8 Types of leading edge devices.

Most modern civil transports have full span leading edge slats whilst older designs normally have inboard Krueger flaps and outboard leading edge slats. Slats work by increasing the camber and chord of the wing. The increment in lift produced is limited by the subsequent stall at the trailing edge flaps. Krueger flaps increase lift in a similar way to slats but without increasing the wing chord. They are sometimes used on the inboard section of wings, with slats used outboard for their favourable pitch characteristics and to help prevent tip stall.

It should be noted that leading edge devices are often interrupted along the span by engine pylon structures which reduce their effectiveness. All leading edge devices increase mechanical complexity of the wing and thereby mass and cost.

Estimation of maximum lift coefficient

The lift increment for a wing with both leading and trailing edge devices may be estimated from two-dimensional section data using the formula below:

$$\Delta C_{L_{max}} = \Delta C_{L_{max}} \, (S_{flapped}/S_{ref}) \cos \Lambda_{HL}$$

where: $\Delta C_{L_{max}}$ = sectional (2D) lift coefficient increment of the device
$S_{flapped}$ = wing area in flowpath of the device (defined in Fig. 8.9)
S_{ref} = aircraft gross wing area, the design reference area
Λ_{HL} = sweepback angle of the hinge line of the device – may be approximated to the wing TE sweep (for flaps) or LE sweep (for slats, etc.) if wing details are unavailable

The sectional two-dimensional lift coefficient increments for various leading edge devices are shown in Table 8.2. The term c'/c refers to the amount of chord extension of the slat. This is typically 1.05–1.10, based on an empirical relationship of the percentage of exposed wing span for c'/c as shown in Fig. 8.10.

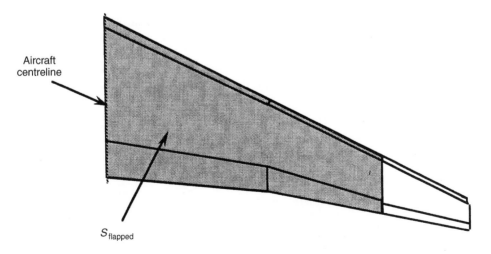

Aircraft
centreline

$S_{flapped}$

Fig. 8.9 Definition of $S_{flapped}$.

Table 8.2

Leading edge device	$\Delta C_{L_{max}}$
Krueger flap	0.3
Slat	$0.4\,c'/c$

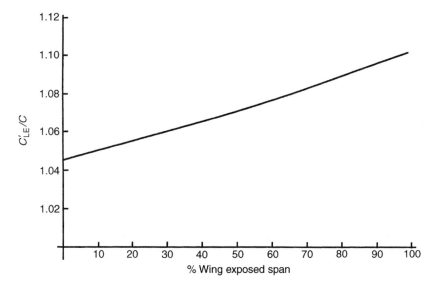

Fig. 8.10 Typical values of effective chord for leading-edge high-lift devices.

The sectional (two-dimensional) lift coefficient increments for various trailing edge devices are shown below (Table 8.3). The values are representative of landing flap settings. For take-off values use 60 to 80% of these values.

For trailing edge flaps the term c'/c represents the amount of chord extension due to Fowler movement. Typical values are 1.25–1.30. A statistical relationship for some common types of trailing edge flaps for c'/c as a function of flap angle is shown in Fig. 8.11.

It can be seen from Table 8.3 that the lift increment increases substantially

Table 8.3

Leading edge device	$\Delta C_{L_{max}}$
Double-slotted flap	1.6
Triple-slotted flap	1.9
Single-slotted Fowler flap	$1.3\,c'/c$
Double-slotted Fowler flap	$1.6\,c'/c$
Triple-slotted Fowler flap	$1.9\,c'/c$

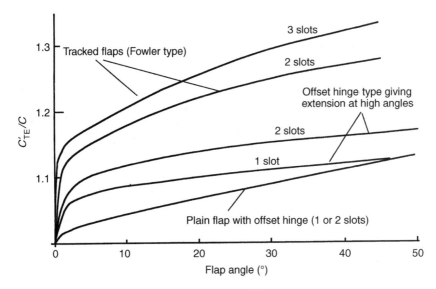

Fig. 8.11 Typical values of effective chord for trailing-edge flaps (fully extended).

moving down the table but at the expense of extra mechanical complexity, mass and cost. Aircraft designers adopt the simplest system consistent with the performance requirement. Again there is a possible trade-off between smaller wing area with complex flap system or vice-versa.

The aircraft data file (Data A) shows the type of high-lift devices used on some past and current civil transport aircraft along with the wing maximum lift coefficient obtained. Some of these are plotted in Fig. 8.12 as a function of the wing sweepback angle (quarter chord). The lines show a cosine relationship.

Comparison with actual aircraft data shows the above estimation method to be reasonably accurate while being simple to apply.

Lift curve slope

The lift curve slope is important as it affects the aircraft angle of attack in the take-off and landing configurations and also the setting of the wing relative to the fuselage. In the take-off case the angle of attack with the take-off flap lowered to an angle which develops sufficient lift for lift-off is required. A check can then be made that the aircraft geometry and undercarriage position are such that this angle of attack can be achieved on rotation. In the landing configuration the angle of attack is needed to establish that the pilot's view meets the appropriate airworthiness criteria. A knowledge of the lift curve slope in cruise is required so that the wing attitude relative to the fuselage can be set such that the fuselage is approximately level and at a low-drag altitude during cruise.

Theoretically the lift curve slope (per radian) for an isolated wing section is equal to 2π. In practice the value is somewhat less than this due to three-dimensional

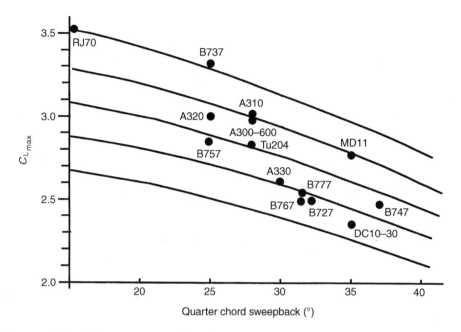

Fig. 8.12 Relationship between maximum lift coefficient and sweepback angle.

flow effects. The lift curve slope of a three-dimensional wing as a function of aspect ratio (A) can be shown to be

$$(dC_L)/(d\alpha) = 2\pi\,[A/(A+2)]$$

In general, flap deflection will only affect the zero-lift angle if the lift curve slope is held approximately constant. Further corrections are also required due to the position of the fuselage relative to the wing configuration (i.e. high-, mid- or low-) and for the effect of sweep. A more accurate method of determining lift curve slope is given by ESDU.[3]

Drag estimation

Each main component of the aircraft (wing, fuselage, tail surfaces, nacelles; and in the low speed flight phases the flaps and undercarriage) must be separately assessed for its contribution to the overall drag of the aircraft.[2] It is not sufficient to consider only the wing effects in the estimation of drag.

For subsonic civil aircraft the overall drag of the aircraft can be considered under three categories:

1. profile drag resulting from the pressure field around the shape and from the surface skin friction effects of the boundary layer;
2. lift induced drag resulting from the changes in pressure due to attitude variations resulting from the generation of lift;

3. wave drag from shock waves as parts of the accelerated flow over the surfaces become supersonic.

These effects result in the following formations for aircraft drag coefficient:

$$C_D = C_{D_0} + C_{D_i} + \Delta C_{DW}$$

where: C_{D_0} = estimated total profile drag coefficient (i.e. the summation of the drag from all the aircraft components appropriate to the flight conditions under investigation)

C_{D_i} = total effect of all the lift dependent components (principally this is a function of C_{D_i}; as the design becomes more established this term may be extended to include a direct C_L term as well as the square term)

ΔC_{DW} = additional drag resulting from the shock waves. As civil aircraft are not intended to be flown past the drag divergence Mach number this term may be assumed to be 0.0005 if no other details are available

The total drag coefficient can be plotted against C_L^2 to give the graph shown in Fig. 8.13 and known as the drag polar. Each of the drag terms will now be assessed.

Profile drag
The profile drag can be estimated using the formula below:

$$C_{D_0} = C_f F Q \, [S_{wet}/S_{ref}]$$

where: C_f = skin friction coefficient which is a function of Reynolds number

F = component form (shape) factor

Q = interference factor

S_{wet} = component wetted area

S_{ref} = reference area used for the calculation of C_D (normally the wing gross area)

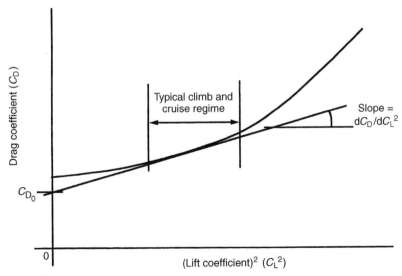

Fig. 8.13 Aircraft drag polar.

We start by calculating the Reynolds number Re for each component:

$$Re = (Vl)/u$$

where: V = aircraft forward speed in the flight case under investigation
u = kinematic viscosity at the speed and height of operation
l = component characteristic length, i.e. fuselage overall length, wing mean chord, tail mean chord, nacelle overall length

The skin friction coefficient for turbulent boundary layer conditions can now be calculated for each component using the Prandtl–Schlichting formula:

$$C_f = [0.455]/[(\log Re_c)^{2.58} (1 + 0.144 Mn^2)^{0.65}]$$

where: M = Mach number at operational conditions under investigation.
Re_c = Reynolds number of component

For any component or area with laminar flow the following equation should be used:

$$C_f = 1.328/(Re_c)^{0.5}$$

For components with both laminar and turbulent flows the value for C_f should be a weighted (by area) average of the two results.

The form factors for each component are calculated from the input geometry using a specific formula for each component.

(i) For the fuselage

$$F = 1 + 2.2/(\lambda)^{1.5} - 0.9/(\lambda)^{3.0}$$

where: $\lambda = l_f/[(4/\pi)A_x]^{0.5}$
l_f = fuselage overall length
A_x = fuselage cross-sectional area
(Note: $[(4/\pi)A_x]^{0.5}$ = fuselage diameter for circular fuselage shapes)
$Q = 1.0$

(ii) For the wing

$$F = (F^* - 1)\cos^2 \Lambda_{0.5c} + 1$$

$$F^* = 1 + 3.3(t/c) - 0.008(t/c)^2 + 27.0(t/c)^3$$

where: $\Lambda_{0.5c}$ = sweepback angle at 50% chord
$Q = 1.0$ for well filleted low/mid wings, $= 1.1$–1.4 for small or no fillet
(a value of 1.0 to 1.2 seems to work for conventional designs)

(iii) For the tail surfaces. F as for wings with

$$F^* = 1 + 3.52\,(t/c)$$

$$Q = 1.2$$

(iv) For the nacelle. Estimating the drag of a nacelle is complicated by the intricate

geometry of many nacelles and the inter-relationship with the definition of engine thrust. For initial estimates you may use $FQ = 1.25$ for wing-mounted engines and a 20% higher value for aft fuselage-mounted installations (to account for the increased interference on the rear of the aircraft).

(v) For the undercarriage. This is influenced by the size of the undercarriage and the number of wheels. At the preliminary design stage much of this data will not be available. The number of wheels and the general undercarriage size will be primarily influenced by the aircraft maximum landing weight. It is suggested that for aircraft with a multi-wheel bogie type undercarriage the following formulae may be used for undercarriage drag:

imperial units $\Delta D/q = 0.0025\,(W_L)^{0.73}$
W_L in lb and $(\Delta D/q)$ in sq ft
metric units $\Delta D/q = 0.00157\,(W_L)^{0.73}$
W_L in kg and $(\Delta D/q)$ in sq m

where $\Delta D/q$ is the increase in drag area (i.e. $S \cdot C_{D_0}$) where S is the wing reference area and W_L is the weight/mass of the undercarriage.

For smaller aircraft in the F100, DC9 and B737 class, which will usually have twin-wheel main undercarriages, the following formula is suggested:

imperial units $\Delta D/q = 0.006\,(W_L)^{0.73}$
W_L in lb and $(\Delta D/q)$ in sq ft
metric units $\Delta D/q = 0.00093\,(W_L)^{0.73}$
W_L in kg and $(\Delta D/q)$ in sq m

For a quicker result just assume $\Delta D/q = 0.02\,S$ (where S is the wing reference area).

(vi) For the flaps. The drag due to the high-lift systems depends upon the types of trailing edge and leading edge flaps envisaged for the aircraft under consideration. As was noted earlier there are several possibilities, from plain flaps to area extending flaps all with one, two or three slots. The method suggested for use at the preliminary design stage covers area extending flaps such as:

- Fowler type with one, two or three slots;
- flaps with offset hinge and a linkage to give some area extension at low flap angles again with one or two slots and plain flaps with offset hinge with one or two slots.

The parameters fundamentally influencing the flap drag increments have been taken as type of flap, flap angle, wing area increase and the sweep angle. The definition of the flap drag increment along with the above parameters is important, particularly the wing area increase. The symbols used are defined in the following list along with the associated figures:

ΔC_D is the flap drag increment = total drag increment for both leading and trailing edge devices extended, or for trailing edge devices only – see Fig. 8.14

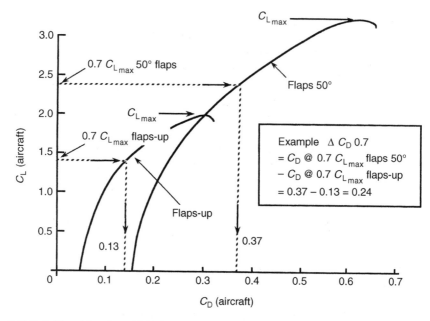

Fig. 8.14 Definition of flap drag at 0.7 $C_{L_{max}}$.

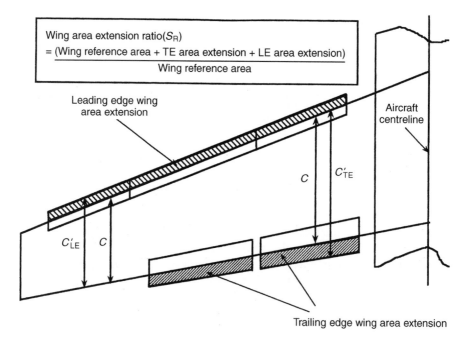

Fig. 8.15 Definition of wing area extension ratio.

$\Lambda_{0.25}$ wing 1/4 chord sweep

S_R extended flap area ratio as defined in Fig. 8.15

N number of slots in trailing edge flap systems

β trailing edge flap angle as defined in Fig. 8.16

C'_{TE} effective wing chord with wing trailing edge flaps extended (LE flaps retracted) as defined in Fig. 8.17

C'_{LE} effective wing chord with leading edge devices extended (TE flaps retracted)

$C_{L_{max}}$ certified maximum lift coefficient at appropriate flap setting

$\Delta C_{D1.21V_S}$ total flap drag increment at $C_{L_{max}}/1.44$, for TO/second segment climb

$\Delta C_{D1.31V_S}$ total flap drag increment at $C_{L_{max}}/1.69$, for landing/approach

The objective of this method has been to produce a quick estimation of the flap drag at take-off, second segment and landing approach conditions. Flap drag has therefore been presented at $1.2V_S$ and $1.3V_S$ (where V_S is the aircraft stall speed with flaps extended) at the appropriate flap condition. Figures 8.18 and 8.19 give the drag increment at $1.2V_S$ and $1.3V_S$ respectively. It is important that note is taken of the definition of this drag increment as shown in Fig. 8.14. By its nature this method of assessing flap drag is crude and more rigorous methods should be applied as soon as the flap and wing geometry have been established.

(vii) Drag of secondary items. The drag of secondary items may be as high as 10% of the profile drag calculated using the above method. The extra drag is typically due to excrescence, surface imperfections and system installations. For initial project design work the following estimates are suggested.

Wing: 6% of wing profile drag
Fuselage and empennage: 7% of fuselage profile drag
Engine installation: 15% of nacelle profile drag
Systems: 3% of total profile drag

In addition to the above items the cockpit windshield will increase fuselage drag by 2–3%.

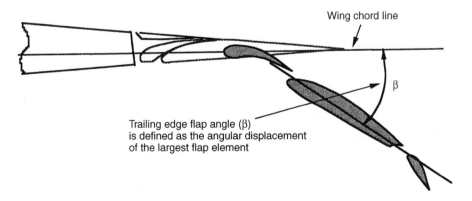

Wing chord line

β

Trailing edge flap angle (β)
is defined as the angular displacement
of the largest flap element

Fig. 8.16 Definition of trailing edge flap angle.

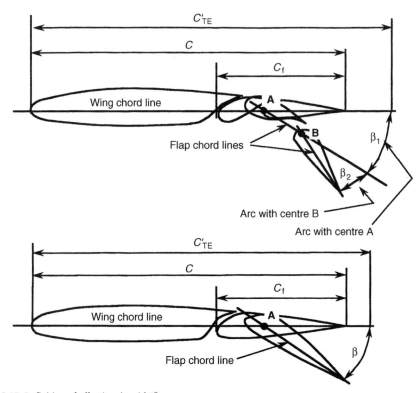

Fig. 8.17 Definition of effective chord (C').

Trim drag may also be a significant contribution to the total aircraft drag. At the initial project design stage, however, insufficient data is available to accurately estimate it. A typical figure for a well designed aircraft is about 5 drag counts (1 count $= 1 \times 10^{-4}$).

There are several published methods for estimating wetted area of the various components but all rely on the detailed definition of the layout. When you know the shape of the component it is relatively easy to make a detailed estimate of the wetted area of the part using your own initiative.

Some performance calculations are done with one engine failed. In such a condition the aircraft will be subjected to drag increases from the flow blockage of the failed engine and the extra drag from the asymmetric flight attitude of the aircraft. The drag increment from the engine, known as the windmill drag, can be estimated from $\Delta C_D = 0.3\, A_f/S$ (where A_f is the area of the fan cross section, and S is the wing reference area). The drag increment from asymmetric flight is difficult to predict quickly. The best that can be done in the project stage is to add a percentage to the overall drag of the aircraft. A value of 5% C_{D_0} seems reasonable for conventional design configurations.

Adding all the component drags together with the corrections detailed above gives the aircraft profile drag coefficient:

$$C_{D_0} = \Sigma(C_{D_{\text{components}}})$$

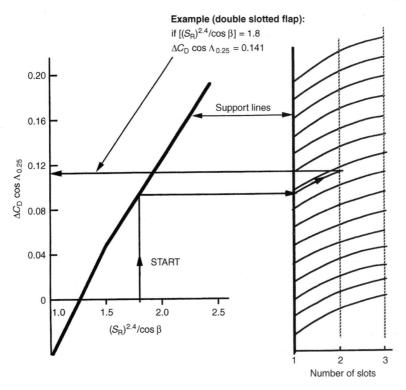

Fig. 8.18 Example of estimation of flap drag at 1.2 V_S.

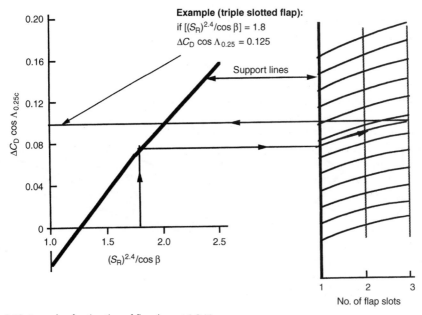

Fig. 8.19 Example of estimation of flap drag at 1.3 V_S.

Lift induced drag

Lift dependent drag arises from three principal effects:

1. a component from the wing planform geometry
2. a contribution from non-optimum wing twist
3. a component due to viscous flow forces

All of these effects are associated with the distribution of lift along the wing span (sometimes called span loading). The best (lowest induced drag) loading consists of a smooth elliptical distribution from wing tip to tip with no discontinuities due to fuselage, nacelles, flaps, etc. Obviously, except for high performance sailplanes, it is not feasible to arrange the aircraft layout to get such a spanwise load distribution. The comments below will help you to estimate the lift dependent drag coefficient for civil aircraft.

The component arising from the planform geometry is derived from classical lifting line theory details of which can be found in good aerodynamic textbooks. In this theory the wing is represented by a series of horseshoe vortices which generate the aerodynamic circulation around and along the wing shape. Figure 8.20 shows the theoretical distribution of the induced drag factor relative to the wing aspect ratio and taper ratio. These values are corrected by the application of an empirically derived factor (C_2) derived from previous aircraft designs (Fig. 8.21).

As might be expected, aircraft with older wing sections are inferior at high aspect ratios. Modern aircraft tend to adopt higher aspect ratios partly because the wings are designed using advanced technology three-dimensional aerodynamic analysis methods.

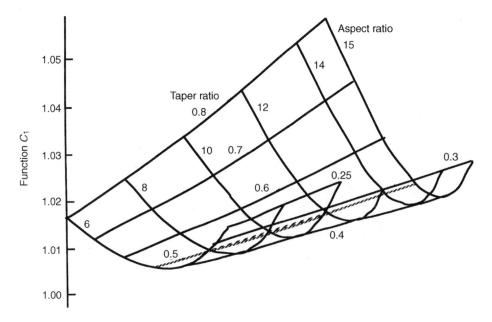

Fig. 8.20 Uncorrected planform factor C_1.

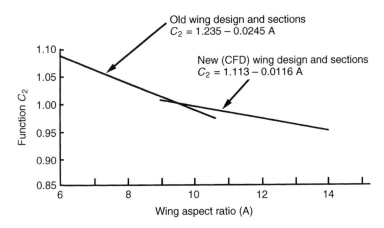

Fig. 8.21 Empirical correction C_2 to planform factor.

Induced drag coefficient is estimated by the equation below:

$$C_{D_i} = [(C_1/C_2)/(\pi A)]C_L^2$$

$$\text{i.e.} \quad dC_{D_i}/dC_L^2 = C_1/C_2/(\pi A)$$

where: C_1 is found from Fig. 8.20

C_2 is found from Fig. 8.21

A is the wing aspect ratio

C_L is the lift coefficient of the aircraft in the flight condition under investigation (i.e. aircraft mass, speed and altitude)

The contribution from the effect of non-optimum wing twist requires a knowledge of the distribution of aerofoil section twist and changes of the sectional lift curve shape, along the span. In the early stages of the project design such intricacies in wing geometry will not have been decided; however, a contribution is appropriate as the final wing shape will include such distributions. The addition of C_{D_i} for such effects will be between 0.0003 to 0.0005 with a value of 0.0004 being suitable for conventional civil turbofan layouts.

The viscous flow effects are significant. These forces manifest themselves mainly in the boundary layer growth arising from changes in wing incidence. Without the assistance of powerful computer fluid flow analysis it is difficult to predict these effects accurately. An empirical analysis of conventional civil aircraft geometry and operating conditions shows that the contribution to dC_D/dC_L^2 is proportional to aircraft profile drag. The following relationship is appropriate for current wing geometry.

$$dC_D/dC_L^2 = 0.35\,C_{D_0} \quad \text{(for older technology designs)}$$

$$dC_D/dC_L^2 = 0.15C_{D_0} \quad \text{(for advanced technology designs)}$$

The older technology value seems to work well for B737–300, B757 and B767, the A320 lies in between the two values and the A330/340 and B777 match the advanced technology value.

Hence the total lift dependent drag coefficient for the aircraft is the sum of the three effects:

$$dC_D/dC_L^2 = C_1 C_2/\pi A + 0.0004 + 0.15 C_{D_0}$$

Wave drag

As civil aircraft are designed to be operated away from the worst effects of wave drag rise it is acceptable to use a simple addition to the aircraft drag when operating at speeds greater than M 0.7 (e.g. in cruise). The additional drag due to compressibility will lie between 5 and 20 drag counts.

Cruise drag

The method above will enable an estimate to be made of the lift to drag ratio (L/D) in the cruise phase. Early generation jet airliners had L/D values in the range 15 to 17. Modern airliners using super-critical wing technology have values around 18 to 22.

References

1. SAWE, Paper No. 2228, May 1994.
2. NASA CR 151971 (Lockheed) 1978, 'A delta method for empirical drag build-up techniques'.
3. ESDU Data Sheet 70011 with amendments A to C, July 1977.

9

Powerplant and installation

Objectives

The main engine characteristics of concern to the aircraft designer are:

- maximum engine thrust available in the various segments of the flight (i.e. take-off, climb and cruise)
- engine fuel consumption
- engine mass
- engine geometry

This chapter describes how the basic engine parameters affect these engine characteristics and thereby enables you to understand the fundamental reasons for the differences between individual engines. At the commencement of an aircraft design study there are usually a number of possible engines available in the appropriate thrust range. The data in this chapter should enable the most appropriate type of engine to be selected.

If an engine is not available at the appropriate thrust, an assessment of the required engine size needs to be made. The aircraft thrust requirements cannot be divorced from the basic engine characteristics as these affect the size of the aircraft required to meet the design specification, and by suitable iteration the thrust requirements. Help is given on how to scale a known set of engine characteristics to enable such an assessment to be carried out. Guidance on how to derive the engine nacelle geometry from the basic engine dimensions and how to install the nacelle under a wing is also provided.

No attempt has been made to show how to calculate detailed engine performance. The main objective of this chapter is to enable you to understand the fundamental reasons for the differences between the engine characteristics of various engine types (for example, you will be able to understand how the specific thrust affects the engine characteristics). A simplified approach to the fundamental principles and the derivation of some formulae used in the engine analysis is given towards the end of the chapter. References to textbooks on gas turbine theories are given at the end of the chapter.

Introduction

Current high bypass ratio turbofan engines are thermodynamically much more efficient than the early turbojet and low bypass types. This has been largely brought about by the introduction of advanced technologies which have enabled turbine blades to withstand high centrifugal loads whilst operating in gas temperatures considerably higher than the melting point of the unprotected blade material.

The Chairman of Rolls Royce once summed this up by saying

> high pressure turbine blades operate at temperatures roughly 300°C above the melting point of the metal from which they are constructed . . . additionally each blade experiences a centrifugal load equivalent to having a double-decker bus hanging on it.

Many technologies are brought to bear to enable an engine to be designed that is safe and has a service life acceptable to the airlines.

In the following paragraphs the factors that influence the choice of engine characteristics are discussed. To illustrate the underlying principles a simplified approach has been adopted. The choice of powerplant will depend upon the duty required of it by the particular aircraft and the associated cost. A cost efficient engine will have:

- low initial price
- low maintenance cost
- low weight
- high reliability
- high fuel efficiency
- assuming it is a podded engine, a low wetted area

This categorisation is somewhat of an oversimplification because there is an interdependency between all these items. The first four items will be largely a function of the mechanical design of the engine whereas the last two items are mainly influenced by the engine cycle and its internal operating efficiency. The engine cycle is defined by the bypass ratio, pressure ratio and turbine entry temperature.

On large aircraft the fuel price accounts for some 30% of the aircraft direct operating costs. High fuel efficiency is therefore a major factor in such aircraft designs. Improving fuel efficiency will not only have a direct saving on aircraft fuel cost but will also have an indirect effect due to the reduced fuel mass to be carried. A more fuel efficient engine will require less fuel to fly a given range and hence will lead to a lower take-off weight and a general scaling down of the aircraft and engine size. This is sometimes referred to as the 'snowball' effect. Due to the reduced take-off weight and smaller engine size, the aircraft first cost, depreciation charges, insurance charges, landing fees and maintenance charges will all be reduced. Thus a very powerful cost saving process stems from improving the engine efficiency.

Engine cycle

The parameters that define the engine cycle have been mentioned above. Before proceeding these parameters need to be defined.

- *Bypass ratio* This is the ratio of the air passing through the bypass duct to the air passing through the gas generator (e.g. a bypass ratio of 5 has a total airflow of 6 units, 5 of which go through the bypass duct and 1 of which goes through the gas generator).
- *Pressure ratio* This is the ratio of the air pressure at the entry to the engine to the pressure at the outlet of the compression system (i.e. as it enters the combustion section).
- *Turbine entry temperature* This is the gas temperature at the entry to the first turbine stage in the engine.
- A parameter which depends on the combination of the above characteristics is called the *specific thrust*. This is defined as the amount of thrust per unit airflow:

$$\text{(thrust) divided by (engine total air mass flow)}$$

Engine thrust

Since reference will be made to engine thrust in the sections describing the various efficiencies of the engine it will be helpful to define this term.

Engine thrust results from the change in momentum of the air through the engine plus any pressure thrust due to the static pressure ratio across the final (exhaust) nozzle.

The change of momentum of the air through the engine is given by:

$$M\,(V_j - V_0)/g$$

where: M = air flow through the engine
V_j = velocity of the exhaust jet
V_0 = velocity of the air entering the engine
g = gravitational acceleration

Pressure thrust is given by:

$$(p - p_0)A$$

where: $(p - p_0)$ = pressure difference at the nozzle
A = the nozzle cross-sectional area

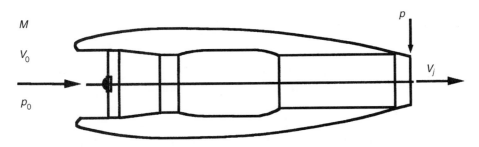

Fig. 9.1 The derivation of nett thrust (source ref. 1).

$$\text{engine thrust} = M(V_j - V_0)/g + (p - p_0)A$$

If the final nozzle is unchoked (i.e. not at supersonic speed) the nozzle pressure term disappears and the thrust becomes:

$$M(V_j - V_0)/g$$

In the case of a choked nozzle, with V_j representing the jet velocity when the flow is fully expanded to atmospheric pressure (p_0), the nozzle pressure term again disappears and the thrust also becomes:

$$M(V_j - V_0)/g$$

The term MV_0/g is often referred to as the **momentum drag**. The term $(MV_j/g) + (p - p_0)A$ is called the **gross thrust**. The difference between the gross thrust and momentum drag is referred to as the **nett thrust**.

Main factors influencing fuel efficiency

Fuel consumption is directly linked to engine efficiency. The higher the overall efficiency, the lower is the fuel consumption of the engine per unit thrust. The overall engine efficiency can be broken down into three main components:

- the power producing component (i.e. the gas generator);
- the transmission system which conveys the power from the gas generator to the propulsive jet;
- the propulsive jet system.

The overall efficiency can be written as:

$$\eta_o = \eta_{th} \times \eta_t \times \eta_p$$

where: η_{th} = thermal efficiency of the gas generator
η_t = transmission efficiency
η_p = propulsive efficiency of the jet

Each of these efficiencies will be discussed.

Thermal efficiency

It can be shown that:

$$\eta_{th} = 1 - \left(\frac{1}{r}\right)^n = [1 - (1/r^n)]$$

where: r = overall pressure ratio of the engine thermodynamic cycle including any intake ram pressure rise
n = function of the gas constant γ

Therefore, the higher the pressure ratio, the higher will be the thermal efficiency. Consequently the specific fuel consumption will be lower.

However, it should be noted that available turbine materials and turbine cooling techniques limit the maximum pressure ratio. Providing cooling air can negate the effect of higher pressure ratio due to the higher temperatures of the cooling air from the compressor. For small engines there will also be a limit due to the practicalities of engineering the blades for the final compressor stages.

Transmission efficiency

Transmission efficiency can be determined from:

$$\eta_T = \frac{(1 + \mu)}{\left(1 + \dfrac{\mu}{\eta_f \mu_t}\right)}$$

where: μ = bypass ratio
η_f = fan efficiency
η_t = turbine efficiency

If fan and turbine efficiencies were 100% then the transmission efficiency would also be 100%. The transmission efficiency therefore depends primarily upon the fan and turbine efficiencies. These are determined by the current material and manufacturing technologies that are available.

Propulsive efficiency

Propulsive efficiency can be determined from:

$$\eta_p = \frac{2V_0}{\dfrac{X_n}{M} + 2V_0}$$

where: V_0 = the aircraft velocity
X_n = the engine thrust
M = the engine airflow
X_n/M is referred to as the specific thrust

Lower specific thrust means higher propulsive efficiency; however, the lower the specific thrust for the same overall pressure ratio and turbine entry temperature the larger will be the fan diameter for a given thrust. Large fan diameters mean more weight for both engine and nacelle and more aircraft drag. However, due to the improved propulsion efficiency, there will be an improvement in the engine specific fuel consumption (SFC). (SFC is the amount of fuel burned per unit time per unit thrust – lb/hr/lb or in SI units kg/hr/Newton.)

Let us assume flight above the tropopause where the ambient temperature is constant (at this height the speed of sound is 295 m/s in ISA conditions [see Data D]). For simplicity, assume that the engine hot and cold jet stream velocities are equal then the equation above gives:

$$\eta_p = \frac{590Mn}{\dfrac{X_n}{M} + 590Mn}$$

where: X_n = the nett thrust in Newtons
M = the engine mass flow in kg/sec

The above expression is sometimes quoted in imperial units in which thrust is in pounds, mass flow in lb/sec, velocity in ft/sec. If W is the mass flow in lb/sec then M in the above formula needs to be replaced by W/g thus giving:

$$\eta_p = \frac{2Mn \cdot a}{\dfrac{X_n}{W} \cdot g + 2Mn \cdot a}$$

where: X_n/W = specific thrust
a = speed of sound

In the tropopause, $a = 968$ ft/sec and $g = 32.2$ ft/sec^2, giving:

$$V_0 = \frac{Mn \cdot 968}{32.2} \approx 30Mn$$

$$\eta_p \approx \frac{60Mn}{\left(\dfrac{X_n}{W} + 60Mn\right)}$$

where thrust (X_n) is in pounds force and 'mass' flow (W) is in lb/sec.

Propulsive efficiency is sometimes shown to be a function of bypass ratio. This is only true for engines of the same overall pressure ratio and turbine entry temperature.

Engines of differing bypass ratio can have the same specific thrust and hence the same propulsive efficiency as shown in Table 9.1.

Nevertheless, in defining engine performance and efficiency, the term specific thrust will be used in this chapter. The above values for specific thrust are those related to the take-off rating. Currently the design point for engines of this type is at the top of climb where the specific thrust will be much lower, for the engines below it will be about 16. Bypass ratio will still be used on occasions as it does describe the flow split between the bypass stream and the gas generator.

Table 9.1

Engine	Tay 650	Trent 772
Bypass ratio	3.06	4.89
Pressure ratio	16.20	36.84
Thrust (lb)	15 100	71 100
Mass flow (lb/sec)	418	1978
Specific thrust (sec)	36.12	35.94

Effect of specific thrust

$$\text{Nett thrust, } X_n = M(V_j - V_0)$$

where: V_j = jet velocity (m/s)
V_0 = flight speed (m/s)
M = air mass flow (kg/sec)

The same static thrust can be generated by a high V_j and a low massflow (as in a pure jet engine, i.e. high specific thrust) or by a low V_j and a high massflow (as in a bypass engine with a low specific thrust). However, as flight speed increases the MV_0 term becomes increasingly important as the rate of fall off of thrust with flight speed becomes worse for lower specific thrust engines, as shown in Fig. 9.2.

Therefore, for a given thrust requirement at flight speed, the low specific thrust engine will have more static thrust available than a high specific thrust engine.

It has been shown that reducing specific thrust improves engine efficiency. However, for the same thrust requirement reducing specific thrust will increase the fan diameter. The mass of the basic engine will therefore increase and the engine nacelle will become larger, heavier and create more drag. The nett result will be a reduced engine efficiency improvement with increased engine size.

Figure 9.3 shows a typical effect of specific thrust for the same cruise thrust.

Each of these engines has the same cruise thrust and has been installed to meet the same noise regulations. An indication of the relative powerplant weights and maximum nacelle diameters is shown.

Making simplified assumptions, Fig. 9.4 shows how specific thrust affects the static thrust if a constant core energy is assumed.

Due to the simplification of the theory used in Fig. 9.4, the increase in static thrust will not be as great in practice as that shown. As specific thrust decreases, the fan diameter will increase. There is a limit to the amount of airflow that can

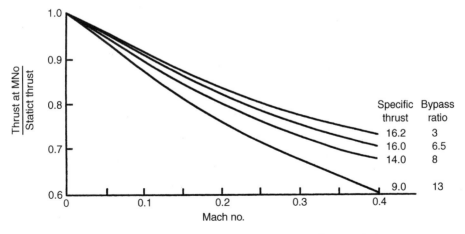

Fig. 9.2 Effect of Mach number and specific thrust on the thrust lapse rate (source ref. 2).

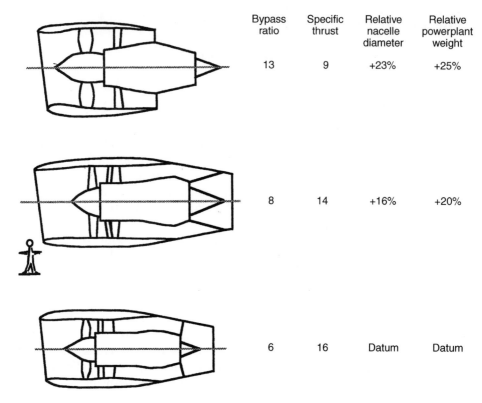

	Bypass ratio	Specific thrust	Relative nacelle diameter	Relative powerplant weight
	13	9	+23%	+25%
	8	14	+16%	+20%
	6	16	Datum	Datum

Fig. 9.3 Effect of specific thrust on engine size and weight at constant cruise thrust (source ref. 2).

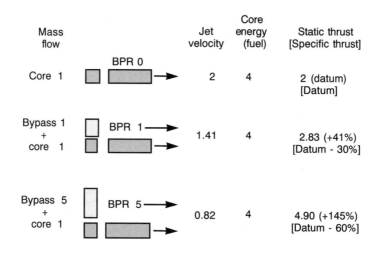

Mass flow		Jet velocity	Core energy (fuel)	Static thrust [Specific thrust]
Core 1	BPR 0	2	4	2 (datum) [Datum]
Bypass 1 + core 1	BPR 1	1.41	4	2.83 (+41%) [Datum - 30%]
Bypass 5 + core 1	BPR 5	0.82	4	4.90 (+145%) [Datum - 60%]

Notes: *(Constant cycle parameters), (Energy $\propto MV^2$), (Thrust $\propto MV$).*

Fig. 9.4 Effect of specific thrust on static thrust for constant core energy (source ref. 2).

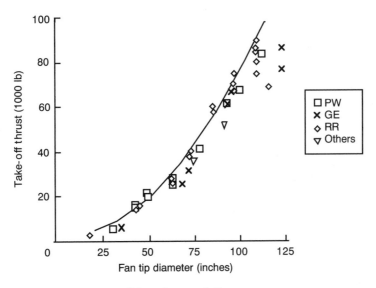

Fig. 9.5 Effect of fan diameter on take-off thrust (source ref. 2).

be passed per unit area of fan. It follows therefore that the maximum take-off thrust is mainly a function of fan diameter. This is shown in Fig. 9.5 for a range of current engines from various manufacturers.

Specific fuel consumption and flight speed

The specific fuel consumption is defined as the rate of fuel consumption per unit of thrust. If we assume a 100% efficient engine then the ideal fuel consumption is:

$$SFC_{\text{ideal}} = (\text{heat equivalent of aircraft work})/(\text{calorific value of fuel} \times \text{thrust})$$

Example

Using fuel with a calorific value of 10 000 CHU/lb and assuming flight above the tropopause in a standard atmosphere (i.e. constant ambient temperature) where the speed of sound is 968 ft/sec (note imperial units) and converting seconds to hours (i.e. 3600), and ft lb/hr to Joule/hr (i.e. 1.35582), then the ideal specific fuel consumption becomes:

$$\text{aircraft work per hour} = \text{thrust} \times \text{velocity}$$
$$= \text{thrust} \times \text{Mach no.} \times 968 \times 3600 \times 1.35582 \, \text{Joule/hr}$$

Converting CHU to Joule (using 1899.11):

$$(\text{calorific value of fuel} \times \text{thrust}) = (10\,000 \times 1899.11 \times \text{thrust})$$

Substituting the values above into the ideal SFC equation we get:

$$SFC = \frac{1.35582 \times 968 \times 3600 \times Mn}{10\,000 \times 1899.11} \approx \frac{Mn}{4}$$

Typically, for an aircraft flying above the tropopause at $Mn = 0.84$, the ideal SFC will be about 0.21.

Actual SFC for an engine with an overall efficiency of η_o becomes:

$$SFC_{\text{actual}} = \frac{SFC_{\text{ideal}}}{\eta_o} = \frac{Mn}{4\eta_o}$$

Modern large high bypass ratio turbofans have SFCs in the range 0.55–0.6. This implies overall efficiency (η_o) in the range 35–40%.

Factors governing choice of engine cycle

The choice of engine cycle (i.e. engine pressure ratio, turbine entry temperature and specific thrust) is determined by a process of arriving at the best compromise between engine specific fuel consumption, engine weight and price. The criteria at the initial design stage will be either in terms of direct operating cost of the aircraft or the capital cost of acquiring the aircraft. The aircraft mission is important in this consideration. Long-range aircraft will be more sensitive to changes in specific fuel consumption due to the 'snowball' effect from fuel mass. On the other hand, short-range aircraft will be more sensitive to engine cost than to the specific fuel consumption. In such an aircraft it may not be appropriate to employ the highest bypass ratio and pressure ratio possible along with the highest turbine entry temperature capability as these lead to increased complexity and hence higher cost and weight. For example, Table 9.2 compares the powerplants for the Fokker F100 (a typical short-range aircraft) and the B-747 with two different engines (a typical long-range aircraft).

On the long-range aircraft fuel efficiency is the main driver. In this case the engine overall efficiency needs to be the highest practicable (within current technology constraints). The best thermal efficiency is obtained first of all by designing for the highest component efficiency (i.e. compressor and turbines); this will then determine the pressure ratio to give the highest efficiency. The turbine entry temperature will be determined by the turbine material properties and the blade cooling technology. The higher the turbine entry temperature, in general, the higher the thrust. So one of the major effects of high turbine temperatures is to keep the overall size and weight of the engine down for a given thrust. The highest

Table 9.2 Comparative engine data

Aircraft	F100	B-747	B-747
Engine	R.R. Tay	RB211–524	CF6–80C2
Bypass ratio	3.07	4.3	5.15
Pressure ratio	16.60	33.0	30.4
Cruise SFC (lb/hr/lb)	0.69	0.57	0.564

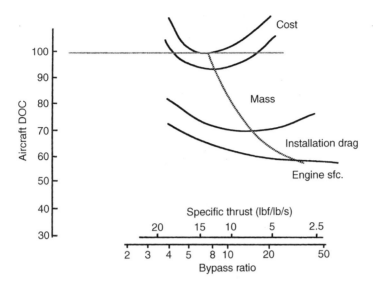

Fig. 9.6 Effect of engine parameters on direct operating cost (source ref. 3).

transmission efficiency is obtained by maximising the fan and turbine efficiencies. The propulsive efficiency is at its highest when the average jet velocity is at its lowest. Low jet velocities imply high bypass ratio or low specific thrust. The choice of bypass ratio or specific thrust is determined by balancing the improvement in propulsive efficiency against the increase in weight, drag and cost as bypass ratio and thereby the size of the engine is increased.

The effect of engine parameters on aircraft direct operating cost (DOC) is shown in Fig. 9.6. Considering the effect of engine SFC alone produces a low value for optimum specific thrust (e.g. 3.0 lbf/lb/sec in the example). However, as specific thrust decreases, installation drag, aircraft mass and aircraft price increase. When all effects are taken into account the optimum specific thrust value is much higher (e.g. 14.0 in the example).

Engine performance

The engine performance data will contain thrust, fuel flow and airflow. Fuel flow is often in the form of a specific fuel consumption (i.e. fuel flow per unit thrust). Thrust and fuel flow are obviously required in order to calculate the aircraft performance. The airflow value is needed to design the engine intake so that the flow requirements of the engine can be met in all the important phases of the flight without excessive penalties (for example, due to spillage, i.e. the intake providing more air than the engine requires).

The engine will also have certain rated thrust levels. These ratings are set such that the engine can pass its certification type test and at the same time give an acceptable engine service life to the airlines.

These ratings are usually set by a turbine entry temperature limit and are summarised as follows.

- *Take-off rating:* is the maximum thrust rating and corresponds to the highest turbine entry temperature. To protect engine service life this thrust level is normally cleared for only five minutes per flight but this can be increased to ten minutes to meet special cases.
- *Maximum continuous:* is the highest thrust rating for the engine cleared for continuous use. It is only used in the event of a serious emergency condition. Although it could be used for normal operations the engine service life would be unacceptably reduced.
- *Maximum climb:* is the maximum thrust used in normal climb
- *Maximum cruise:* is the maximum thrust used in normal cruise

Take-off and maximum continuous ratings are determined by the engine certification type test.

The maximum climb and maximum cruise ratings are set in conjunction with the required take-off rating to achieve a desired service life. The term required take-off rating has been used quite deliberately as this may be less than the maximum take-off rating set by the type test. It is possible that the selected ratings may not match specific aircraft operating requirements. For example, an aircraft may need less take-off thrust but more climb and cruise thrust. It is therefore possible to re-rate the engine to accommodate different combinations of rating while still achieving approximately the same engine service life.

This interplay between the engine ratings and the aircraft requirements is quite important in optimising the engine size and hence the powerplant weight and costs. This compromise can only be achieved through a dialogue between the airframe manufacturer and the engine supplier. The selection of the engine and its ratings plays an important role in the project design discussions between the engine and airframe companies.

Engines operating at a constant turbine entry temperature (TET) will lose thrust as the ambient temperature increases. The engine control system for an engine which is 'flat rated' will progressively increase TET as the ambient temperature rises to maintain constant thrust. Aircraft are often designed to meet specified hot day requirements (typically ISA + 15°C at take-off and ISA + 10°C at climb and cruise). At day temperatures lower than these the engine thrust will increase and hence give the aircraft a performance superior to that at the design point. This is not usually necessary so the engine is flat rated to increase the service life. An example at take-off is shown in Fig. 9.7.

Flat rating also improves service life as only on a comparatively small number of occasions does the aircraft operate at or above the hot day design point.

The development of electronic control systems for engines has allowed airlines to de-rate the engines from the maximum ratings established above for both take-off and climb depending upon the aircraft operational requirements. For example, the aircraft take-off weight may be less than the maximum, the runway available may be longer than that for which the aircraft has been designed, and the ambient temperature may be lower than the specification critical figure. The required take-off thrust under these conditions, observing all the necessary safety margins, may

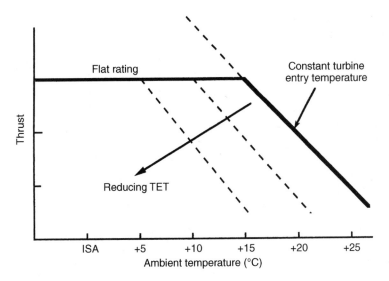

Fig. 9.7 Flat rating.

only be 90% of the maximum rating. For aircraft with high bypass engines the excess take-off power available makes this option commonplace.

This is illustrated in Fig. 9.8. In this way reducing thrust from the maximum permitted rating will lower turbine entry temperatures and thereby increase the service life of the engine. In the case of climb, an aircraft operating at less than its maximum take-off mass may not need the maximum permitted climb rating to give an acceptable climb performance. Changing the engine ratings will affect engine service life due to the changed operating temperatures in the engine.

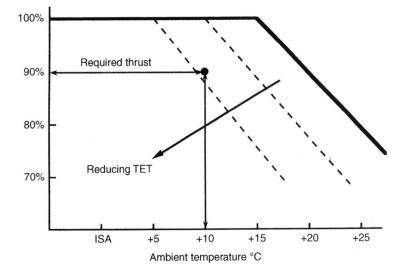

Fig. 9.8 Effect of de-rating.

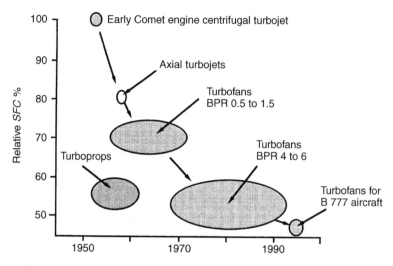

Fig. 9.9 Historical trends in cruise fuel consumption (source ref. 2).

Historical and future engine developments

Engine development has shown a steady increase in cruise fuel efficiency over the years. This is shown on Fig. 9.9 in terms of the improvement in specific fuel consumption.

It should be noted that the De Havilland Ghost engine used in the first jet-powered airline (the De Havilland Comet) paid a large price in cruise fuel efficiency compared to the then current piston and turboprop engines to gain the much higher cruise speed. It was not until the introduction of the high bypass ratio turbofans with bypass ratios in the 4–6 range (e.g. RB211, the JT9 and the CFF–6) that the lost fuel efficiency was regained.

One of the future engine developments which could give further improvements in cruise fuel efficiency would be the advanced turboprop or propfan. However, despite considerable amount of research and development a propfan installation has not been produced that would show a significant benefit in direct operating costs to make it attractive to the airlines at current fuel prices. There are still problems associated with noise and installation (particularly when wing-mounted) and the maximum cruise speed will be restricted to about $Mn = 0.8$ as shown in Fig. 9.10.

The immediate future developments would appear to be vested in the ducted turbofan, exploiting new materials and manufacturing technology in materials, turbine blade cooling techniques and component efficiency improvements.

Engine installation

The object of the engine installation designer is to take the engine shown in Fig. 9.11 enclose it in a nacelle and install it on the aircraft as shown in Fig. 9.12 for the lowest possible weight, drag and cost.

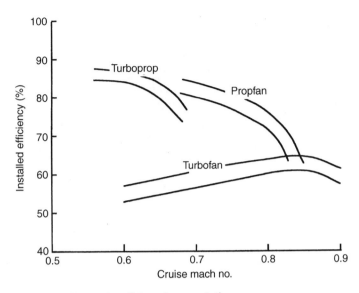

Fig. 9.10 Effect of flight speed on engine efficiency (source ref. 3).

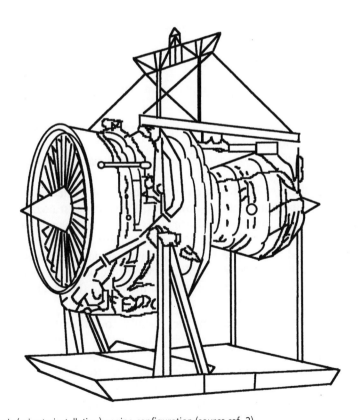

Fig. 9.11 Basic (prior to installation) engine configuration (source ref. 2).

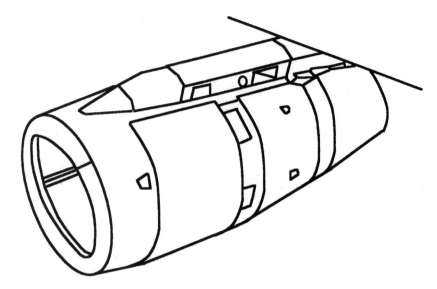

Fig. 9.12 Typical wing installation (source ref. 2).

Although Fig. 9.12 shows an under-wing layout the remarks below apply equally to rear fuselage installations. The majority of new aircraft have externally podded engines installed under the wing and therefore this installation is considered in detail.

There are many issues involved in putting a nacelle around an engine. These are listed below.

- Air must be delivered at conditions acceptable to the engine compressor system throughout the aircraft flight envelope.
- The nacelle must efficiently separate the air which passes through the engine from that which passes outside.
- The external nacelle flow must be such as to minimise the external drag.
- There must be an acceptable fairing down to the final exhaust nozzle.
- Sufficient space must be provided for the engine mechanical systems and accessories while at the same time minimising the impact on the nacelle profile.
- Provision is required for a thrust reverser and its associated systems.
- Sufficient acoustic lining area must be provided to enable the aircraft to meet the required noise rules, again with minimum impact on the nacelle profiles.
- Safety considerations require:
   1. ventilation of various areas in the nacelle to prevent the build-up of dangerous gases and to control temperatures within ranges acceptable to the accessories;
   2. fire detection and extinguishing systems in separate zones subject to a fire hazard;
   3. fireproof bulkheads to enclose zones considered to be fire hazards.
- Accessibility for maintenance purposes is also required.

At the preliminary design stage it is not possible to cater for all these considerations in detail.

Derivation of pod geometry

The geometry described here for the nacelle installation has been based on the analysis of existing aircraft installations covering the major aircraft manufacturers. It covers two different types of installation (separated and mixed jets):

- engines with separate jets installed in a pod with separate nozzles for the fan flow and core flow as shown in Fig. 9.13;
- engine with mixed flows installed in a fully cowled pod as shown in Fig. 9.14.

The statistical data used for the analysis shown in Tables 9.3(a) and (b) was restricted mainly to wing-mounted powerplants. However, some limited information on rear fuselage-mounted engines indicated that basic pod geometries may be derived using the data for lower bypass ratio engines.

Nacelle geometry is influenced by the position of accessories. At the preliminary design stage it may not have been established whether the engine accessories should be mounted on the fan casing or on the core structure. The nacelle geometry shown in this chapter will give typical nacelle geometry and will enable the effect of engines of different designs to be compared without consideration of mounting of accessories. Figs 9.13 and 9.14 define the geometric parameters referred to in the following tables (Tables 9.3(a) and (b)). To arrive at a nacelle geometry the following input data is required:

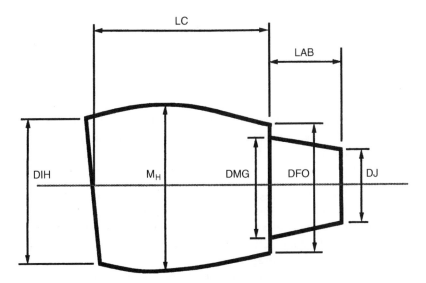

Fig. 9.13 Typical engine nacelle dimensions for turbofan with separate jets (variables defined in Table 9.3) (source ref. 2).

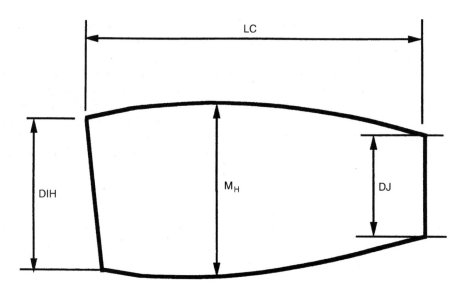

Fig. 9.14 Typical engine nacelle dimensions for mixed flow configuration (variables defined in Table 9.3) (source ref. 2).

- fan diameter (D_F)
- bypass ratio (μ)
- overall compression ratio (OPR)
- total airflow at sea level static (SLS), ISA and take-off rating (W_a)
- aircraft maximum operating Mach number (M_{Mo})

If a plug nozzle is fitted then area required at hot nozzle exit is A_J, where:

$$A_J = \frac{\pi}{4}(D_J)^2$$

A_J is then the required annulus area and (D_P/D_J) plug can be taken as equal to 0.65.

Therefore with plug nozzle it can be shown that: D_J plug $= 1.25\ D_J$ (no plug).

Having defined the engine nacelle, it is now possible to install it on the airframe. There are again many issues to be resolved. These are listed below and shown in Figs 9.15 and 9.16.

- Gully depth. This is the necessary clearance between the wing and the nacelle to ensure efficient wing aerodynamics with a low nacelle drag.
- Penetration. This is the overlap of the final nozzle exhaust plane relative to the wing leading edge. In conjunction with the gully depth it is chosen to minimise the powerplant interference drag.
- Intake highlight to ground clearance must be set high enough to avoid the formation of ground effect vortices which could cause losses due to fan flow distortion and injection of debris from the ground.

- The direction of the thrust reverser efflux must be located to ensure that it does not interfere with the wing flaps and fuselage aerodynamics and also avoid hot gas re-ingestion by the engine.
- The relationship of the intake to the undercarriage must be such that any foreign object or excessive spray thrown up does not impact on the engine safety.
- In the unlikely event of a fan disk or turbine disk disintegrating the likely path of the debris must present no extra safety hazard to the aircraft; in particular it should not present a hazard to any other engine. This requirement has implication on the routing of aircraft and engine services and the siting of fuel tanks.

Table 9.3(a) Typical geometrical relationships for separate jets (source ref. 2)

Dimension	Symbol	Estimating relationship	Input required
Intake highlight diameter	DIH	$DIH = 0.037W_a + 32.2$	W_a (lb/sec)
Maximum height of main cowl	M_H	$M_H = 1.21D_F$	D_F (in)
Main cowl length	LC	$LC = [2.36D_F - 0.01(D_F M_{Mo})^2]$	$D_F\ M_{Mo}$
Main cowl diameter at fan exit	DFO	$DFO = (0.00036\,\mu W_a + 5.84)^2$	μ ratio W_a (lb/sec)
Gas generator cowl diameter at fan exit	DMG	$DMG = (0.000475\,\mu W_a + 4.5)^2$	μ, W_a (lb/sec)
Gas generator cowl diameter at 'hot' nozzle exit	DJ	$DJ = (18 - 55 \times K)^{0.5}$ where $$K = \left\{ \ln\left(\frac{1}{\mu+1}\right)\left(\frac{W_a}{OPR}\right) \right\}^{2.2}$$	W_a (lb/sec) μ, OPR
Length of gas generator afterbody	LAB	$LAB = (DMG - DJ)0.23$	DMG (in) DJ (in)

Units in inches unless stated otherwise.

Table 9.3(b) Typical geometrical relationships for mixed flow (source ref. 2)

Dimension	Symbol	Estimating relationship	Input required
Intake highlight diameter	DIH	$DIH = 0.037W_a + 32.2$	W_a (lb/sec)
Maximum height of cowl	M_H	$M_H = 1.21D_F$	D_F (in)
Main cowl length	LC	$LC = [3.51D_F - 0.21(D_F M_{Mo})^2]$	$D_F\ M_{Mo}$
Gas generator cowl diameter at 'hot' nozzle exit	DJ	$DJ = (18 - 55 \times K)^{0.5}$ where $$K = \left\{ \ln\left(\frac{W_a}{OPR}\right) \right\}^{2.2}$$	W_a (lb/sec) μ, OPR

Units in inches unless stated otherwise.

- The ground clearance must be such as not to be a hazard to the powerplant in the event of either a nose-wheel collapse on landing, or if the aircraft lands with excessive role or pitch.

The installation of the pod on the wing represents a very complex technical problem. However, the major airframe manufacturers have defined some simple rules for positioning the wing-mounted pods.

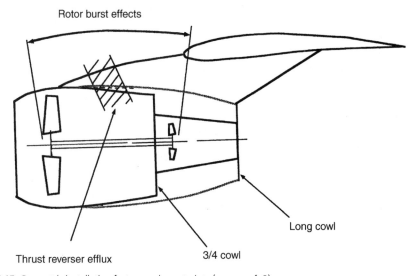

Fig. 9.15 Geometric installation factors and constraints (source ref. 2).

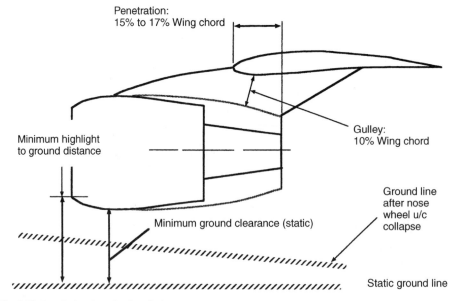

Fig. 9.16 A typical under-wing installation.

A reasonable installation is shown in Fig. 9.16. It should be possible to install a wing-mounted engine for little or no interference drag. Rear fuselage-mounted engines are much more difficult because of the disturbed flow around the rear fuselage with its adverse pressure gradients. Interference drags can be as high as 60% or even 80% of the isolated pod drag.

Rules for scaling an engine

There are limits on the degree of scaling that can be sensibly carried out. It would be impractical and inaccurate to scale an engine configuration from its baseline to 0.5 thrust scale. For example, the size of the last stages of the compressor may be too small to be manufactured. A crude rule of thumb suggests that thrust scaling should be limited to 20%.

If engine thrust scale is x {i.e. $x =$ (required engine thrust)/(known engine thrust)} fuel flow scales by the same ratio (i.e. SFC remains constant). Dimensions scale by the square root of x, and engine and nacelle mass scale by x. {i.e. (engine thrust/engine mass) and (engine thrust/nacelle mass) ratios remain constant.}

These rules are only approximate; for example the combustion system dimensions do not depend solely upon the engine size but involve other parameters. However, at the preliminary design stage these rules will give a reasonable first assessment of the weight and performance of a scaled engine.

Engine price

Estimating the price of the engine is always regarded as a difficult issue because of the many variables involved. For an initial guess the graph below (Fig. 9.17)

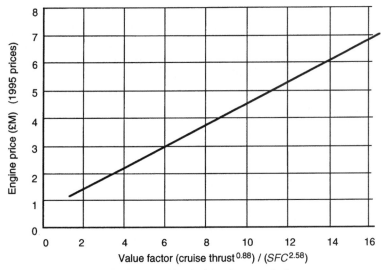

Fig. 9.17 Civil engine price estimation based on historical data (source ref. 4).

based on existing (1995) engine prices may be used. Both thrust and SFC in the value factor are at the maximum cruise rating at $Mn = 0.8$ at 35 000 ft.

Maintenance costs

These costs are even more difficult to generalise than engine price. Estimation methods usually need detailed information from the engine manufacturers. Maintenance costs will usually be quoted in two parts:

1. maintenance man hours per flight hour
2. engine parts cost per flight hour.

To convert 1. to a cost per flight hour a value for the labour rate is required. For more details refer to Chapter 12 on aircraft cost estimation.

Engine data

Data B contains the principal data for all current turbofan engines. On the following pages of this chapter are a series of graphs (Figs 9.18–9.21) showing engine performance for four types of engine:

- low bypass ratio
- medium bypass ratio
- high bypass ratio
- ultra high bypass ratio

The principal performance and geometry for each engine type is listed in Table 9.4.

In the early project stages this data can be scaled and used in association with Data B to provide engine performance. It should be noted that the detailed data is in a non-dimensional form. Scaling from this data would neglect the effect of Reynolds number change on thrust and SFC. At the project stage this is acceptable. As the design process evolves specific engine data, which takes these effects into account, must be used.

Table 9.4 Comparative engine data

Bypass ratio	3.0	6.5	8.0	13.0
Maximum SL static thrust $(E_N{}^*)$ (lb)	15 100	95 000	99 300	97 979
Thrust/engine weight	5.12	4.03	3.83	3.17
Nacelle length (m)	2.405	5.94	9.14*	5.63
Nacelle diameter (m)	–	3.81	4.11	4.74
Fan diameter (m)	1.14	3.02	3.38	–

* Full length nacelle.

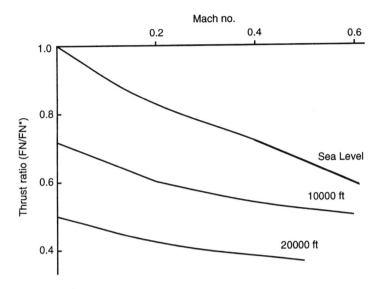

Fig. 9.18(a) Bypass ratio 3.0–take-off thrust.

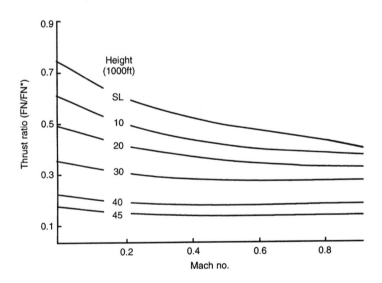

Fig. 9.18(b) Bypass ratio 3.0–maximum climb thrust.

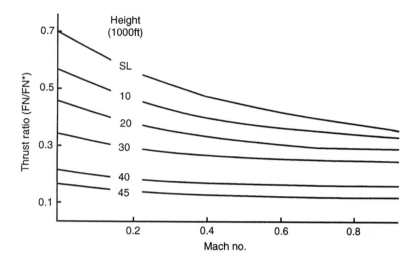

Fig. 9.18(c) Bypass ratio 3.0–maximum cruise thrust.

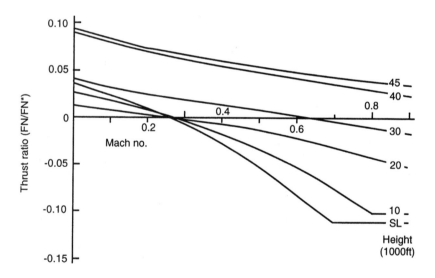

Fig. 9.18(d) Bypass ratio 3.0–descent thrust.

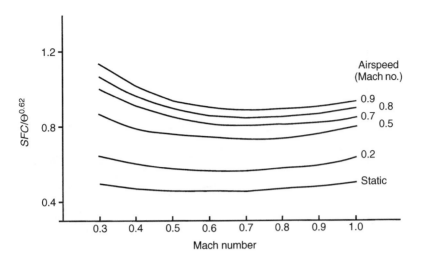

Fig. 9.18(e) Bypass ratio 3.0–*SFC* loops.

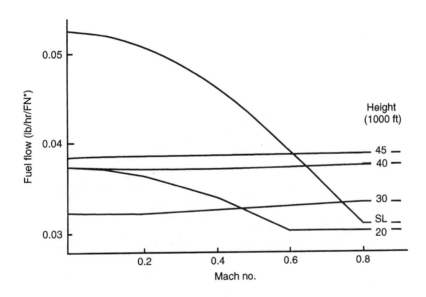

Fig. 9.18(f) Bypass ratio 3.0–descent fuel flow.

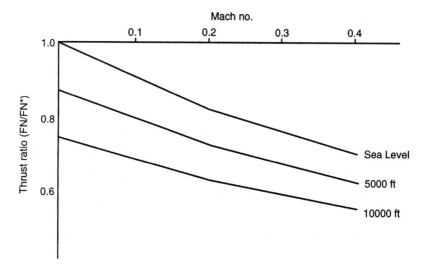

Fig. 9.19(a) Bypass ratio 6.5–take-off thrust.

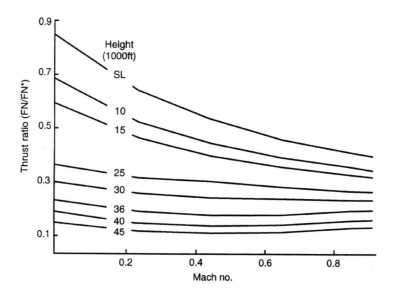

Fig. 9.19(b) Bypass ratio 6.5–maximum climb thrust.

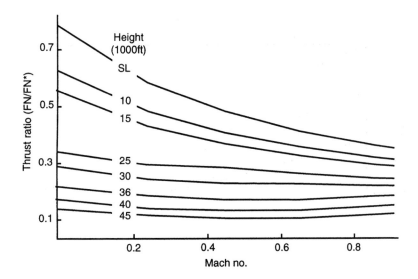

Fig. 9.19(c) Bypass ratio 6.5–maximum cruise thrust.

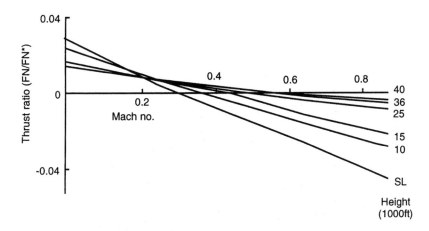

Fig. 9.19(d) Bypass ratio 6.5–descent thrust.

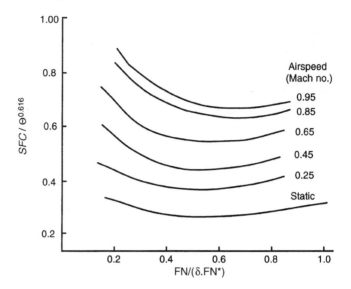

Fig. 9.19(e) Bypass ratio 6.5–*SFC* loops.

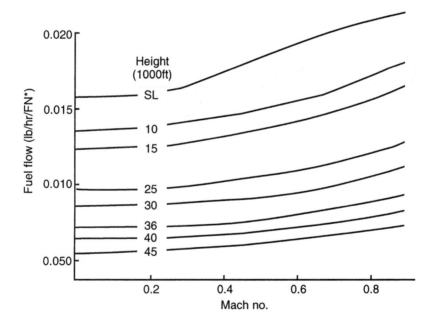

Fig. 9.19(f) Bypass ratio 6.5–descent fuel flow.

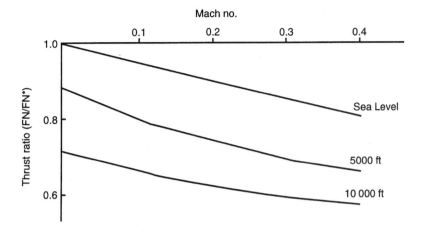

Fig. 9.20(a) Bypass ratio 8.0–take-off thrust.

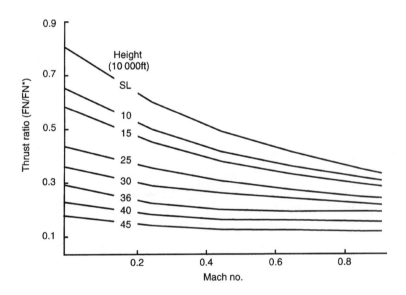

Fig. 9.20(b) Bypass ratio 8.0–maximum climb thrust.

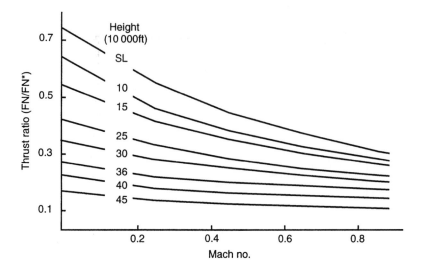

Fig. 9.20(c) Bypass ratio 8.0–maximum cruise thrust.

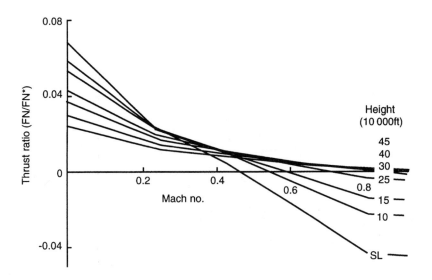

Fig. 9.20(d) Bypass ratio 8.0–descent thrust.

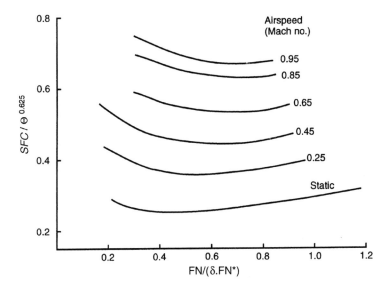

Fig. 9.20(e) Bypass ratio 8.0–*SFC* loops.

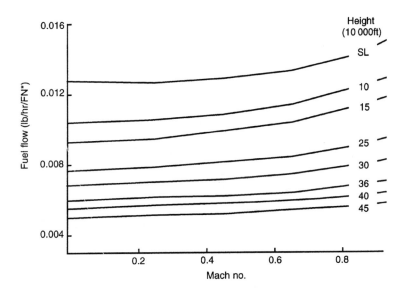

Fig. 9.20(f) Bypass ratio 8.0–descent fuel flow.

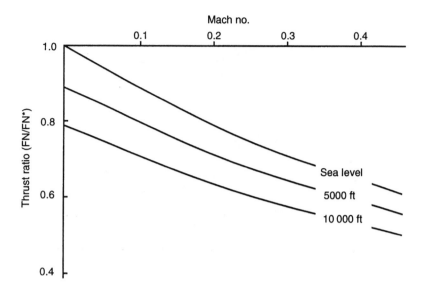

Fig. 9.21(a) Bypass ratio 13.0–take-off thrust.

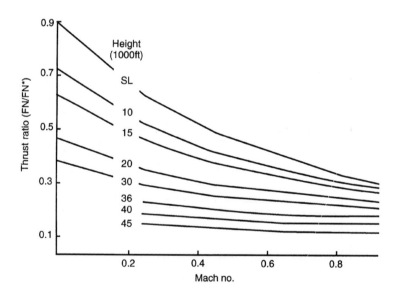

Fig. 9.21(b) Bypass ratio 13.0–maximum climb thrust.

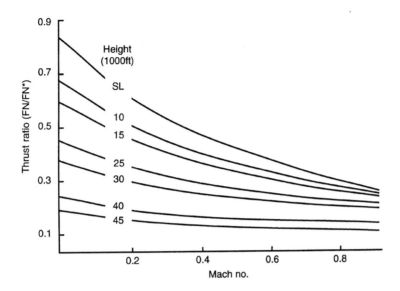

Fig. 9.21(c) Bypass ratio 13.0–maximum cruise thrust.

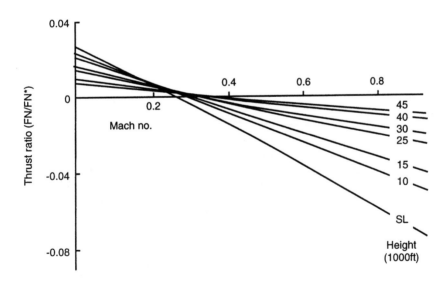

Fig. 9.21(d) Bypass ratio 13.0–descent thrust.

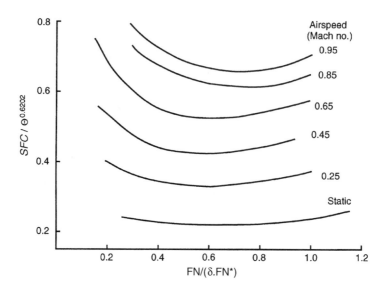

Fig. 9.21(e) Bypass ratio 13.0–*SFC* loops.

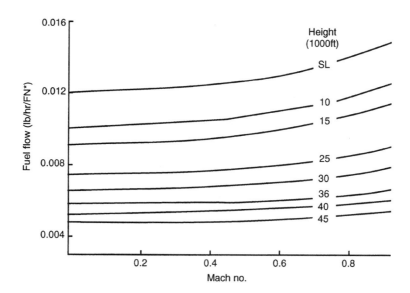

Fig. 9.21(f) Bypass ratio 13.0–descent fuel flow.

Derivation of formulae

The derivation of the formulae given in the early part of this chapter for the thermal, propulsive and transmission efficiencies are given below.

Thermal efficiency

Gas turbines operate to a Joule (or Brayton) cycle. It is this cycle that controls the thermal efficiency. A simple cycle is shown in Fig. 9.22 which also defines the temperature T and pressure p at various sections.

Assuming ideal components (i.e. 100% compressor and turbine efficiencies, zero loss in the combustion system and an ideal gas) the thermal efficiency can be expressed as the ratio (nett work output)/(heat supplied):

$$\text{nett work output} = C_p(T_3 - T_4) - C_p(T_2 - T_1)$$

and

$$\text{heat supplied} = C_p(T_3 - T_2)$$

where C_p is the specific heat at constant pressure with units of Joule/kg Kelvin. Therefore:

$$\eta_{th} = \frac{(T_3 - T_4) - (T_2 - T_1)}{(T_3 - T_2)}$$

From the isentropic relation between pressure and temperature we have:

$$\frac{T_2}{T_1} = r^{[(\gamma-1)/\gamma]} = \frac{T_3}{T_4}$$

where γ is the gas constant and r is the pressure ratio evaluated as:

$$r = \frac{p_2}{p_1} = \frac{p_3}{p_4}$$

Defining the exponential in the above equation as:

$$n = [(\gamma - 1)/\gamma]$$

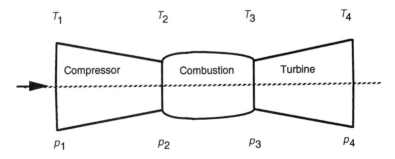

Fig. 9.22 A simple cycle.

Then thermal efficiency

$$\eta_{th} = \frac{T_4(r^n - 1) - T_1(r^n - 1)}{(T_3 - T_2)}$$

$$\eta_{th} = (r^n - 1)\left[\frac{T_4 - T_1}{T_3 - T_2}\right]$$

Defining

$$(T_3 - T_2) \text{ as } T_4 T_1 \left[\frac{T_3}{T_4} \cdot \frac{1}{T_1} - \frac{T_2}{T_1} \frac{1}{T_4}\right]$$

Then

$$(T_3 - T_2) = T_4 T_1 \left[\frac{r^n}{T_1} - \frac{r^n}{T_4}\right]$$

$$= r^n T_4 T_1 \left(\frac{1}{T_1} - \frac{1}{T_4}\right)$$

$$\eta_{th} = (r^n - 1)\left[\frac{T_4 - T_1}{r^n T_4 T_1(1/T_1) - r^n T_4 T_1(1/T_4)}\right]$$

$$= \frac{r^n - 1}{r^n}\left[\frac{T_4 - T_1}{T_4 - T_1}\right]$$

$$= 1 - \left(\frac{1}{r}\right)^n [= 1 - (1/r^n)]$$

where r is the pressure ratio of the engine thermodynamic cycle including any intake ram pressure rise at flight conditions and n is a function of the gas constant γ.

Note that with ideal components the turbine operating temperature has no influence on the thermal efficiency. This does not hold true when realistic component efficiencies are introduced. Figure 9.23 shows the effect of turbine entry temperature on engine thermal efficiency when realistic component efficiencies are considered.

Transmission efficiency

Transmission efficiency for a bypass engine is a function of the fan turbine efficiency and fan efficiency and the bypass ratio.

Assume an engine with mixed exhausts and with the following characteristics:

bypass ratio $= \mu$
core mass flow $= M$
velocity achieved if the available core energy was expanded down to ambient
    pressure $= V_C$
jet velocity $= V_j$
fan turbine efficiency $= \eta_t$
fan efficiency $= \eta_f$
freestream velocity $= V_0$

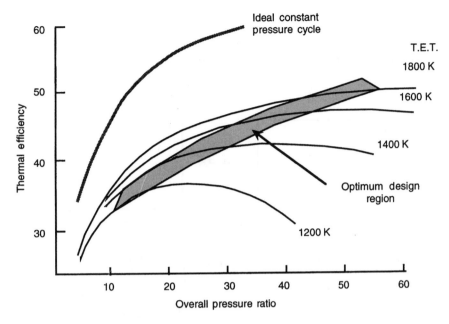

Fig. 9.23 Core thermal efficiency (source ref. 5).

These parameters are shown diagrammatically in Fig. 9.24.

We define engine transfer efficiency as:

$$\eta_t = \frac{\text{kinetic energy available from the core}}{\text{kinetic energy at the propelling nozzle}}$$

Kinetic energy available from core is given by

$$\frac{M}{2}(V_C^2 - V_0^2)$$

and kinetic energy available at the propelling nozzle by

$$\mu \frac{M}{2}(V_j^2 - V_0^2) + \frac{M}{2}(V_j^2 - V_0^2) = \frac{M}{2}(V_j^2 - V_0^2)(\mu + 1)$$

Hence the engine transfer efficiency is:

$$\eta_T = \frac{\dfrac{M}{2}(V_m^2 - V_0^2)(\mu + 1)}{\dfrac{M}{2}(V_C^2 - V_0^2)}$$

$$= \frac{(V_j^2 - V_0^2)(\mu + 1)}{(V_C^2 - V_0^2)}$$

Now consider the overall energy transfer. The energy transferred from the core is given by:

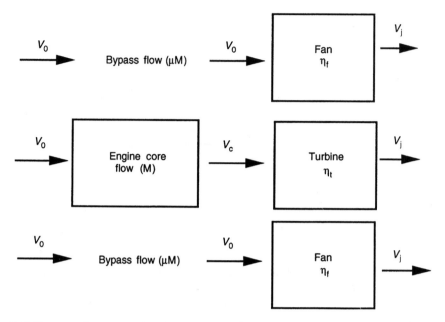

Fig. 9.24 Diagrammatic representation of a bypass engine (source ref. 2).

$$\eta_t \eta_F \frac{M}{2}(V_C^2 - V_j^2)$$

Energy in the bypass stream is given by:

$$\frac{\mu M}{2}(V_j^2 - V_0^2)$$

These must be equal. Therefore

$$\eta_f \eta_t M(V_C^2 - V_j^2) = \mu M(V_j^2 - V_0^2)$$

and so

$$V_C^2 = \frac{\mu}{\eta_f \eta_t}(V_j^2 - V_0^2) + V_j^2$$

and

$$(V_C^2 - V_0^2) = \frac{\mu}{\eta_f \eta_t}(V_j^2 - V_0^2) + (V_j^2 - V_0^2)$$

Substituting $(V_C^2 - V_0^2)$ in the equation of η_T:

$$\eta_T = \frac{(1+\mu)}{\left(1 + \dfrac{\mu}{\eta_f \eta_t}\right)}$$

As stated earlier, η_T is a function of the turbine and compressor efficiencies. Apart from the transmission efficiency these component efficiencies have other effects as described in the next section.

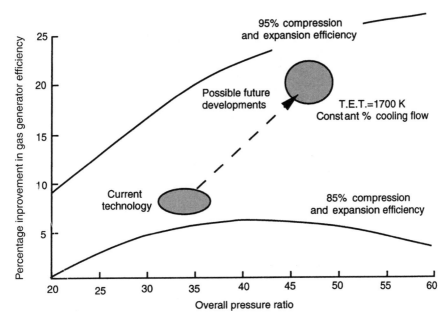

Fig. 9.25 Influence of component efficiency on cycle pressure ratio (source ref. 3).

The influence of component efficiency

Increasing component efficiency levels has three basic effects.

- Thermal efficiency is improved by the direct effect of the improved component efficiencies.
- The optimum pressure ratio increases as shown in Fig. 9.25. The higher pressure ratio in turn improves the thermal efficiency.
- The best match of turbine entry temperature and pressure ratio changes.

However, higher pressure ratio will result in higher cooling air temperatures from the compressor and therefore more cooling air will be required, unless improved turbine materials or much better cooling techniques are employed. It is possible that in the absence of such improvements the increased cooling air requirement could negate the advantage of the higher pressure ratio. Figure 9.26 shows the effect of high temperature material technology and improved cooling techniques on the thermal efficiency. Constant core energy is assumed; therefore as turbine entry temperature is increased and/or the cooling airflow reduced the gas generator air mass flow can be reduced.

Propulsive efficiency

Propulsive efficiency (η_p) is defined as:

$$\frac{\text{work done on aircraft}}{\text{work done on aircraft} + \text{kinetic energy imparted to jet}}$$

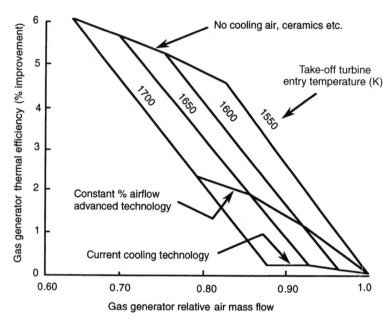

Fig. 9.26 Gas generator efficiency (source ref. 3).

Work done on the aircraft is the engine nett thrust multiplied by the aircraft speed.

As shown earlier the nett thrust is evaluated as:

$$M(V_j - V_0)$$

Then the work done on the aircraft is

$$V_0[M(V_j - V_0)]$$

The kinetic energy imparted to the jet (defined as the work wasted in the exhaust) can be expressed as:

$$\frac{M(V_j - V_0)^2}{2}$$

Propulsive efficiency (η_p) then becomes:

$$\eta_p = \frac{V_0[M(V_j - V_0)]}{V_0[M(V_j - V_0)] + \frac{1}{2}M(V_j - V_0)^2}$$

$$= \frac{MV_0(V_j - V_0)}{MV_0(V_j - V_0) + \frac{1}{2}M(V_j - V_0)^2}$$

$$= \frac{V_0}{V_0 + \frac{1}{2}(V_j - V_0)} = \frac{2V_0}{2V_0 + (V_j - V_0)} = \frac{2V_0}{V_j + V_0}$$

where: η_p = propulsive efficiency
V_0 = flight speed
V_j = jet velocity

For a given flight speed the jet velocity will determine the propulsive efficiency. The choice of jet velocity will determine the mechanical configuration of the engine from the high jet velocity of a pure jet engine to the modest jet velocity of a turbofan engine and ultimately to the low jet velocity of propeller engines. In other words, as bypass ratio increases the jet velocity drops and the propulsive efficiency increases. However, as shown earlier the nett thrust (X_n) is a function of jet velocity:

$$X_n = M(V_j - V_0)$$

where: M = airflow mass flow
  V_j = jet velocity
  V_0 = aircraft velocity

Therefore, substituting this definition of V_j into the equation for η_p gives:

$$V_j = \frac{X_n}{M} + V_0$$

and

$$\eta_p = \frac{2V_0}{\dfrac{X_n}{M} + 2V_0}$$

Aircraft flight Mach number $(Mn) = V_0/a$ where a is the speed of sound at the ambient temperature. Then

$$\eta_p = \frac{2Mn \cdot a}{\dfrac{X_n}{M} + 2Mn \cdot a}$$

If we assume flight above the tropopause where the ambient temperature is constant (at this height the speed of sound is 295 m/s in ISA conditions [see Data D]). For simplicity, assume that the engine hot and cold jet stream velocities are equal; then the equation above gives:

$$\eta_p = \frac{590Mn}{\dfrac{X_n}{M} + 590Mn}$$

where: X_n = nett thrust in Newtons
  M = engine mass flow in kg/sec

The above expression is sometimes quoted in imperial units in which thrust is in pounds, mass flow in lb/sec, velocity in ft/sec. Replacing M in the above formula with W/g gives:

$$\eta_p = \frac{2Mn \cdot a}{\dfrac{X_n \cdot g}{W} + 2Mn \cdot a}$$

where: X_n/W = specific thrust

In the tropopause, $a = 968$ ft/sec and $g = 32.2$ ft/sec^2, giving:

$$V_0 = \frac{Mn \cdot 968}{32.2} \approx 30Mn$$

and

$$\eta_p \approx \frac{60Mn}{\left(\dfrac{X_n}{M} + 60Mn\right)}$$

where thrust X_n is in pounds force and 'mass' flow M is in lb/sec.

Comments

As mentioned earlier, in this chapter a simplified approach has been adopted to more easily illustrate the underlying principles. For example, the turbofan engines have been treated as having a single stream with average jet velocities and pressures whereas in reality the bypass and the gas generator flow must be treated separately. No distinction has been made between the fully cowled engine with a single nozzle with mixed exhaust flows and the separate jets of the three-quarter cowled engine. Component efficiencies have been mentioned but no attempt has been made to quantify them or to demonstrate the matching of turbines to compressors or how to choose the working point on the compressors to give an acceptable surge margin.

Disregarding the simplifying assumptions mentioned above you should now have an underlying appreciation of the matching of engine performance with aircraft requirements and be able to make reasonable choices in the initial design stages.

To gain more understanding of the subject there are many textbooks on gas turbine theory and design[6,7] which deal with engine detail design in much more depth than is appropriate here.

References

1. Rolls Royce Ltd. publication, *The Jet Engine*, 1986.
2. Rolls Royce Ltd. unpublished data.
3. Bennett H. W., Proceedings. 'Aero Engine Development for the Future', 1983, **197**, No. 48.
4. Loughborough University, 'An investigation into aircraft development strategies', May 1996.
5. Wilde, G. L., 'Future large civil turbofans and powerplants', *Aeronautical Journal*, 1978.
6. Harman, R. T. C., *Gas Turbine Engineering*, Macmillan Press Ltd., UK, 1981.
7. Cohen, H., Rogers, G. F. C., Saravanamuttoo, H.I.M., *Gas Turbine Theory*, Longman UK, 1996.

10

Aircraft performance

Objectives

Aircraft performance is a fundamental part of the design process in that it will determine the minimum engine thrust and wing area. The selection of these two items is important as it is very expensive to make radical changes at a later stage in the design process. This is particularly true of the engine which currently takes longer to develop than the airframe. Both airframe and engine manufacturers are working hard to reduce the development time.

Aircraft performance predictions are usually produced by computer programs. This chapter introduces the methods underlying these programs. After completing this chapter it will be possible to understand the principles behind aircraft performance and when required you will be able to carry out calculations independent of the computer. A specimen performance calculation is included at the end of the chapter to show how this may be done.

Introduction

All civil transport aircraft must meet specified airworthiness regulations with regard to flying qualities and aircraft performance. Aircraft have to be designed to meet all these criteria. They must also meet operational requirements such as speed, range, payload, etc. The flight profile may be split in to several phases as shown in Fig. 10.1(a) for the main mission profile and 10.1(b) for the reserves profile.

The main operational profile can be considered in the following segments:

1. start-up and taxi-out
2. take-off and initial climb to 1500 ft
3. climb from 1500 to initial cruise altitude
4. cruise at selected speed and altitude including any stepped climb required
5. descent to 1500 ft
6. approach and landing
7. taxi-in

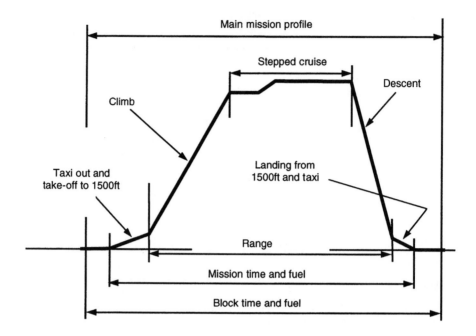

Fig. 10.1(a) Main mission flight profile definition.

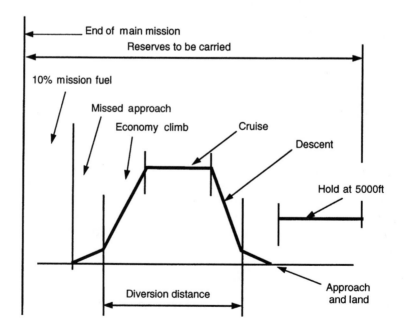

Fig. 10.1(b) Reserve fuel flight profile definition.

Table 10.1

	Domestic flights	Typical international flights
Allowances		
Start and taxi out	9 min	12 min
Approach and landing	6 min	6 min
Taxi in (from reserves)	5 min	6 min
Reserves		
Block fuel	45 min	Allowance of 5%
Diversion	200 nm	200 nm
Hold		30 min at 1500 ft
Diversion approach and landing overshoot	6 min	6 min

Assumed en-route flight conditions: temperature ISA, wind zero.

The diversion distance will be specified. In addition the following reserves are usually specified:

8. hold for a specified time at speed for minimum fuel consumption
9. climb to diversion cruise altitude
10. cruise at speed for maximum range
11. descent to 1000 ft
12. contingency (estimated as a percentage (e.g. 10%) of the fuel used in the main mission profile segments (1–6))

The allowances for taxi, diversion distances and hold times will vary with each route to be flown by the aircraft. To enable a comparison to be made between various aircraft a standard set of allowances forms part of the aircraft initial specification. A typical set of allowances and reserves is shown in Table 10.1.

Performance calculations

The performance calculations for each phase of the flight and their impact on the aircraft range are described below.

Start-up and taxi-out

This phase does not contribute to the range but it does impact on the fuel burn. The customer will usually define the fuel required for this phase in terms of minutes at the engine setting for taxi. Typical times are given in Table 10.1.

Take-off and initial climb to 1500 ft

The take-off and initial climb are two of the most critical safety phases of the aircraft flight. The combined analysis of both phases is referred to as the aircraft take-off performance. A detailed definition of the airfield limits for the aircraft will be held as part of the Aircraft Flight Manual. In the project design phase the field

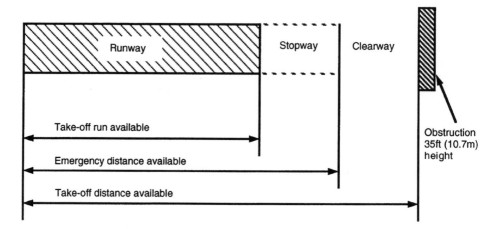

Fig. 10.2 Available take-off distances.

requirement is regarded as an essential part of the initial specification and as such will directly influence the aircraft configuration (e.g. it may dictate the size of engine required).

Again this phase does not contribute to the range but it does impact upon the fuel burn. The fuel required is defined as minutes of fuel burn at maximum take-off rating, with a typical time of two minutes.

Take-off distance

The distances available at an airfield are given in the following form.

- Take-off run available – this is the length of the paved runway.
- Emergency distance available – this is the sum of the runway length plus stopway length. The stopway is an area at the end of the runway, which is clear of obstructions.
- Take-off distance available – this is the sum of the runway, stopway and clearway lengths. The clearway distance is the distance from the end of the stopway to the first obstruction with a height of 35 ft or more.

These distances are shown diagrammatically in Fig. 10.2.

Take-off distances required by the aircraft are related to aircraft performance.

- Take-off run required – this is the distance from the start of the take-off run to the point of lift-off, plus one third of the airborne distance from lift-off to 35 ft [10.7 m] (known as the screen height). For the case when all engines are operating, this distance is multiplied by 1.15. For safe operation the length of take-off run required must not be greater than the take-off run available.
- Take-off distance required with all engines operating – this is the distance from start of the take-off run to an altitude of 35 ft multiplied by 1.15. This length must not be greater than the take-off distance available.
- In the event of the failure of one engine the distances are determined for two possible actions.

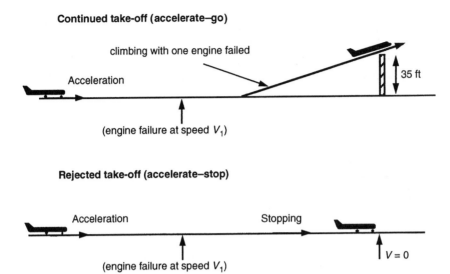

Fig. 10.3 Take-off options with engine failure.

1. Continued take-off after failure – this is the distance from the start of take-off run to the speed at which the critical engine failed plus the distance from this speed to a height of 35 ft. The total distance (accelerate–go) must not be greater than the take-off distance available.
2. Rejected take-off – this is the distance from the start of the take-off run to the speed at which the critical engine failed plus the distance to come to a stop from this speed using wheel brakes alone. The accelerate–stop distance must not be greater than the emergency distance available.

The two definitions for engine-out calculations are shown in Fig. 10.3.

Take-off performance – all engines operating

The take-off performance of the aircraft with all engines operating is relatively straightforward. The aircraft, starting at rest, accelerates along the runway at a low angle of attack altitude (nose-wheel on the ground) and passes a speed (V_S) which equals the aircraft stall speed in the take-off configuration (e.g. undercarriage down and flaps deflected to the take-off angle). At a speed (V_R) which is greater than V_S, the pilot rotates the aircraft by lifting the nose-wheel off the ground. During this manoeuvre the aircraft angle of attack may be limited to avoid the rear fuselage touching the runway. The aircraft will continue to move along the runway until the speed increases to the lift-off value (V_{LOF}). At this point the aircraft starts to climb away from the runway slowly gaining height until the screen height specified in the requirements (usually 35 ft; 10.7 m) is achieved. At this point the aircraft speed should be equal to or greater than speed V_2 (see definition below). Speed V_2 is obviously a critical parameter in the estimation of take-off performance. The airworthiness requirements prescribe various criteria that have to be met in determining V_2. The speeds defined for take-off calculations are illustrated and described in Fig. 10.4.

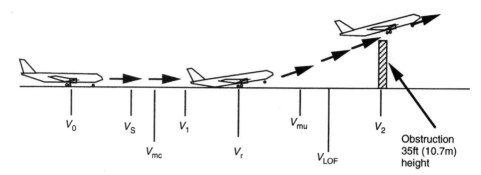

Fig. 10.4 Take-off speed definitions.

V_S Stalling speed (in take-off configuration).

V_{mc} Minimum control speed. At this speed if a critical engine fails the aircraft can be flown with zero yaw and a bank (roll angle of less than 5°).

V_1 Critical power failure speed.

V_R Rotation speed. This can equal V_1 but must be at least $1.05\,V_{mc}$

V_{mu} Minimum unstick speed. Due to tail interference with the ground there may be insufficient incidence available to lift off the runway at V_R.

V_{LOF} Lift-off speed. The aircraft becomes airborne. With all engines operating $V_{LOF} > 1.1\,V_{mu}$ and with one engine failed $V_{LOF} > 1.05\,V_{mu}$.

V_2 Take-off climb speed. This is the speed to be achieved at the 35 ft screen height. It has to have the following margins over V_{mc} and V_S:

$V_2 > 1.1\,V_{mc}$

$V_2 > 1.2\,V_S$ or $> 1.15\,V_S$ provided that with one engine failed there is a significant effect on the stall speed with the remaining engine(s) at maximum power.

Estimation of take-off distance – all engines operating

The take-off analysis consists of three parts:

1. ground roll
2. transition to climb
3. climb

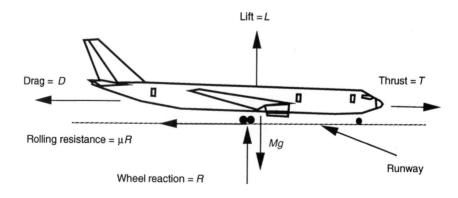

1. Ground roll. Resolving the forces acting on the aircraft during the ground roll we have:

(a) horizontally:

$$T = D + M \, (dV/dt) + \mu R$$

(b) vertically:

$$R = Mg - L$$

Combining gives:

$$M \, (dV/dt) = T - D - \mu \, (Mg - L)$$

where: M = aircraft mass
dV/dt = aircraft acceleration
μ = runway coefficient of friction
R = undercarriage reaction force = $Mg - L$
g = gravitational acceleration

Noting that, from the aircraft drag polar in the take-off configuration,

$$C_D = a + bC_L^2$$

where: C_D = aircraft drag coefficient = $a + bC_L^2$
C_L = aircraft lift coefficient

combining with the equation for drag:

$$D = 0.5 \, \rho V^2 S C_D$$

where: ρ = air density
V = aircraft velocity
S = wing reference area

and writing aircraft weight and lift in this form:

$$W = Mg \text{ and } L = 0.5 \, \rho V^2 S C_L$$

gives:

$$(dV/dt) = [(T/W - \mu) + (\rho/(2W/S))(-a - bC_L^2 - \mu C_L)V^2]g \qquad (10.1)$$

The ground run (s_G) is the integral of $[(1/2a) \, dV^2]$ from zero to V_{LOF} $\qquad$ (10.2)

where a is the acceleration (i.e. (dV/dt) as defined in equation (10.1). The aircraft incidence will be constant during the majority of the ground run so that for all practical purposes C_L can be assumed to be constant and lift will be proportional to V^2.

An examination of the variation of take-off thrust with speed shows that it is approximately linear. As the ground run integration is with respect to V^2, a mean value of thrust at speed $0.707 \, V$ would be sufficiently accurate for the calculation.

Therefore with T and C_L constant, two further constants K_T and K_A can be defined as follows:

$$K_T = (T/W) - \mu \tag{10.3}$$

$$K_A = \rho(-a - bC_L^2 - \mu C_L)(2W/S) \tag{10.4}$$

Substituting K_T and K_A in equation (10.1) gives:

$$(dV/dt) = [K_T + K_A \cdot V^2]g \tag{10.5}$$

and combining equations (10.2) and (10.5) and integrating between zero and lift-off speed gives:

$$s_G = 1/(2\,gK_A) \cdot \ln\,[(K_T + K_A \cdot V_{LOF}^2)/K_T]$$

Values of the rolling friction (μ) for dry conditions for several pavement types are shown below:

paved runway	$\mu = 0.02$
hard turf, gravel	$\mu = 0.04$
short dry grass	$\mu = 0.05$
long grass	$\mu = 0.10$
soft ground	$\mu = 0.10$ to 0.30

Civil jet transports will only operate from a paved runway; the other surfaces are included for interest or if unconventional operations are to be analysed.

2. Transition to climb. During transition the aircraft accelerates from V_{LOF} to the take-off climb speed V_2. It is difficult to accurately predict the ground distance covered during transition. A simple method will be presented, but where possible it should be calibrated against the measured performance of current aircraft. During transition we assume that the aircraft flies at $0.9C_{L\max}$ and that the average speed in the manoeuvre is given by:

$$V_{TRANS} = (V_{LOF} + V_2)/2$$

We also assume that the aircraft is flying along an arc, as shown in Fig. 10.5. Resolving normal to the flight path gives:

$$L = Mg\cos\theta + MV_{TRANS}^2/r \tag{10.6}$$

where r is the radius of the arc. The load factor n, is defined as:

$$n = L/Mg \tag{10.7}$$

with

$$L = 0.5 \cdot \rho \cdot V_{TRANS}^2 \cdot S \cdot 0.9\,C_{L\max}$$

For small angles $\cos\theta = 1.0$. Substituting equation (10.6) into (10.7) gives:

$$n = 1 + (V_{TRANS}^2/rg)$$

or

$$r = V_{TRANS}^2/[g(n-1)] \tag{10.8}$$

Hence, the ground distance for transition, s_T, is given by:

$$s_T = r \cdot \gamma \tag{10.9}$$

where γ is the final climb gradient.

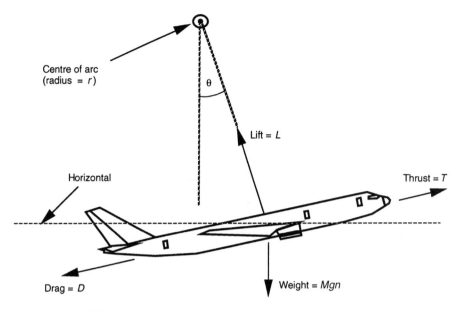

Fig. 10.5 Transition flight path definition.

The height at the end of transition is:

$$h_T = r \cdot \theta \cdot \theta / 2$$

where θ is the angle of the flight path arc as shown in Fig. 10.6.

The screen height may be exceeded during the transition manoeuvre.
The distance to the screen height, s_S, can be approximated to:

$$s_S = [(r + h_S)^2 - r^2]^{0.5} \tag{10.10}$$

where: h_S = screen height

Referring to equations (10.8) and (10.9) it can be seen that the transition manoeuvre is governed by the values of n and V_{TRANS}. Typical values for these parameters are $n = 1.2$, $V_{TRANS} = 1.15 \, V_S$.

3. Climb. The ground distance from the end of transition to the screen height, s_C, is given by:

$$s_C = (\text{screen height} - h_t) / \tan \gamma_C \tag{10.11}$$

where: h_t = height at end of transition
γ_C = best climb angle

For small climb angles $\tan \gamma_C = \sin \gamma_C = \gamma$.

The total distance is then calculated by summing the individual components (groundroll + transition + climb). For JAR/FAR certification rules the take-off field length is then multiplied by 1.15 to allow for operational variabilities.

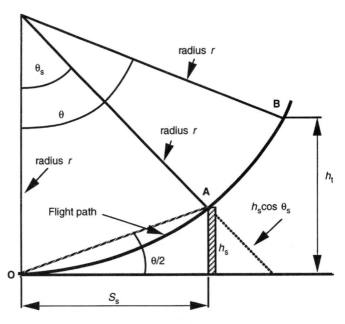

Fig. 10.6 Transition flight path geometry.

Estimation of take-off distance – with one engine failed

As described previously, the pilot has two options in the event of an engine failure during the take-off run:

- continue the take-off and fly away on the remaining engines;
- apply emergency braking and bring the aircraft to rest further down the runway.

These two options were shown on Fig. 10.3. They are often referred to as 'accelerate–go' and 'accelerate–stop'. If the engine failure occurs early in the take-off run it would be preferable to stop the aircraft. If the failure was later (when the aircraft had achieved a large amount of kinetic energy) it may be better to continue the take-off and fly away. For a given aircraft weight it is possible to calculate the total distance required to either stop or fly (to the obstacle height) for a range of speeds at which the engine fails. These calculations enable the failure speed at which the 'accelerate–go' distance is equal to the 'accelerate–stop' distance to be identified. This common distance is known as the 'balanced field length' (BFL). The output of this type of calculation is shown in Fig. 10.7.

When calculating the balanced field length the equations developed for the all engines operating case can be used. Up to the point of engine failure the calculations are exactly the same. At and after the point of engine failure the following changes have to be made.

- 'Pilot's reaction time' – this is a time delay between the engine failure and the pilot recognising it and initiating some action. A delay time of two seconds is currently prescribed during which the aircraft speed is assumed to be constant at the engine failure speed.

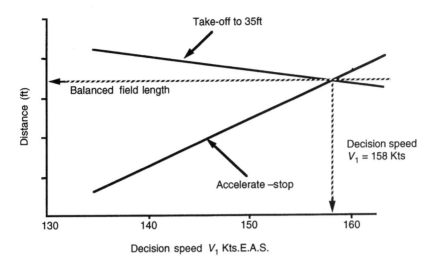

Fig. 10.7 Definition of balanced field length.

- In the 'accelerate–go' case, allowance has to be made for:
  1. the increased drag due to the failed engine and the effect on the aircraft drag of the resulting aircraft yaw altitude;
  2. any allowable emergency power for the live engine(s).
- In the 'accelerate–stop' case, the performance and limitations of the braking system are important.

Balanced field length is a function of aircraft weight; therefore it is possible to operate an aircraft out of an airport with restricted runway length by reducing aircraft weight (removing payload and/or fuel). The take-off field length required for a given take-off weight and ambient condition (temperature and airfield height) is the greater of the balanced field length and 1.15 times the all engines operating distance to 35 ft.

Simplified empirical take-off performance estimation

The equation for acceleration during the ground roll is:

$$(dV/dt) = (T/Mg) - (D/Mg) - (\mu R/Mg) \tag{10.12}$$

where: D = aerodynamic drag (N)
μR = force due to the rolling friction (N)
T = engine thrust (N)
M = aircraft mass (kg)

Compared to thrust, the forces from drag and rolling friction are very small. Figure 10.8 shows the magnitude of the forces during take-off.

Neglecting aerodynamic drag and rolling friction in equation (10.12) we find that acceleration, a, is given by:

$$a = (dV/dt) = (T/Mg)$$

In general, the distance is equal to $(V^2/2a)$.

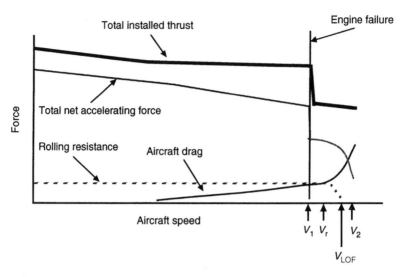

Fig. 10.8 Forces on the aircraft during take-off.

As discussed earlier, we can take the thrust at the mean energy speed ($0.707\,V_2$). Therefore distance, s, is given by:

$$s = V_2^2/[2(T/Mg)]$$

where: V^2 = true airspeed = $V_{2EAS}/\sigma^{0.5}$
EAS = equivalent airspeed
σ = relative density = ρ/ρ_0
$C_L @ V_2 = Mg/S_w \cdot q$

Using the definitions above:

$$s = kM^2g^2/(S_w \cdot T \cdot C_{LV_2} \cdot \sigma) \tag{10.13}$$

where: k = constant
$C_{LV_2} = Mg/S_w \cdot q$
S_w = reference wing area
q = dynamic pressure = $0.5\,\rho V_{TAS}$
TAS = true air speed
ρ_0 = sea level air density

An analysis of the certified take-off performance of current aircraft will enable the above relationship to be calibrated and the values of the constant k to be established. The calibration will depend upon the number of engines. It has been found that in general, twin-engined aircraft take-off performance will be determined by balanced field length and four-engined aircraft by the all engines operating case.

Take-off climb
The take-off climb begins at lift-off and finishes at 1500 ft above the runway. For analysis, the climb profile is split into four segments. Airworthiness requirements

Table 10.2 Climb segment definitions

Segment	Height	Flap	U/C	Rating	Gradient (%) number of engines		
					2	3	4
1st	0–35 ft	TO	Down	TO	+'ve	0.3	0.5
2nd	400 ft	TO	Up	TO	2.4	2.7	3.0
3rd	400+ ft	Variable	Up	Max. Cont.	Level flight acceleration		
4th	1500 ft	En-route	Up	Max. Cont.	1.2	1.5	1.7

demand that, with one engine failed, certain climb gradients are met in each segment. The definition of the segments along with the required gradients and associated operating conditions are summarised in Table 10.2.

The take-off climb performance of civil transport aircraft is usually defined by second-segment climb requirements. The second segment starts with 'the under-carriage up' and ends 400 ft above ground. During this segment a minimum climb rate must be demonstrated with an aircraft speed at least 15% above the stall speed and with one engine inoperative.

Consider the aircraft shown in Fig. 10.9, which is climbing at constant forward speed V:

$$T = D + Mg \sin \gamma$$

where: T = total engine thrust = $N \cdot F_n$
N = number of engines
F_n = thrust from each engine
Mg = aircraft weight = W

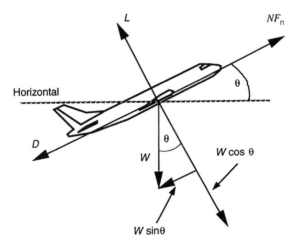

Resolving along and normal to the flight path:–
$NF_n - D = W \sin\theta$
$L = W \cos \theta$ (= W when θ is small)
Hence, climb gradient (θ) = tan θ = $(NF_n - D) / W$

Fig. 10.9 Forces on the aircraft during climb.

Hence

$$\text{climb gradient} = (T - D)/W \tag{10.14}$$

The aircraft drag during second-segment climb can be split into two parts:

- the drag of the basic configuration
- the asymmetric drag due to a failed engine.

The drag of the basic configuration may be estimated by using methods discussed in Chapter 8. The asymmetric drag consists of the windmilling drag of the failed engine plus the additional trim drag from the rudder/aileron deflection required to counteract the asymmetric flight geometry. For initial project work the windmilling drag of the failed engine may be estimated:

$$C_D S = 0.3\, A_f$$

where A_f is the fan cross-sectional area.

During the second-segment climb the additional trim drag may be assumed initially to be 5% of the basic profile drag. If the aircraft is take-off climb critical it would be appropriate to use a more detailed method as described in Chapter 8.

The engines will be at maximum take-off power setting. However, the thrust produced will be lower than the sea-level static value due to the effect of forward speed. This must be taken into account as the thrust at V_2 can be around 20% lower than the static value.

Combining the take-off climb performance with the take-off distance data gives the overall take-off capability of the aircraft. A typical chart is shown in Fig. 10.10.

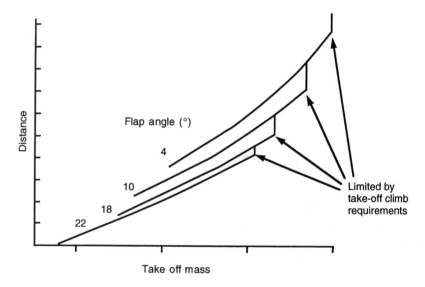

Fig. 10.10 Effect of flap deployment on take-off distance.

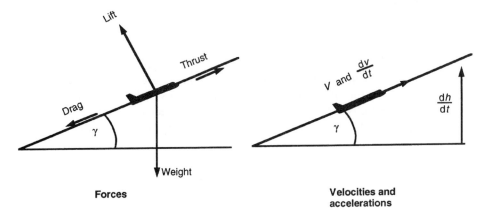

Fig. 10.11 Climb analysis.

En-route performance

In normal flight the aircraft can be flying in any of the following conditions:

- constant altitude, constant speed
- climbing or descending at constant speed
- accelerating or decelerating at constant altitude
- climbing or descending and accelerating or decelerating

Figure 10.11 shows the forces diagram and the velocity acceleration diagram for the general case with a climb gradient γ.

Assuming small angles:

$$T = D + W\,(dV/dt) + W\sin\gamma$$

Since

$$\sin\gamma = (dh/dt)/V$$

$$T = D + W\,(dV/dt) + W\,(dh/dt)/V$$

Therefore

$$V\,(T - D)/W = V\,(dV/dt) + (dh/dt) \qquad (10.15)$$

The thrust available (T) has to overcome the drag (D) and at the same time provide the necessary acceleration (dV/dt) and/or rate of climb (dh/dt).

Climb from 1500 ft to initial cruise altitude

This is split into two parts:

1. 1500 ft to 10 000 ft where the speed is restricted to 250 kt in the USA (but not necessarily in the rest of the world);
2. from 10 000 ft to the initial cruise altitude.

The distance flown in both of these phases contributes to the range. The climb speed above 10 000 ft is at constant EAS followed by a constant Mach number. Various criteria can be applied to determine these climb speeds.

The two most common are:

- speed for minimum time to height
- speed for minimum direct operating cost.

In normal operation the second of these two options is used. The speed is higher than that required for minimum time to height. This reduces the block time, improves the productivity and hence improves the direct operating cost (DOC). The Mach number at the top of climb in these circumstances is usually the cruise Mach number.

Climbing at constant equivalent airspeed (EAS) implies an increasing true airspeed as altitude increases. Climbing at constant Mach number below the tropopause implies a reduction in the true air speed as altitude increases. Hence when calculating climb performance these effects need to be taken into account. From equation (10.15) the rate of climb with no acceleration is given by:

$$(RoC)_0 = [(T - D)/W]V$$

and the rate of climb with acceleration by

$$(RoC)_a = [T - D - (W/g)(\mathrm{d}V/\mathrm{d}t)](V/W)$$
$$= [(T - D)/W]V - (V/g)(\mathrm{d}V/\mathrm{d}t) \qquad (10.16)$$

Now

$$(\mathrm{d}V/\mathrm{d}t)/(\mathrm{d}h/\mathrm{d}t) = (\mathrm{d}V/\mathrm{d}h)$$

Dividing equation (10.16) by $(\mathrm{d}h/\mathrm{d}t)$ and substituting for $(RoC)_0$ we get:

$$(RoC)_a = (RoC)_0/[1 + (V/g)(\mathrm{d}V/\mathrm{d}h)]$$

The climb can now be computed using a step by step procedure using the appropriate engine climb rating as specified by the engine manufacturer.

Cruise performance

An aircraft in cruise is neither accelerating or climbing. It will normally be flying at a constant speed (Mach number or EAS) at a constant altitude. An examination of equation (10.15) for the cruise case shows that with zero acceleration and zero rate of climb, thrust equals drag (as would be expected). Before looking at cruise performance and its implication on range the aircraft must be capable of climbing to the desired initial cruise altitude in an acceptable time. This requires a reasonable rate of climb at the initial cruise altitude. The initial cruise altitude will often define the climb thrust required and the aircraft wing area, as discussed below.

At the initial cruise altitude a rate of climb of 300 ft/min is normally required. Therefore the thrust available at this height must equal the aircraft drag plus the thrust to produce the required climb rate. The climb gradient from equation (10.14) may be written as:

$$\gamma = V_c/V = (T - D)/W$$

V_c is the rate of climb (300 ft/min = 1.52 m/s) and V is the aircraft speed at the initial cruise altitude (i.e. the aircraft cruise speed). The drag (D) may be estimated using the methods outlined in Chapter 8. Thus rearranging the required increase in thrust to give V_c ft/min rate of climb:

$$\Delta T = (V_c/V)W$$

Total thrust required now becomes:

$$T = (V_c/V)W + D$$

Note that the thrust calculated above is at a climb rating and must be related back to an equivalent static sea-level take-off thrust to determine the installed thrust required for the aircraft design.

The aircraft at the initial cruise condition should be operating at a lift coefficient to give a lift/drag ratio near the maximum. If the lift coefficient is too high it implies that the wing area is too small or conversely if it is too low the wing is too large.

In the cruise phase the range flown is calculated using the appropriate aircraft and engine parameters along with the fuel available for cruise. Assume that the total aircraft thrust at cruise rating is T, and that at this condition the engine has a specific fuel consumption (SFC) denoted by c [measured in (N/sN)]. The decrease in aircraft weight with time due to fuel burn is:

$$(dW/dt) = -c \cdot T$$

For straight and level flight, we have $(T = D)$ and $(L = W)$, therefore:

$$(dW/dt) = -cD(W/L)$$

and

$$dt = (1/c)(L/D)(dW/W) \qquad (10.17)$$

Integrating between an initial cruise weight W_1 and the final cruise weight W_2 gives:

$$t = (1/c)(L/D)\ln(W_1/W_2) \qquad (10.18)$$

If the aircraft is flying at a speed V then the range flown is:

$$R = (t_2 - t_1)V \ln(W_1/W_2) \qquad (10.19)$$

This is the classical Breguet Range Equation. The equation assumes the lift to drag ratio (L/D) remains constant throughout the cruise phase. This implies that the aircraft is constantly climbing to maintain the same lift coefficient but this height change is very gradual and can be ignored in initial project calculations.

In practice, air traffic control does not allow cruise-climb profiles and insists on airlines adopting a stepped climb procedure as shown in Fig. 10.12.

Currently the steps must be 4000 ft, but with the introduction of satellite navigation this may be reduced to 2000 ft. The maximum cruise range for a given speed can be derived as follows. The range covered in time dt is given by:

$$dR = V_T dt$$

where V_T = cruise speed

Noting that $V_T = [2W/\rho S C_L]^{0.5}$ and recalling equation (10.17) for (dt) gives:

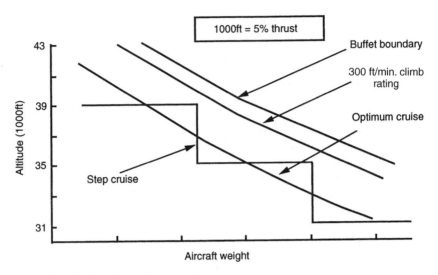

Fig. 10.12 Cruise altitude options and constants.

$$dR = -[2W/\rho S C_L]^{0.5}(1/c)(L/D)(dW/W)$$

Noting that $(L/D) = C_L/C_D$ and rearranging the terms gives:

$$dR = -[2/\rho S]^{0.5}(1/c)(C_L^{0.5}/C_D)(dW/W^{0.5})$$

Integrating between W_1 and W_2 gives:

$$R = (2/c)[2/\rho S]^{0.5}(C_L^{0.5}/C_D)(W_1^{0.5} - W_2^{0.5})$$

This equation shows that the maximum range will occur when $(C_L)^{0.5}/C_D$ is a maximum, from which it can be shown that:

$$C_L \text{ (max range)} = (a/3b)^{0.5} \tag{10.20}$$

$$C_D \text{ (max range)} = (4/3)a = (4/3)C_{D0} \tag{10.21}$$

where: a and b are determined from the drag polar $(C_D = a + bC_L^2)$.

In the initial project design phase the range is often specified and the mass of fuel must be estimated to fly this range. Rearranging the Breguet Range Equation we get:

$$(W_f/W) = 1 - \exp\{R \cdot c/[V \cdot (L/D)]\} \tag{10.22}$$

where: (W_f/W) = fuel weight ratio
c = engine specific fuel consumption (SFC)
R = still air range

In the cruise phase there are various options including these three:

1. cruise at speed and altitude to give minimum fuel burn;
2. cruise at speed and altitude to give minimum direct operating cost;
3. cruise at speed and altitude to give maximum endurance.

Option 1. The fuel burn can be obtained by applying equations (10.20), (10.21) and (10.22) over a range of altitude.

Option 2. This is usually flown at the aircraft design Mach number and at an altitude to maximise the range with the given fuel load. The speed in this case is usually higher than for case 1 in order to reduce block time thereby increasing the aircraft productivity and hence minimising direct operating cost.

Option 3. Is applicable to the holding/diversion conditions. Equation (10.17) shows that the endurance is a function of the engine specific fuel consumption (c) and the aircraft lift drag ratio (L/D). As the holding altitude is usually specified, to minimise the holding fuel the aircraft needs to be flown at the highest L/D possible (i.e. as close to the minimum drag speed as is practical to give acceptable aircraft speed stability).

Descent performance

The descent calculation is similar to the climb except that in this case the thrust is less than the drag.

$$\text{Rate of descent} = [(D - T)/W] \cdot V \cdot (1 + V/g \cdot dV/dh)^{-1}$$

The descent speeds are usually M_{mo} and V_{mo} down to 10 000 ft and 250 kt below 10 000 ft. The thrust is at a setting which enables the cabin air conditioning to be maintained. There is one other criterion that must be observed: the rate of descent must be such that the maximum rate of change of pressure in the cabin does not exceed 300 ft/min. Once these conditions are met the descent fuel, distance and time can be calculated on a step by step basis similar to the climb. At the initial project stage in aircraft design there are usually insufficient data to establish the thrust setting; it is therefore acceptable to extend the cruise to the destination and ignore the descent manoeuvre. This is a pessimistic assumption as the air miles/kg of fuel is much higher in the descent than in the cruise. Often a first approximation of the thrust and fuel flow on the descent is possible and can be used to give an allowance for the time, distance and fuel used in the descent.

Landing

The landing distance is predicted in a similar manner to take-off. The landing phase is split into several segments as shown in Fig. 10.13. The distance covered during each segment is calculated and then summed to obtain the total distance. This value is then factored to account for pilot and operational uncertainties.

Each of the separate phases will now be described:

1. Approach. For civil transport aircraft the approach flight path is typically made at 3° to the horizontal. For FAR and JAR certification the approach starts at a screen height of 50 ft and ends at the flare height, h_F. The value of h_F is given by:

$$h_F = r \cdot \gamma_A \cdot \gamma_A/2$$

where γ_A is the approach gradient. The approach distance, s_A, is given by:

$$s_A = (\text{obstacle height} - h_F)/\tan \gamma_A \tag{10.23}$$

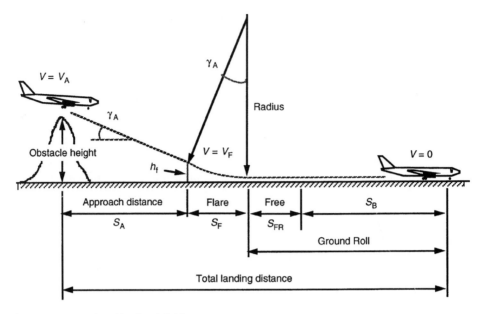

Fig. 10.13 Approach and landing definitions.

2. Flare. During the flare manoeuvre, the aircraft decelerates from the approach speed (V_A), to the touch-down speed (V_{TD}). The average speed during the flare, V_F, is given by:

$$V_F = (V_A + V_{TD})/2$$

Typically $V_{TD} = 1.15\,V_S$.

The radius of the flare manoeuvre, r, is given by:

$$r = V_F^2/[g(n-1)]$$

Typically $n = 1.2$.

The ground distance covered during the flare, s_F, is given by:

$$s_F = r \cdot \gamma_A \tag{10.24}$$

3. Ground roll. After touch-down the aircraft will roll for a few seconds before the pilot applies the brakes and spoilers. The distance covered is called the free-roll distance. This may be approximated using:

$$s_{FR} = t \cdot V_{TD}$$

where t is duration of free roll.

The braking distance (s_B) is calculated using the following equation with speed from V_{TD} to rest:

$$s_B = [1/(2gK_A)]\ln[(K_T + K_A V_{TD}^2)/K_T] \tag{10.25}$$

where: $K_A = -(\rho a)(2W/S)$
$\qquad\quad K_T = -\mu$

and in general thrust = zero.

The thrust and lift coefficient must be evaluated at the average energy speed (i.e. $0.707\,V_{TD}$). Typically thrust will be zero or idle and with spoilers deployed C_L will be zero. The effects of braking are included by modifying the value of μ. Maximum values under various conditions are shown below:

dry paved runway $\mu = 0.5$
wet paved runway $\mu = 0.3$
icy paved runway $\mu = 0.1$

Many modern civil jet transports use coefficients less than 0.5 in normal operating conditions to reduce brake and tyre wear. The maximum coefficient is dependent on wheel size which is limited in order to fit the brake discs and systems. Note that there is weight transfer during braking on to the nose undercarriage. A typical value of μ would be 0.3.

The total landing distance calculated by the equations above is factored by 1.66 for JAR/FAR certification rules to account for variability in pilot and operational uncertainties during the landing manoeuvre.

Simplified empirical landing calculations

If we examine the above equations we note that:

- the approach distance is independent of the aircraft speed and is only a function of the approach angle;
- the flare distance is a function of V^2 and the normal acceleration;
- the ground roll is a function of V^2 and the available braking force.

The total landing distance is primarily a function of V^2, the normal acceleration in the flare and the braking force. The normal acceleration in the flare must lie in a small band for passenger comfort. The braking force is a function of the braking coefficient of friction on the runway surface and the amount of aircraft mass reacted by the braked wheels. This later item depends upon the mass distribution between the nose-wheels and the main undercarriage when in the braking mode and also the amount of residual lift from the wings. The residual lift from the wings depends upon how efficient the lift spoilers are and how quickly they are deployed after touch-down (the effect of thrust reversers is not allowed).

If we assume that all aircraft pull a similar amount of 'g' in the flare, the mass distribution between nose and main wheels is similar and the spoilers have roughly the same efficiency; then the landing performance is purely a function of the approach speed, V_A, squared. Airworthiness requirements specify that $V_A = 1.3\,V_S$ where V_S is the stalling speed in the landing configuration.

$$V_S^2 = \text{constant} \times M/(S_w \cdot C_{L_{max}} \cdot \sigma)$$

$$\text{landing distance (approx.)} = M/(S_w \cdot C_{L_{max}} \cdot \sigma) \tag{10.26}$$

An analysis of the landing performance of current aircraft will enable this relationship to be calibrated. One might expect more scatter for landing distance than take-off as the major decelerating force from the wheel brakes is subject to more variability than the accelerating force from the engines during take-off.

Fuel for approach and landing

This phase of the flight does not contribute to the range but it does affect the fuel burn and hence the fuel available for the other flight segments. The customer will usually define the fuel required for this phase in terms of minutes at the approach fuel flow. A typical time would be six minutes.

Taxi-in after landing

Again this does not contribute to the range but it does impact on the total fuel burn. The fuel allowance for this item would come out of the reserves. A typical time would be six minutes at the taxi fuel flow.

Reserves

This is extra fuel that has to be carried to cater for unforeseen events (e.g. air traffic control may prevent the aircraft flying at its optimum altitude, head-winds may be more severe than forecast, thunderstorms may force the aircraft to depart from the planned flight plan, the destination airfield may have a traffic problem and the aircraft may have to join a holding pattern, the destination airfield may be closed for weather or other operational reasons and the aircraft may have to divert). It is necessary to consider the worst case scenario, with the defined hold and diversion requirements.

The fuel allowed for the overshoot at the end of the normal operational mission is typically 1–1.5 min at take-off power. The fuel for hold is calculated at the speed and fuel flow for maximum endurance. The diversion fuel is calculated using climb, cruise and descent techniques to give the minimum fuel for the required diversion distance.

Payload range

The aircraft mission analysis enables the fuel required for various range and take-off weights to be calculated and a payload–range diagram to be constructed. This shows the ability of the aircraft to perform different missions. A typical payload–range diagram is shown in Fig. 10.14. A full explanation of the diagram is given in Chapter 7. Three points define the payload–range diagram.

1. Range with maximum payload

This point is defined as the range flown with maximum payload. The maximum payload of aircraft is defined as either the maximum volumetric payload or the maximum zero fuel weight limited payload.

- The maximum volumetric payload includes:
  1. mass of passengers in high-density layout, assuming standard passenger weights;
  2. mass of passengers baggage, assuming standard baggage weights;
  3. remaining hold volume used for cargo, assuming standard cargo densities.

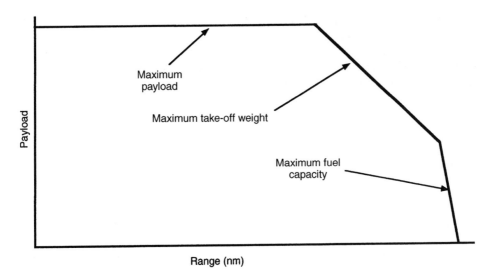

Fig. 10.14 Payload–range diagram.

- The maximum zero fuel mass limited payload is:

   zero fuel mass limited payload = maximum zero fuel mass − operating mass empty

The maximum zero fuel weight is normally based on structural requirements. In the absence of any other criteria at the conceptual stage it is acceptable to assume that the maximum zero fuel weight is equal to the operating weight empty plus the maximum volumetric payload.

2. Maximum fuel range

This point is defined as the range flown with maximum fuel and the associated payload to maintain the aircraft maximum take-off weight. The maximum fuel weight corresponds to the fuel tank volume available. Most civil transport aircraft store fuel in the wing. Additional fuel for long-range aircraft may be stored in the wing centre section and in some exceptional cases also in the tail structures. The payload with maximum fuel is then calculated as:

payload = maximum take-off mass − maximum fuel mass − operating mass empty

3. Ferry range

This point corresponds to the range flown with maximum fuel and zero payload. Thus, in this case the aircraft is no longer at its maximum take-off weight and the aircraft cruise lift to drag ratio (L/D) will be different from that in cases 1 and 2. The take-off weight for the segment is calculated as:

take-off weight = operating weight empty + maximum fuel weight

The aircraft specification will define the design range and corresponding design payload. The design payload on a civil airliner is usually less than the maximum payload. It normally consists of the maximum passenger load plus their baggage and in addition some airlines require a certain amount of cargo. This will usually place the aircraft design point on the payload–range diagram between cases 1 and 2.

Example calculations for performance estimation

The performance of an aircraft can now be calculated. A typical example is given covering take-off, landing and payload–range calculations. The specimen aircraft has the characteristics shown in Table 10.3.

Table 10.3

Maximum take-off mass	230 000 kg (507 055 lb)
Operating mass empty	130 000 kg (286 600 lb)
Maximum landing mass	184 000 kg (405 646 lb)
Payload	
maximum passengers (3 class) 305	29 050 kg.
+ cargo (to be determined)	
Wing area	376.4 m^2

Aerodynamic data	Drag	$C_{L\,max}$
Clean undercarriage	$C_D = 0.015 + 0.05\,C_L^2$	
TO flap undercarriage down	$C_D = 0.035 + 0.05\,C_L^2$	1.75
TO flap undercarriage up	$C_D = 0.025 + 0.05\,C_L^2$	
Landing flap, undercarriage down	$C_D = 0.055 + 0.05\,C_L^2$	2.50

Engine thrust: sea-level static = 660 970 N

Take-off calculation – field length

The engine performance used in this example is taken from Fig. 9.19 for the bypass ratio 6.5 engine.

The lift coefficient during the take-off ground roll is assumed to be 0.7 with a $C_{L_{max}}$ of 1.715. Higher values of $C_{L_{max}}$ can be used provided that the take-off climb requirements can be met. These higher values are achieved by increasing the trailing edge flap angle as described in Chapter 8. The effect of increasing the trailing edge flap angle is to increase the C_L for a given incidence, thereby increasing the lift coefficient at the aircraft incidence during the ground roll.

The speeds associated with the lift coefficients at take-off weight (i.e. $C_{L_{max}} = 1.715$ and $C_{L_{V_2}} = 1.215$) are shown in Table 10.4. The drag polar is given in Table 10.3.

Table 10.4 Aircraft speeds

Speeds	(m/s)	Mach no
V_{Stall}	74.70	0.2195
V_{LOF}	82.20	0.2416
V_{TRANS}	85.90	0.2524
V_2	89.66	0.2634

Calculations for all engines operating

The coefficients, K_T and K_A are calculated using equations (10.3) and (10.4) respectively. The thrust term is factored to account for the decrease in thrust with increasing aircraft forward speed. The value of 0.864 seems appropriate for modern turbofan engines. A more accurate value should be used if detailed engine data are available.

$$K_T = (0.864 \cdot 660\,970)/(230\,000 \cdot 9.81) - 0.02$$

$$= 0.2331$$

$$K_A = 1.225/[2(230\,000 \cdot 9.81)/376.4] \times (0.02 \cdot 0.7 - 0.035 - 0.05 \cdot 0.7^2)$$

$$= -4.649 \times 10^{-6}$$

Stall speed (as quoted in Table 10.4)

$$V_S^2 = (230\,000 \cdot 9.81)/(0.5 \cdot 1.225 \cdot 376.4 \cdot 1.75)$$

$$= 74.7\,\text{m/s}$$

Lift-off speed (as quoted in Table 10.3)

$$V_{\text{LOF}} = 1.1 \times V_S = 82.2\,\text{m/s}$$

Ground distance:

$$s_G = (2 \cdot 9.81 \cdot 4.649 \cdot 10^{-6})^{-1}$$
$$\times \ln\{[0.2331 + (-4.649 \cdot 10^{-6} \cdot 82.2^2)]/0.2331\}$$

$$= 1586\,\text{m}$$

Transition to climb:

$$V_{\text{TRANS}} = 1.15 \times V_S$$

$$= 85.9\,\text{m/s}$$

$$R = 7380/[9.81(1.2 - 1.0)]$$

$$= 3761\,\text{m}$$

Final climb gradient:

$$\text{speed} = V_2 = 1.2 \times V_S$$

$$\text{Assume: } C_D = 0.025 + 0.05\, C_L^2 \text{ (take-off flap, gear up)}$$

$$C_L = 1.75/(1.2)^2 = 1.215$$

$$C_D = 0.0988$$

$$\text{Drag} = 0.5 \cdot 1.225 \cdot (1.2 \cdot 74.7)^2 \cdot 376.4 \cdot 0.0988$$

$$= 183\,028$$

$$\text{Engine lapse rate} = 0.803$$

Climb gradient:

$$\gamma = (T - D)/W$$

$$= (0.803 \cdot 660\,970 - 183\,028)/(230\,000 \cdot 9.81)$$

$$= 0.154$$

Ground distance covered during transition:

$$s_T = R \times \gamma$$

$$= 3761 \times 0.154$$

$$= 580$$

Height at end of transition:

$$h_T = R \cdot \theta \cdot (\theta/2)$$

$$= 3761 \cdot 0.154 \cdot (0.154/2)$$

$$= 44.6$$

The screen height is exceeded during the transition manoeuvre, so equation (10.10) must be used to determine the ground distance to the screen height:

$$s_S = [(3761 + 10.67)^2 - 3761]^{0.5}$$

$$= 284\,\text{m}$$

Therefore the take-off distance to the screen height is:

$$1586 + 284 = 1870\,\text{m}$$

Finally, for FAR/JAR regulations this value is factored by 1.15 to allow for pilot and operational variations, giving FAR take-off distance $= 2150\,\text{m}$.

Calculation for balanced field length

As before $C_{L_{max}} = 1.75$ and $C_{L_{V_2}} = 1.215$. Speeds are given in Table 10.5. Detailed analysis is shown in Tables 10.6 to 10.9.

Table 10.5

Speeds	(m/s)	Mach no
V_{Stall}	74.70	0.2195
V_{LOF}	82.20	0.2416
V_{TRANS}	85.90	0.2524
V_2	89.66	0.2634

Table 10.6 Calculation of ground run

Engine failure speed (V_1)	60	73.96	78.85	82.2
Mach number (Mn)	0.176	0.217	0.232	0.242
0.707 M	0.124	0.154	0.164	0.1710
Thrust lapse rate	0.9	0.877	0.87	0.864
Thrust (N)	594 873	579 671	575 044	571 078
K_T	0.2436	0.2369	0.2349	0.2331
K_A (constant)		−0.000 004 649		
Distance (m) to engine failure speed S_g	780	1244	1438	1586
Distance (m) travelled in two seconds reaction time at V_1	120	148	158	164

Table 10.7 Distance from end of reaction time to V_{LOF}

Mean speed during transition is given by: $V_m = (V_1 + V_{LOF})/2$

V_m (m/s)	71.1	78.1	80.5
M mean	0.209	0.23	0.82
Thrust lapse rate	0.835	0.822	0.818
Thrust (engine failed)	275 955	271 659	270 337
ΔC_D windmilling drag(constant)		0.003486	
ΔC_D asymmetric(constant)		0.00125	
$(1) = T/W - \mu$	0.1023	0.1004	0.09981
$(2) = (\mu C_L - a - b C_L^2)$(constant)		−0.05024	
$(3) = (V_m^2)[\rho/2(W/S)]$	0.5161	0.6225	0.662
$(4) = (2) \cdot (3)$	−0.02593	−0.03127	−0.03326
$(5) = (1) - (4)$	0.07637	0.06913	0.06655
Acceleration $[dV/dt](m/s^2)$	0.749	0.678	0.653
Δt (secs)	29.6	12.15	5.13
Δs (m)	2107	949	413

Transition to climb

$$V_m = [(V_2 - V_{LOF})/2] = 85.9 \, \text{m/s}$$

with $n = 1.2$ $$R = (V_m^2)/[g(n-1)] = 3761$$

Table 10.8 Final climb gradient

V_2	89.66 m/s (174 kt)
Thrust lapse rate	0.803
Thrust	265 379 N
C_{D_0}	0.10574
Drag	195 884
Gradient	0.0308
Transition distance $(r \cdot \gamma)$	116 m
Height at end of transition	1.78 m
Δh_t to screen	8.89 m

Distance from end of transition to screen height = 266.5 m

Engine failure speed (m/sec)	60	73.96	78.85	82.2
Total distance to screen (m)	3411	2745	2413	2154

Table 10.9 Stop distance

Failure speed (m/s)	60	73.96	78.85	82.2
K_T		−0.3		
K_A		+0.0000051		
S_{stop} (m)	594	889	1004	1087
Total distance (m)	1494	2281	2600	2837
V_1^2	3600	5470	6217	6757

It is convenient to plot the distances as a function of failure speed squared $(V_1)^2$ as shown in Fig. 10.15.

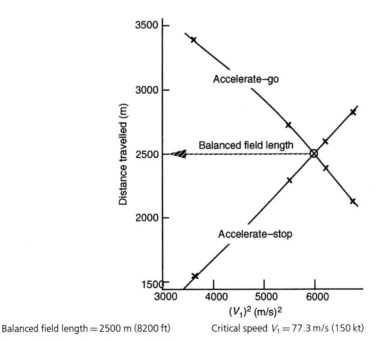

Balanced field length = 2500 m (8200 ft)   Critical speed V_1 = 77.3 m/s (150 kt)

Fig. 10.15 Calculation of balanced field length.

Calculation for second-segment climb

From the previous example we have:

$$C_D = 0.025 + 0.05C_L^2 \quad \text{(take-off flap, under carriage up)}$$

Lift coefficient at V_2 (from take-off example) $= 1.215$, $C_D = 0.09881$.
Drag coefficient due to failed engine:

$$C_D = 0.3\, A_f/S$$

For a 330 kN engine, a reasonable fan diameter would be 2.50 m.

$$C_D = (0.3 \times 2.50)/376.4 = 0.00199$$

Trim drag:

$$C_D = 0.05 \times 0.0988 = 0.00494$$

Total drag coefficient:

$$C_D = 0.09881 + 0.00199 + 0.00494 = 0.10574$$

Drag:

$$D = 0.5 \times 1.225 \times (1.2 \times 74.7)^2 \times 376.4 \times 0.10574 = 195\,884\,\text{N}$$

Climb gradient, one engine inoperative:

$$\gamma = (0.5 \times 0.803 \times 660\,970) - 195\,884/(230\,000 \times 9.81) = 0.0308$$

The minimum airworthiness gradient for a twin-engined aircraft with engine failure in the second segment is 0.024. Hence this aircraft meets the requirements. In the event that the second-segment climb was not met there are a number of steps that could be taken.

1. Increased engine thrust. If the engine is already operating at its maximum permissible take-off rating this can only be achieved by scaling the engine to a higher thrust. This would increase the engine weight and powerplant drag thus increasing the aircraft take-off weight or altering the payload–range capability.

2. Increase the number of engines. If the number of engines is increased from two to three or four, the thrust remaining with one engine inoperative will be much greater. Re-evaluating the climb gradient above for a three engine aircraft gives:

$$\gamma = [((3-1)/3 \times 0.803 \times 660\,970) - 195\,884]/(230\,000 \times 9.81)$$
$$= 0.07$$

i.e. 7.94% against a requirement of 2.7%. Re-evaluating for a four-engine aircraft gives:

$$\gamma = [((4-1)/4 \times 0.803 \times 660\,970) - 195\,884]/(230\,000 \times 9.81)$$
$$= 0.089$$

i.e. 8.9% against a requirement of 3.0%. As can be seen, in each case there is a large increase in the gradient. The disadvantage of increasing the number of engines is generally accepted to be increased maintenance costs and often increased propulsion system weight.

3. Reduce the take-off lift coefficient ($C_{L_{max}}$). If the take-off distance is not critical for the design, the take-off lift coefficient may be reduced by retracting the flaps thus reducing drag and increasing the one engine inoperative climb gradient.

4. Taking off at speeds higher than V_2 **(overspeeding).** Even if the take-off distance is not critical the climb gradient can be improved by taking off at speeds higher than V_2. This procedure is known as overspeeding. At V_2 the aircraft is below its minimum drag speed; therefore by climbing out at speeds between V_2 and the minimum drag speed the climb gradient can be improved by virtue of the reduced drag. It should be noted that there will be a reduction in thrust due to the increased momentum drag at the higher speeds, but this will, in general, not be sufficient to nullify the drag improvement.

Payload–range

The payload–range diagram is the next calculation to be done.

There are normally three cases to be considered:

1. maximum take-off weight, maximum design payload
2. maximum take-off weight, maximum full load
3. maximum full load, zero payload.

Case 1. will be used as an example.

Before a payload–range calculation can be attempted several parameters must be determined. These include the allowances, the distance travelled and the associated fuel burn. These must be calculated for the various segments of the flight profile.

Taxi fuel

At the maximum take-off weight and with a rolling coefficient of friction of 0.2 the thrust required to taxi the aircraft can be established. With take-off weight = 507 055 lb (230 000 kg) and runway friction $\mu = 0.02$:

$$\text{thrust required} = 0.02 \times 507\,055 = 10\,141\,\text{lb} \ (45.1\,\text{kN})$$

Hence thrust required per engine = 5070 lb (22.6 kN).

From the engine data a plot of thrust against fuel flow can be obtained as shown in Fig. 10.16.

The fuel flow at a thrust of 5070 lb = 1700 lb/hr (771 kg/hr) per engine. With two engines the fuel flow per aircraft = 3400 lb/hr (1542 kg/hr). This gives a taxi fuel flow of 60 lb/min (27.2 kg/min). Start-up and taxi-out = 6 min. Hence the fuel allowance for taxi, etc. = 6 × 60 = 360 lb (163 kg).

Except for the full tanks case this is extra to the take-off mass. Aircraft will have a maximum ramp weight which is greater than the maximum take-off weight to allow for this extra fuel.

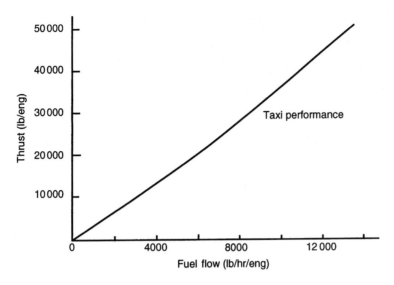

Fig. 10.16 Engine fuel flow during taxi.

Take-off allowance
This is usually taken as two minutes at take-off power. The calculation is shown in Table 10.10.

Table 10.10

take-off thrust sea-level static	74 315 lb (330 · b kN)
specific fuel consumption	0.32 lb/hr/lb
fuel flow	23 780 lb/hr (10 787 kg/hr)
fuel flow per aircraft	47 560 lb/hr (21 574 kg/hr)
	792 lb/min (359 kg/min)
take-off allowance	1580 lb (717 kg)

Approach
V_S in the approach configuration $\geqslant 1.1\ V_S$ in the landing configuration. For this aircraft the landing $C_{L_{max}} = 2.5$. Assume that the approach speed (V_A) is $1.3\ V_S$ (approach configuration). Then:

$$V_A = 1.43\ V_S \text{ (landing configuration)}$$
$$C_{LA} = 2.5/1.43^2 = 1.223$$
$$M_A = 0.235$$
$$C_D = 0.025 + 0.05\ C_L^2$$
$$L/D = 12.26$$
$$\text{maximum landing weight} = 405\,000\,\text{lb } (184\,000\,\text{kg})$$
$$\text{thrust required per engine} = 16\,517\,\text{lb } (73.5\,\text{kN})$$

Figure 10.17, from engine data shows thrust versus fuel flow in the approach condition.

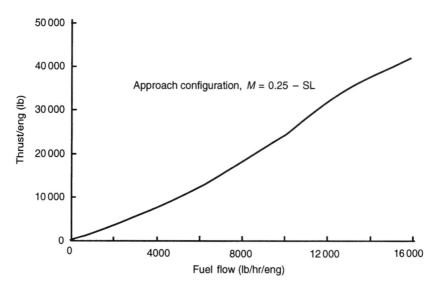

Fig. 10.17 Engine fuel flow on approach.

Total fuel flow $= 14\,400$ lb/hr (6532 kg/hr) per aircraft $= 240$ lb/min (109 kg/min). Assuming a five minute approach allowance the fuel used is 1200 lb (545 kg).

Taxi-in

Time allowance six minutes, fuel allowance at 60 lb/min $= 360$ lb.

The various fuel allowances which do not contribute to the range have been calculated and are summarised below:

taxi-out 360 lb (163 kg) (extra to take-off mass)
take-off 1580 lb (717 kg)
approach 1200 lb (345 kg)
taxi-in 360 lb (163 kg) (taken from reserves)

Climb performance

The next stage in the calculation is to compute the climb performance. The first consideration is the determination of the installed engine data at the various climb speeds and altitudes. Installed engine data is defined as the engine performance after allowance has been made for: (1) air and power off takes and (2) intake losses. The installed engine data for the climb is shown on Fig. 10.18.

For this example the climb speeds are 250 kt EAS up to $10\,000$ ft, accelerate to 320 kt EAS at $10\,000$ ft, then climb at 320 kt EAS until $Mn = 0.82$ is reached and then maintain $Mn = 0.82$ for the remainder of the climb (i.e. assuming operation in the USA). A shorthand method of showing these speeds is:

$$(250/320)\,\text{kt EAS} = Mn0.82$$

A step by step calculation can now be carried out to calculate the time, fuel used and distance travelled from the start of climb to the initial cruise altitude. A typical calculation for the maximum take-off weight case is shown in Table 10.21 at the

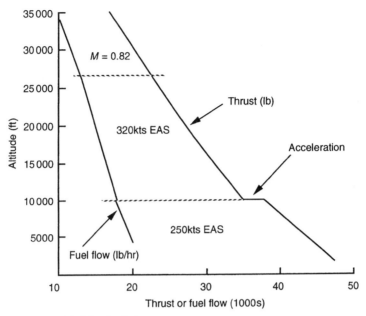

Fig. 10.18 Engine thrust and fuel flow in climb.

end of this chapter. The results are plotted in Fig. 10.19. Note that this calculation has been done in Imperial units.

Climb performance to 35 000 ft from 1500 ft is summarised below:

time   19.5 min
distance   137 nm (254 km)
fuel   9573 lb (439 kg)

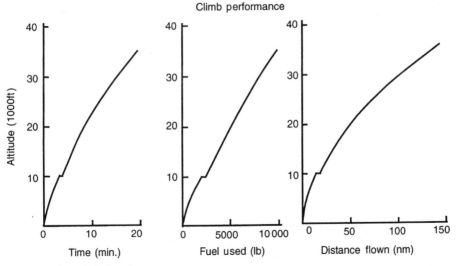

Fig. 10.19 Estimated climb performance.

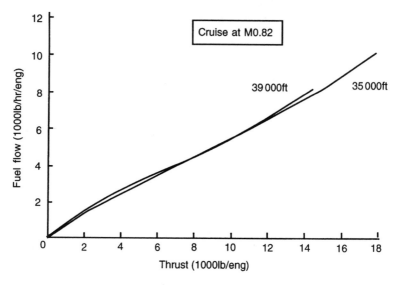

Fig. 10.20 Estimated engine performance in cruise.

Cruise performance

A stepped cruise technique will be used, with an initial cruise altitude of 35 000 ft and one step to 39 000 ft. A plot of engine fuel flow against engine thrust at the cruise Mach number of $Mn = 0.82$ is required for these two altitudes. This is shown in Fig. 10.20 (note, this shows installed engine data).

The calculation is done for a range of aircraft weights at each altitude as shown in Table 10.11. The results are plotted on Fig. 10.21.

Descent performance

This can be calculated in a similar way to the climb. The appropriate engine setting for the descent is, as mentioned earlier, not always available. However, often the engine manufacturer will give flight idle data and this will enable a first calculation of the descent performance to be made. As the design progresses and more information on the environmental control system becomes available some

Table 10.11

Altitude (ft)	350 000			390 000		
Weight (kg)	224 970	210 730	196 494	195 890	182 050	168 210
(lb)	495 967	464 578	433 190	431 855	401 345	370 835
CL	0.522	0.489	0.456	0.551	0.512	0.473
Drag (kn)	121	114	107	105	98	91
(lb)	27 199	25 668	24 124	23 653	22 034	20 533
Fuel flow (kg/hr)	6574	6197	5838	5755	5341	5000
(lb/hr)	14 494	13 662	12 870	12 688	11 774	11 023
TAS (kt)	472.7			470.3		
NAMPP (kg)	0.0719	0.0763	0.081	0.0817	0.088	0.094
(lb)	0.0323	0.0343	0.0364	0.0365	0.0392	0.0420

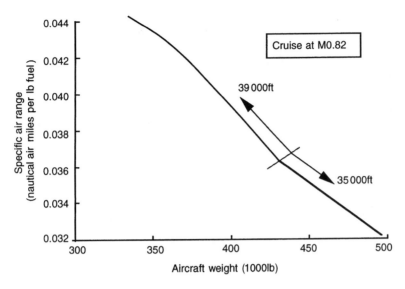

Fig. 10.21 Estimated specific air range in cruise.

adjustment to the engine settings on the descent may be required. In our example the engine performance at flight idle and 250 kt CAS is shown in Fig. 10.22.

A step by step calculation can now be carried out to calculate the time distance and fuel from the final cruise altitude. A calculation for a typical end of cruise weight is shown in Table 10.22 at the end of the chapter. Figure 10.23 shows the time distance and fuel plotted against altitude extracted from this table. Note that this calculation has been done in Imperial units.

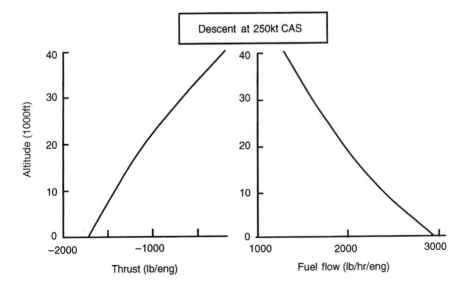

Fig. 10.22 Engine performance in descent.

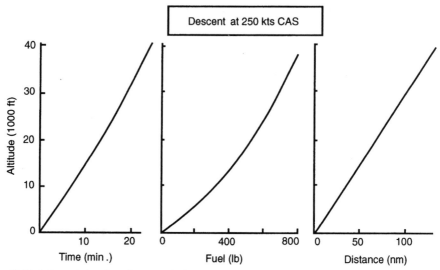

Fig. 10.23 Estimated aircraft performance in descent.

Descent performance from 39 000 ft to sea-level is summarised below:

time   23.2 min
distance   129.5 nm (240 km)
fuel   814 lb (369 kg)

Diversion and hold performance
We now need to calculate the fuel necessary for diversion and holding.

Diversion of 200 nm. To allow for diversion, hold, approach and landing, assume 200 nm diversion, 30 min hold and 6 min on approach. The diversion manoeuvre will include a climb, cruise at speed for maximum range and a descent. The climb is computed at the weight appropriate to that at the end of the main profile. The descent calculation already computed will be sufficiently accurate at this stage. The cruise altitude will usually be determined by an airline operating requirement that the distance covered in the diversion cruise must not be less than half the total diversion distance. The climb and descent calculations that have already been made will enable an approximate assessment to be made.

The cruise altitude is estimated to be 20 000 ft. The climb performance to this altitude at a typical diversion weight is as follows:

time   5.8 min
distance   34 nm (63 km)
fuel   3460 lb (1569 kg)

From the descent calculation the corresponding figures are:

time   12.5 min
distance   62 nm (115 km)
fuel   530 lb (240 kg)

This leaves a cruise distance of 104 nm.

The diversion cruise is normally flown at the speed for maximum range, i.e. at a speed appropriate to

$$C_L = [(C_{D_0} \cdot \pi \cdot e \cdot AR)/3]^{0.5}$$

where e is the Oswald efficiency factor.

Using the drag polar $(C_D = 0.015 + 0.05\,C_L^2)$ gives C_L for maximum range $= 0.316$.

At 20 000 ft and a weight at the start of diversion of 37 500 lb (17 000 kg) gives a cruise Mach number of 0.65. This assumes that the engine specific fuel consumption is constant as speed is varied. In fact it will increase as speed increases and hence will decrease the speed for maximum range from that calculated above.

The effect on the nautical air miles per pound of fuel (NAMPP) is estimated to be about 2–3%. At the project stage this is sufficiently accurate; however as the design progresses more detailed calculations will need to be done.

At 20 000 ft with a typical cruise diversion weight, the C_L of 0.316 gives a Mach number of 0.65. Engine thrust and fuel flow appropriate to $Mn = 0.65$ and 20 000 ft is shown in Fig. 10.24.

This data is used to plot (NAMPP) versus aircraft weight (Fig. 10.25).

As can be seen NAMPP is insensitive to weight over the typical diversion weight range having a typical value of 0.0305.

$$\text{Fuel for 104 nm diversion} = 3410 \text{ lb (1547 kg)}$$

Hence, total diversion fuel is the sum of climb, descent and cruise:

$$3460 + 530 + 3410 = 7400 \text{ lb (3350 kg)}$$

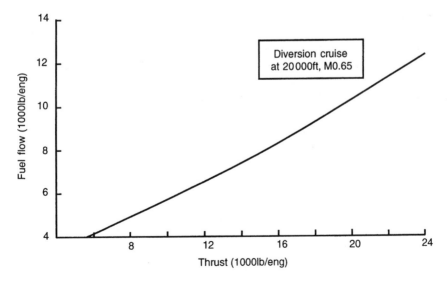

Fig. 10.24 Engine performance during diversion cruise.

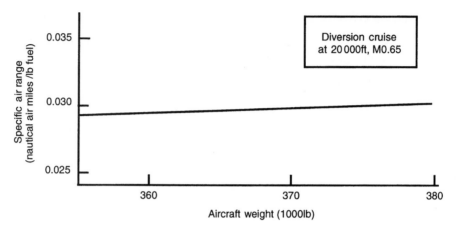

Fig. 10.25 Estimated specific air range in diversion cruise.

Table 10.12 Hold at 5000 ft

Typical weight at hold	355 000 lb (161 027 kg)
Hold at C_L for minimum drag	$C_L = 0.55$
Mach number at 5000 ft	0.36
L/D at 5000 ft	18.26
Drag	19 441 lb (86.5 kN)
Thrust/engine	9720 lb (43.2 kN)

From the engine data at $Mn = 0.36$ and 5000 ft a fuel flow of 5400 lb/hr (2449 kg/hr) per engine can be found from Fig. 10.26. Total fuel needed for 30 min hold = 5400 lb (2449 kg).

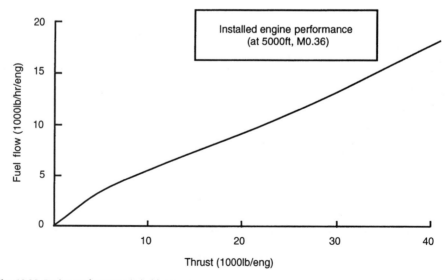

Fig. 10.26 Engine performance in hold manoeuvre.

Payload–range calculations

Case 1. Maximum take-off weight, maximum passenger payload. In this condition the aircraft parameters are shown in Table 10.13.

Table 10.13

take-off weight	507 055 lb (230 000 kg)
operating weight empty	286 596 lb (130 000 kg)
passenger payload	64 050 lb (29 050 kg)
fuel carried	156 409 lb (70 947 kg)
hold + diversion fuel	12 800 lb (5806 kg)

The total reserve fuel consists of 5% block fuel for contingencies + hold + diversion fuel is shown in Table 10.14.

Table 10.14

block fuel	136 770 lb (62 038 kg) less taxi-out fuel
5% block fuel	6838 lb (3102 kg)
total reserve fuel	19 638 lb (8908 kg)

The mission fuel is calculated as shown in Table 10.15.

Table 10.15

	Time (min)	Distance (nm) (km)	Fuel (lb) (kg)	Weight (lb) (kg)
Taxi-out (extra to $MTOM$)	6		360 (163)	
Take-off	2		1580 (717)	
				505 475 (229 282)
Climb to 35 000 ft	19.5	137 (254)	9573 (4342)	
				495 902 (224 940)
Cruise at 35 000 ft	282.9	2229 (4128)	64 902 (29 439) (NAMPP = 0.03435)	
				431 000 (195 800)
Climb to 39 000 ft	4.65	36 (67)	1380 (626)	
Cruise at 39 000 ft	286.7	2247 (4161)	57 325 (26 002) (NAMPP = 0.0392)	
Descent	21.6	130 (240)	810 (367)	
Approach	5		1200 (545)	
Block totals	10.47 hr	4779 (8850)	137 130 (62 202)	

Case 2. Maximum take-off weight, maximum fuel. In this condition the aircraft parameters are shown in Table 10.16.

Table 10.16

maximum take-off weight	507 055 lb (230 000 kg)
maximum fuel capacity	176 370 lb (80 000 kg)
operating weight empty	286 596 lb (13 000 kg)
payload	44 089 lb (20 000 kg)
hold + diversion fuel	12 800 lb (5806 kg)

Reserves = 0.05 block fuel + hold + diversion fuel

block fuel	155 780 lb (70 661 kg)
5% block fuel	7790 lb (3534 kg)
total reserve fuel	20 590 lb (9340 kg)

The mission fuel is calculated as shown in Table 10.17.

Table 10.17

	Time (min)	Distance (nm) (km)	Fuel (lb) (kg)	Weight (lb) (kg)
Taxi-out	6		360 (163)	
Take-off	2		1580 (717)	
				505 115 (229 119)
Climb to 35 000 ft	19.5	137 (254)	9573 (4342)	
				495 542 (224 776)
Cruise at 35 000 ft	281.4	2217 (4106)	64 542 (29 276) (NAMPP = 0.03435)	
				431 000 (195 500)
Climb to 39 000 ft	4.65	36 (67)	1380 (626)	
Cruise at 39 000 ft	391.5	3068 (5682)	76 335 (34 625) (NAMPP = 0.0402)	
Descent	21.6	130 (240)	810 (367)	
Approach	5		1200 (545)	
Block totals	12.2 hr	5588 (10 349)	155 780 (70 661)	

Case 3. Maximum fuel, zero payload. In this condition the aircraft parameters are shown in Table 10.18.

Table 10.18

operating weight empty	286 596 lb (130 000 kg)
maximum fuel capacity	176 370 lb (80 000 kg)
take-off weight	462 966 lb (210 000 kg)
hold + diversion fuel	12 800 lb (5806 kg)
block fuel (as for Case 2)	155 780 lb (70 661 kg)
total reserve fuel	20 590 lb (9340 kg)

The mission fuel is calculated as shown in Table 10.19.

Table 10.19

	Time (min)	Distance (nm) (km)	Fuel (lb) (kg)	Weight (lb) (kg)
Taxi-out	6		360 (163)	
Take-off	2		1580 (717)	
				461 026 (209 120)
Climb(++) to 35 000 ft	15.7	107 (198)	7550 (3425)	
				453 476 (205 695)
Cruise at 35 000 ft	101.8	802 (1485)	22 476 (10 195) (NAMPP = 0.0357)	
				431 000 (195 500)
Climb to 39 000 ft	4.65	36 (67)	1380 (626)	
Cruise at 39 000 ft	645.3	5058 (9367)	120 424 (54 624) (NAMPP = 0.042)	
Descent	21.6	130 (240)	810 (367)	
Approach	5		1200 (545)	
Block totals	13.4 hr	6133 (11 357)	155 780 (70 661)	

(++) Calculated at a start of climb weight of 461 000 lb (209 108 kg).

The resulting payload–range diagram is shown in Fig. 10.27.

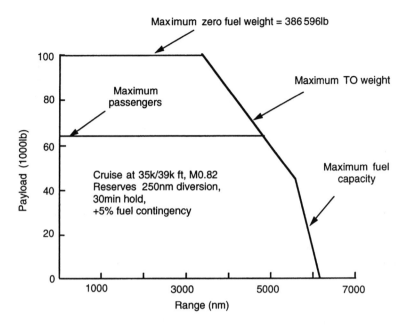

Fig. 10.27 Estimated aircraft payload–range diagram.

Landing performance

The appropriate aircraft parameters are shown in Table 10.20.

Table 10.20

Maximum landing weight	184 000 kg
Reference wing area (S)	376.4 m²
C_D (clean configuration)	$0.015 + 0.05\,C_L^2$
C_D (landing flap, gear down)	$0.055 + 0.05\,C_L^2$
$C_{L_{max}}$ (landing)	2.5
Coefficient of friction (μ)	0.03

We can use these values to calculate the performance:

$$V_S = [(184\,000 \times 9.81)/(0.5 \times 1.225 \times 376.4 \times 2.5)]^{0.5}$$
$$= 55.96\,\text{m/s}$$

Approach phase:

$$V_F = (V_{TD} + V_A)/2$$
$$= (1.15 + 1.30)V_S/2$$
$$= 1.225\,V_S$$

$$r = (1.225 \times 55.96)^2/[9.81(1.2 - 1)]$$
$$= 2395.1\,\text{m}$$

$$h_F = r \cdot \gamma \cdot \gamma/2$$

Civil transport aircraft normally fly the approach angle (γ) at 3 degree. This is equivalent to $3 \times 2\pi/360 = 0.0524$ radians.

$$h_F = 2395.1 \times 0.0524 \times 0.0524/2 = 3.28 \, \text{m}$$

Approach distance:

$$s_A = (15.24 - 3.28)/0.0524 = 228.4 \, \text{m}$$

Flare phase:

$$s_F = 2395.1 \times 0.0524 = 125.4 \, \text{m}$$

Ground roll is calculated in two parts.

(a) Free roll. (*2 seconds at V_{TD}–airworthiness requirement*)

$$s_{FR} = 2 \times V_{TD} = 2 \times 1.15 \times 55.96 = 128.7 \, \text{m}$$

(b) Braked roll

$$s_B = (1/2g \cdot K_A) \ln [(K_T + K_A V_{TD}^2)/K_T]$$

where: $K_T = (T/W) - \mu$

$$T = 0$$

Hence $K_T = -0.03$.

$$K_A = [\rho/2(W/S)](\mu C_L - a - bC_L^2)$$

C_L is assumed to be zero:

$$K_A = [1.225/2(184\,000 \times 9.81/376.4)](0 - 0.055 - 0)$$
$$= -7.02 \times 10^{-6}$$

Thus:

$$s_B = [1/(2 \times 9.81) \times (-7.02 \times 10^{-6})]$$
$$\cdot \ln [(-0.03 + (-7.02 \times 10^{-6} \times (1.15 \times 55.96)^2))/ - 0.03]$$
$$= -672 \, \text{m (sign reversed due to deceleration)}$$

Therefore the landing distance is the sum of the approach, flare, free roll and braked roll phases:

$$228 + 125 + 129 + 672 = 1154 \, \text{m}$$

The FAR/JAR landing distance is factored by 1.66 to account for operational variances:

$$\text{FAR/JAR landing distance} = 1.66 \times 1154 = 1916 \text{m}$$

Airframe/engine matching

The descriptions and methods in previous chapters have provided the tools to calculate the aircraft drag, weight and performance in the various flight phases. At the project stage a definition of the aircraft geometry weight and engine thrust are required to enable these methods to be applied. In this section a method is outlined whereby an engine size and aircraft weight can be determined to allow the aircraft design process to proceed.

The engine size is obviously a function of the geometry and weight of the aircraft. First of all, an outline of the various requirements that size the engine will be given. In Chapter 11 a method of arriving at aircraft take-off weight, using area and engine size, will be given. There are both operational and certification requirements to be met. The thrust required to cruise is an example of an operational case.

The airframe therefore puts certain thrust demands on the engine. The engine has prescribed ratings to meet these demands. These same ratings have to pass a certification type test and also give an acceptable service life for the airlines. The choice of ratings for the engine is an important decision area in the matching of aircraft and engine performance.

In this section the matching of the various rated thrusts to those required by the aircraft are considered. The specification will determine which of these requirements size the engine. At this point the wing area and take-off weight have still to be determined. It should be noted that once the seating arrangement and associated comfort standard have been determined an initial outline of the fuselage can be drawn. The remaining unknowns are therefore the wing area and engine size. The thrust requirements are therefore put in terms of thrust/weight ratio as a function of wing loading.

Take-off field length

In the early stages of project design insufficient detail is known about the aircraft characteristics to carry out a detailed take-off calculation. Hence the project designers have to resort to an empirical approach using equation (10.13). The example shown in the previous section can be used to calibrate equation (10.13) which becomes:

$$\text{take-off field length} = 19.9 \times \lambda W^2/(S \cdot T \cdot C_{L_{V_2}} \cdot \sigma)\,\text{ft}$$

for a twin-engined aircraft. For a four-engined aircraft the constant 19.9 should be changed to 17.0. This expression is in Imperial (British) units (i.e. the weight will be in lb, wing area in sq ft and the thrust in lb).

For a given take-off field length a relationship between wing loading (W/S), thrust loading $(\lambda T/W)$ can be derived for a given $C_{L_{V_2}}$ as shown in Fig. 10.28. For the purpose of this exercise a $C_{L_{\max}}$ (i.e. with take-off flap) of 1.75 has been assumed and the effect of a $C_{L_{\max}}$ of 2.0 indicated. The thrust loading will be at $0.707 \times V_2$ speed and allowance must be made for this when finally determining engine size. Current modern turbofan engines will lose between 10% and 15% of their thrust at $0.707\,V_2$ from that in the static condition.

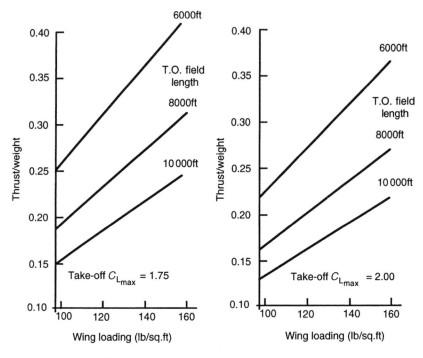

Fig. 10.28 Estimated take-off field length.

Take-off climb-out

The climb-out requirements have been discussed earlier in the chapter.

It was shown that in the various flight segments of the mission a minimum gradient is required in the event of engine failure.

$$\text{Gradient} = [(T/W) - (D/W)]$$

We can write, with the small angles involved, (D/W) as $[1/(L/D)]$. Therefore

$$(T/W) = \text{Gradient} + [1/(L/D)]$$

For a given wing and type of flap the (L/D) will be a function of the flap angle. The $C_{L_{max}}$ will also change with flap angle. A typical relationship of (L/D) versus $C_{L_{max}}$ is shown on Fig. 10.29 where (L/D) is that appropriate to the second-segment configuration and hence includes the additional drag due to the asymmetry in flight and the inoperative engine.

The thrust loading (T/W) will be that associated with a particular forward speed; in the case of second segment it will be at V_2. Due regard needs to be taken of the thrust fall off with speed when sizing the engine. It can be seen from Fig. 10.10 that the choice of flap angle, i.e. $C_{L_{max}}$, effects the thrust requirements for both field length and climb-out. Increasing the aircraft $C_{L_{max}}$ reduces the thrust requirement for a given field length but increases thrust required for climb-out. The choice of flap and flap angle is therefore important. The critical climb-out requirement is usually in the second segment. In this phase the aircraft has take-off flap deployed,

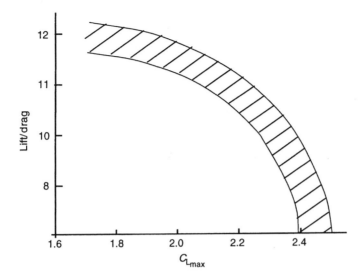

Fig. 10.29 Typical lift/drag ratio and maximum lift coefficient variation.

undercarriage up, one engine failed and the remaining engines at take-off power. For a twin-engined aircraft (reference Table 10.2) in the second segment:

$$(T/W) = 0.024 + [1/(L/D)]$$

Obstacle clearance

There is one issue which can affect the aircraft take-off thrust requirement, and that is obstacle clearance. Many of the major airfields have runways with obstacles which affect the ability of some aircraft to take full advantage of the runway length available. Gatwick is a typical example; on one of the runway directions there is 10 500 ft available but due to an obstacle in the flight path the effective runway available is reduced to approximately 10 000 ft.

The aircraft nett flight path with one engine inoperative must clear any obstacle by at least 35 ft. The nett flight path is the gross flight path reduced by the following gradients:

two-engined aircraft   0.8%
three-engined aircraft   0.9%
four-engined aircraft   1.0%

More details of this requirement can be found in CAP395 – Air Navigation: The Order and the Regulations Section 3 General Regulations.

It is difficult to deal with this type of requirement at the project stage as we do not usually design the aircraft for one specific operation. Obstacle clearance is best allowed for by examining the effect of the obstacle on existing aircraft performance and deciding on an effective runway length to clear the obstacle.

En-route climb performance

There are two issues here that effect the engine: the initial cruise altitude capability and the time taken to arrive at this altitude.

Initial cruise altitude capability

This definition is the condition of the aircraft at the top of climb. By the time the aircraft arrives at this point it has lost about 2–3% of its take-off weight because of the fuel used. Initial cruise altitude capability is defined as the ability to sustain a certain rate of climb at this altitude, at typical cruise speed and with the engines operating at maximum climb rating. A 300 ft/min rate of climb is the usual requirement although airlines may specify other levels (e.g. 500 ft/min). In this calculation we will assume a 300 ft/min requirement. The initial cruise altitude is set at a minimum of 2000 ft above the optimum cruise altitude. This margin is required because in the current air traffic control environment it is not possible to follow the optimum cruise altitude schedule (i.e. approximately a constant C_L technique). Such a procedure would lead to a gradual increase in altitude as fuel is burnt and the aircraft becomes lighter. Air traffic control requires aircraft travelling in the same direction to fly at discrete flight levels separated by 4000 ft in altitude. This means that a minimum of 2000 ft margin above the optimum is required to enable the aircraft to fly a stepped cruise technique. This is shown on Fig. 10.30. Development in navigational aids may enable this separation to be reduced to 2000 ft in the near future.

At the initial cruise altitude required:

climb thrust = drag at cruise speed at 98% take-off weight

+ thrust required for 300 ft/min rate of climb

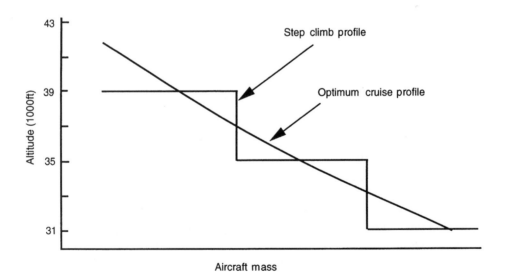

Fig. 10.30 Step cruise profile.

Thrust required for 300 ft/min rate of climb:

$$\Delta X_n = (300 \cdot 0.98 \, MTOW)/(V_{TAS} \cdot 101.3)$$

where:

$$V_{TAS} = \text{true airspeed (kt)}$$
$$\Delta X_n = \text{additional thrust (lb)}$$

$$\text{Total thrust required} = [0.98 \, MTOW/(L/D)]$$
$$+ [(300 \cdot 0.98 \, MTOW)/(V_{TAS} \cdot 101.3)]$$

Time to height

The initial cruise altitude capability will fix the thrust level at the top of climb, but the time to height will also be affected by the thrust ratings at the lower altitudes. This can be an important issue from operational considerations. The ability to climb to altitude quickly may enable the aircraft to take advantage of an air traffic control slot which would be missed by an aircraft with a low climb capability. Such aircraft with limited performance may therefore be subject to delay or be forced to cruise at a lower than optimum altitude. An acceptable climb time can be the subject of some debate but on a long-range aircraft such as the B747–400 is often considered as a typical design criterion. This aircraft climbs to 31 000 ft in about 20 min at $MTOM$. Typical climb performance is shown in Fig. 10.31.

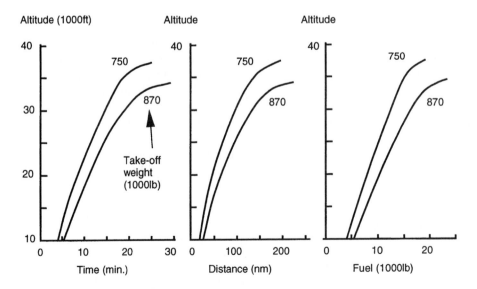

Fig. 10.31 Estimated climb performance.

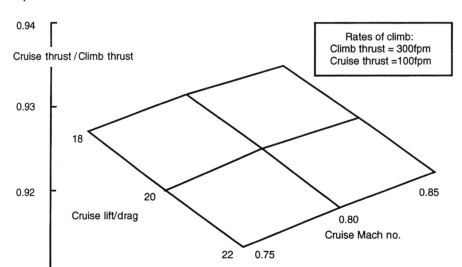

Fig. 10.32 Estimated ratio of cruise to climb thrust.

Cruise performance

Cruise thrust required is around 7–8% less than the climb thrust. As shown in the calculation above climb thrust may be fixed by a rate of climb of 300 ft/min at cruise Mach number and altitude whereas cruise thrust is fixed by a rate of climb of 100 ft/min. Figure 10.32 shows how the ratio of cruise to climb thrust varies with cruise lift/drag ratio and the cruise Mach number.

En-route engine-out ceilings

These requirements can determine the maximum continuous rating for the engine. The regulations say that in the event of an en-route engine failure all terrain must be cleared by a minimum of 2000 ft with a gradient which varies with the number of engines. This means, for example, over the Rocky Mountains an engine-out ceiling of 16 000 ft is desirable, a lower ceiling than this may result in the aircraft having to fly a non-optimum (less direct) route leading to increased fuel requirements.

The aircraft specification should allow for these considerations, if known at the project stage, and set a ceiling appropriate to the envisaged use of the aircraft. The gradients specified by airworthiness requirements for the engine-out ceilings are associated with the number of engines on the aircraft configuration:

number of engines	2	3	4
gradient %	1.1	1.4	1.6

The effect of en-route gradient requirements can be alleviated by using a drift down technique as shown in Fig. 10.33. However, for this purpose no advantage can be taken of any cruise altitude above the engine re-light altitude.

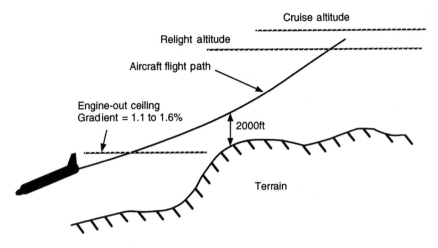

Fig. 10.33 En-route requirement with one engine failed.

Descent

The descent is governed by the rule that the rate of descent in the cabin should not exceed 300 ft/min and the requirements of the air conditioning system. Hence the idle setting on the descent will be set by the requirements to supply the appropriate amount of air at the right pressure to the air conditioning system. The relationship between aircraft altitude and cabin altitude for a typical pressure differential of 8 psi is shown on Fig. 10.34.

Another noteworthy aspect of descent performance is that an aircraft at

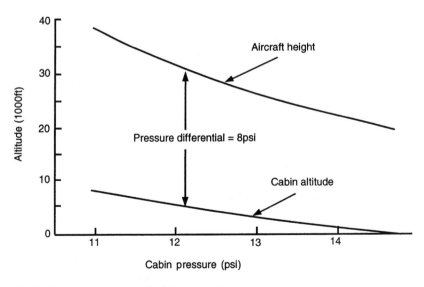

Fig. 10.34 Relationship between aircraft altitude and cabin pressure.

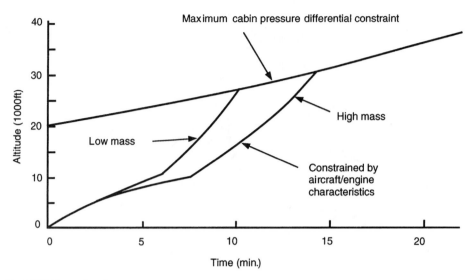

Fig. 10.35 Descent performance.

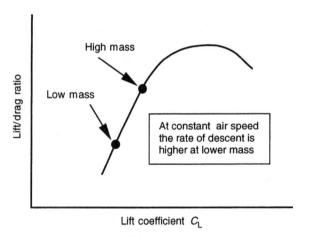

Fig. 10.36 Variation of aircraft lift/drag ratio in descent.

low mass can have a higher rate of descent than an aircraft at high mass (see Fig. 10.35).

The reason for this variation lies in the change of lift to drag ratio with aircraft mass, as shown in Fig. 10.36.

Landing performance

There are two requirements in landing which affect the engine in terms of thrust:

- approach climb
- baulked landing climb

These requirements are not usually critical and need only be examined on short-range aircraft where the maximum landing weight is either the same as or close to the maximum, take-off weight.

Approach climb

With the aircraft in the approach condition, with approach flap, undercarriage retracted, maximum landing weight with one engine inoperative and the remaining engines at take-off power the gradients required by the airworthiness regulations vary with the number of engines in the aircraft configuration as follows:

number of engines	2	3	4
gradient %	2.1	2.4	2.7

The aircraft speed should not exceed $1.5 \times$ stall speed in the approach condition and the stall speed in the approach condition should not exceed $1.1 \times$ the stall speed for the all-engines operating case with landing flap.

Baulked landing climb

With the aircraft in the landing configuration, with all engines operating at a thrust level which is available eight seconds after initiation of movement of the throttle, a gradient of 3.2% is required to meet airworthiness regulations. The associated aircraft speeds in this condition are specified as:

number of engines	2 and 3	4
speed	$1.2 \, V_S$	$1.15 \, V_S$

Concluding comment

With the information in this and earlier chapters it is now possible to embark upon the process of an aircraft project design. The next chapter shows how the initial estimates are made in this process.

Table 10.21

Aircraft climb calculation: tabular method

1. Climb to 10 000 ft at constant (EAS) speed of 250 kt. Accelerate at 10 000 ft to 320 kt. After this the speed is held constant up to the height at which the cruise Mach number (0.82) is achieved (i.e. 26 500 ft). Thereafter airspeed is kept at constant Mach number up to 35 000 ft.
2. Aircraft weight reduces due to fuel burn (an iterative calculation, started with a guessed value).
3. Drag determined from aircraft drag polar (clean).
4. Thrust determined from engine data at maximum climb rating.
5. Rate of climb initially calculated assuming no aircraft acceleration.
6. Rate of climb subsequently corrected (RC corr) to allow for aircraft acceleration.
7. Mean fuel flow (Mean FF) from engine data at climb rating, height and airspeed.
8. Time elapsed in segment (Δt), fuel used in segment (ΔF), distance flown in segment (Δs).
9. Summation of the above in subsequent rows.

Weight at the start of climb $= 507\,055 - 1486 = 505\,569$ lb

Drag polar $C_D = 0.015 + 0.05\,C_L^2$

Altitude (ft)	1500	5000	10 000	10 000	15 000	20 000	26 500	30 000	35 000
Weight (lb)	505 569	504 769	503 645	50 300	501 656	500 230	498 000	497 054	496 000
EAS (kt/Mach)	250	250	250	320	320	320	320	0.82	0.82	0.82
C_L	0.588	0.587	0.586	0.359	0.358	0.357	0.355	0.415	0.520
Drag (lb)	27 748	27 683	27 645	30 064	30 014	29 969	29 890	28 287	27 149
Thrust (lb)	94 800	87 000	75 966	70 070	61 930	54 960	46 736	41 226	34 846
RC (ft/min)	3435	3208	2828	2998	2600	2219	1681	1274	743
Acceleration corr	1.0883	1.1000	1.1120	1.2150	1.2150	1.2610	1.3300 / 0.9100	0.9100	0.9170
RC corr (ft/min)	3156	2916	2543	2468	2140	1760	1264 / 1847	1400	810

Segment data

RC mean (ft/min)	3036	2730	2304	1950	1512		1624	1105
Δt (min)	1.15	1.83	2.17	2.56	4.3		2.16	4.52
$\Sigma \Delta t$ (min)	1.15	2.98	3.54	5.96	8.52	12.82	14.98	19.50
Mean FF (lb/hr)	40 987	37 912	35 560	34 495	31 441	28 430	25 035	21 470
ΔF (lb)	788	1160	480	1248	1343	2037	899	1618
$\Sigma \Delta F$ (lb)	788	1948	2428	3676	5019	7056	7955	9573
TAS mean (kt)	262.5	280	331	388	421	464.5	487	478
Δs (nm)	5.0	8.5	4.5	14	18	33.3	17.5	36
$\Sigma \Delta s$	5.00	13.50	18.00	32.00	50.00	83.30	100.80	137.00
Acceleration (ft/sec)^2	2.825	3.090	2.56					
Mean acceleration	0.81	0.81						
Δt (min)								

Table 10.22

Aircraft descent calculation: tabular method

1. Initiate descent at Cruise Mach number reducing to 250 kt EAS.
2. Weight reduces due to fuel burn (on iterative calculations make sensible guess initially).
3. Drag determined from aircraft drag polar.
4. Thrust determined from engine data at descent rating (flight idle).
5. Initial note of descent (RD) assuming no energy.
6. Subsequent RD corrected to account for energy gained in descent (due to reducing true air speed).
7. Elapsed time in segment (Δt).
8. Fuel flow in segment (Mean FF) from engine data.
9. Fuel burnt in segment (ΔF), summed in subsequent row.
10. Distance travelled in segment (Δs), summed in last row.

Weight at the start of descent 380 000 lb

Altitude (ft)	39 000	35 000	30 000	25 000	20 000	15 000	10 000	5000	sea level
TAS (kt)	462	430	394	365	336	310	289	268	250
Weight (lb)	380 000	379 950	379 850	379 802	379 703	379 592	379 476	379 345	379 186
C_L	0.5030	0.4830	0.4760	0.4640	0.4620	0.4580	0.4550	0.4480	0.4425
Drag (lb)	20 892	20 987	21 020	21 106	21 116	21 136	21 148	21 217	21 262
Thrust (lb)	−300	−450	−676	−750	−1072	−1270	−1400	−1550	−1680
RD (fpm)	2610	2457	2279	2127	1985	1852	1736	1628	1531
Acceleration corr	1.372	1.295	1.23	1.189	1.157	1.13	1.11	1.092	1.08
RD corr (fpm)	1902	1898	1853	1789	1716	1639	1564	1491	1417
RD mean (fpm)		1900	1875	1821	1753	1678	1601	1527	1454
$\Delta \tau$ (min)		2.11	2.67	2.75	2.85	2.98	3.12	3.27	3.44
$\Sigma \Delta t$ (min)		2.11	4.78	7.53	10.38	13.36	16.48	19.75	23.19
Mean FF (lb/hr)		1400	1540	1720	1920	2130	2330	2550	2800
ΔF (lb)		49.00	68.50	79.00	91.20	106.00	121.00	139.00	160.50
$\Sigma \Delta F$ (lb)		49.00	117.50	196.50	287.70	393.70	514.70	653.70	814.20
TAS mean (kt)		445	410	378	350	323	298	278	263
Δs (nm)		15.60	18.20	17.30	16.60	16.00	15.50	15.20	15.10
$\Sigma \Delta s$ (nm)		15.60	33.80	51.10	67.70	83.70	99.20	114.40	129.50

11

Initial estimates

Objectives

This chapter provides a framework which will enable the design process to be approached in a logical manner. An indication is given on how to establish a starting point and then how to proceed to establish a preliminary aircraft design. The method also enables the critical requirements to be established and shows the effect of alleviating any that might have an undesirably large influence on the design.

A specimen calculation is included at the end of this section. On completing this section you should be able to tackle the full design process including iterative and parametric studies. The aircraft design case studies that follow Chapter 15 further expand the methods shown here.

The values and relationships used and further developed for a particular study may only be appropriate for that study. It is risky to use such values in other work without conducting a thorough validation of the data.

Introduction

The aircraft specification lists the aircraft requirements in terms of payload, range and performance (en-route and field). The design process involves selecting an aircraft layout and size to meet these requirements against some overall criteria such as minimum take-off weight or minimum direct operating cost (DOC) so that a baseline configuration can be established.

Apart from information and data for the aircraft, engine data including performance, weight and costs will also be required. To assist in the early stages some typical engine data is given in Chapter 9 (Figs 9.18–9.21). This data is in a non-dimensional form to make it easier to scale engine size in the initial design studies.

The only part of the aircraft that the specification immediately sizes, once the passenger comfort standard has been decided, is the fuselage. There are therefore three major unknown aircraft parameters:

wing area S
engine thrust T
take-off weight W

All three of these parameters are interdependent. The wing area affects the engine size and both of these impact on the empty weight, fuel weight and hence the maximum take-off weight. Our analysis needs to take into account variations of these three major parameters.

It is necessary before starting the analysis to establish approximate values for each of these three parameters. This is done by assessing the characteristics of current aircraft and engines with similar specification. It is also useful to determine some simple weight and performance relationships for these aircraft. This will enable an appropriate range of values of wing area and engine size to be selected and allow an approximation of the take-off weight to be made for the initial estimates. The basic wing geometric characteristics (thickness chord and sweepback) will be dictated by the specified cruise Mach number and can therefore be fixed at an early stage.

The interdependence of wing area, engine thrust and take-off weight makes it impossible to consider them as independent variables. However, an assessment can be made of wing loading (W/S) and thrust loading (T/W) using the performance equations from Chapter 10. The simplified mass equations from Chapter 7 allow for each of the unknowns to be estimated. Guidance on the input required for the performance and mass equations can be obtained from the characteristics of existing aircraft with a similar specification. Typical values for these characteristics are given in the aircraft data files. Care must be taken to ensure that the effect of any new technology under consideration for the design (e.g. carbon fibre structure and laminar flow) is taken into account.

Wing loading

Aircraft wing loading is usually determined by either the landing field length or the approach speed depending upon which of these is specified. For a range of current aircraft the distance from 50 ft is plotted in Fig. 11.1 as a function of:

$$(W/S)/(C_{L_{max}} \cdot \sigma)$$

where: $\sigma = \rho/\rho_0$
       $\rho =$ air density

To comply with airworthiness requirements, the estimated distance from 50 ft is factored by the ratio $(1.0/0.6)$ to obtain landing field length. Care must be taken when using published figures to ascertain if the value is for the factored or actual distance.

To use the data on Fig. 11.1 requires the assessment of a $C_{L_{max}}$ in the landing configuration. An assumption based on the wing thickness/chord ratio and sweepback required for the specification along with current aircraft data should give a reasonable estimate at this stage. For turbofan-powered aircraft cruising at

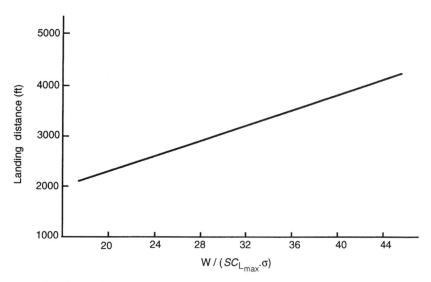

Fig. 11.1 Landing distance from 50 ft.

M0.75–0.85, $C_{L_{max}}$ in the landing configuration would be expected to lie in the range 2.4 to 3.4.

If an approach speed is specified the wing loading can be obtained from the following equation:

$$W/S = q \cdot C_L$$

where: q = dynamic pressure = $0.5\,\rho V^2$
V = aircraft speed
C_L = lift coefficient in the approach condition

The flap setting for approach is not usually the same as used for landing. The airworthiness regulations specify the approach speed as follows:

$$V_{approach} < 1.5 \times V_{stall} \text{ approach flap}$$

$$V_{stall} \text{ approach flap} > 1.1 \times V_{stall} \text{ landing flap}$$

$$V_{approach} > 1.3\, V_{stall} \text{ approach flap}$$

The approach speed must therefore lie between 1.43 and 1.65 V_{stall}, where V_{stall} is determined with landing flap. Or in terms of $C_{L_{max}}$ landing flap, the approach C_L will lie between $(C_{L_{max}}/2.04)$ and $(C_{L_{max}}/2.72)$.

At some airports during busy times air traffic control sets the approach speed. For some aircraft this means that landing flap is set and landing gear is down on approach.

In each of the above cases W/S will be that appropriate to the aircraft maximum landing weight. This can be corrected to W/S at the maximum take-off weight by the ratio of maximum landing weight to maximum take-off weight. This can be obtained from consideration of current aircraft with a similar specification.

An initial estimate of the wing loading has now been made. Two further checks are required. Firstly, that at the initial cruise altitude the aircraft is operating at a reasonable lift coefficient to give $Mn(L/D)$ somewhere near the maximum, and secondly that the ride (response to gusts) is acceptable.

Cruise check

The cruise Mach number and the initial cruise altitude are given in the specification. The dynamic pressure $(0.5\rho V^2)$, denoted as q can then be computed:

$$q = 0.7\,pMn^2 \text{ (in British units)}$$

where: $p =$ ambient pressure (lb/sq ft) at the initial cruise altitude
$Mn =$ cruise Mach number

The wing loading (W/S) at start of cruise is approximately $(0.98 \times W/S)_{\text{take-off}}$ due to the use of fuel during taxi, take-off and climb. Then

$$\text{cruise } C_{\text{L}} = 0.98\,(W/S)\cdot q^{-1}$$

This can be compared with $C_{L_{\text{des}}}$ obtained from Fig. 8.3. A typical curve of lift to drag ratio versus C_{L} at the cruise Mach number is shown in Fig. 11.2 which also shows the $C_{L_{\text{des}}}$ from Fig. 8.3. It should be noted that the curve is flat around the maximum L/D value. In the example shown 99% of the maximum L/D value covers the C_{L} range from 0.43 to 0.50. It is therefore not necessary to precisely match the $C_{L_{\text{des}}}$ value as the effect on aircraft performance of not operating at the maximum value is minimal.

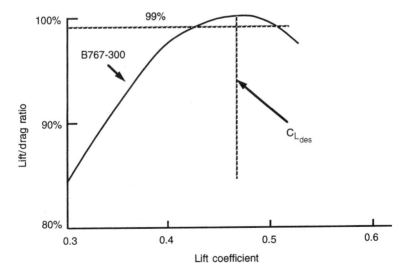

Fig. 11.2 Lift coefficient for optimum cruise.

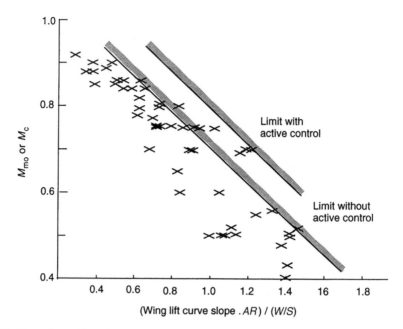

Fig. 11.3 Ride and speed limitations.

Ride check

It is necessary to check whether the chosen wing characteristics and the loading will give a satisfactory gust response at the maximum cruise Mach number. The ability of a given wing planform and wing loading to give a satisfactory gust response is assessed by referring to the boundary conditions of Fig. 11.3 where limiting speed is shown as a function of a wing gust response parameter.

Two boundary conditions are shown, one with active controls giving the aircraft a gust alleviation capability, and the other without active control.

The gust response parameter is defined by the equation:

$$[\alpha_{1wb} \cdot AR/(W/S)]$$

where: α_{1wb} = wing body lift curve slope as calculated from the EDSU (Engineering Science Data Units) data sheets
AR = aspect ratio
(W/S) is the wing loading at the start of the cruise

The wing aspect ratio is the term that makes allowance for aeroelastic effects. The higher the aspect ratio the more flexible the wing will be. On a swept wing, as the wing flexes, the ideal incidence will change, hence the local lift will change and this will affect the gust response. Provided the wing loading (W/S) is greater than about 340 kg/sq m (70 lb/sq ft) this is not usually a critical item.

Engine size

Engine size is usually determined by either the take-off performance or the climb performance at the initial cruise altitude. The particular requirement which determines the engine size will be dictated by both engine and airframe characteristics.

As noted in Chapter 9 short-range aircraft do not require such a high cruise efficiency engine as long-range aircraft. Engine bypass ratio and overall pressure ratio are the main parameters to affect engine efficiency. However, high efficiency engines lose thrust with speed at a much higher rate than low efficiency engines. Long-range aircraft usually have a fairly relaxed field requirement (10 000 ft+). The engine for these aircraft will be sized for the top of climb requirement.

Conversely the short-range aircraft has a much more severe field length requirement (5000–7000 ft). With a lower bypass ratio and pressure ratio it would be expected that the engine would be sized at take-off. These of course are generalisations and there are engine airframe combinations which do not fit neatly into either of these categories. If there is any doubt it is advisable to check both of these criteria to establish which is critical.

Take-off

The initial estimate for the take-off wing loading along with the specification field performance and equation (10.13) will enable an assessment of the thrust/weight ratio to be made as a function of the lift coefficient at V_2. The $C_{L_{max}}$ at take-off will lie between 1.7 and 2.2, and $C_{L_{V_2}}$ will be between 1.18 and 1.53. Statistical analysis of existing aircraft along with any proposed new technology will enable a first assessment of $C_{L_{V_2}}$ to be made and hence an initial thrust/weight ratio. As was noted earlier this thrust/weight ratio will be at $0.707 \times V_2$ and hence needs correcting to static conditions:

$$(\text{thrust/weight})_{\text{static}} = (\text{thrust/weight})_{0.707 V_2} \cdot (\text{static thrust/thrust}_{0.707 V_2})$$

Initial cruise altitude capability

At the initial cruise altitude the aircraft will require a rate of climb of 300 ft/min when flying at the cruise Mach number with engines set to maximum climb rating. This enables the aircraft at the initial cruise altitude with engines at cruise thrust to have a margin over the maximum cruise thrust. This margin is required to enable the flight control system to maintain height and speed under gusty conditions.

The equation for the thrust required at the top of climb (as shown in Chapter 10) can be rearranged to account for British units as follows:

$$F_n/M_{cl} = (L/D)^{-1} + (300/V \cdot 101.3)$$

where: M_{cl} = mass at the top of climb (lb) and is approximately $0.98 \times MTOM$

F_n = climb thrust required at top of climb (lb)

(F_n needs to be referred back to the corresponding sea-level static value of thrust/weight ratio by using the engine characteristic which relates the maximum climb thrust to the engine sea-level static take-off thrust.)

L/D = lift drag ratio at cruise speed

V = cruise speed TAS (kt)

$(F_{n_{climb}}/F_{n_{TO}})$ can be obtained from the engine data

$(F_{n_{TO}}/MTOM) = (F_n/M_{cl})\,0.98/(F_{n_{climb}}/F_{n_{TO}})$

A first estimate of the cruise L/D is required. A basis for this can be obtained from statistics of current aircraft and the wing geometric characteristics. The L/D is primarily a function of the span loading, the aircraft profile drag and the wing loading. Assuming a parabolic drag polar:

$$C_D = C_{Do} + C_{Di} \qquad C_{Do} = C_f\,(S_{wett}/S_{wing}) \qquad C_{Di} = C_L^2/(\pi A e)$$

$$C_D = C_f\,(S_{wett}/S_{wing}) + C_L^2/(\pi A e)$$

$$C_L \text{ for } (L/D)_{max} = [C_f\,(S_{wett}/S_{wing}) \cdot (\pi A e)]^{0.5}$$

$$C_D \text{ for } (L/D)_{max} = 2 \cdot C_f\,(S_{wett}/S_{wing})$$

$$(L/D)_{max} = \{[C_f\,(S_{wett}/S_{wing}) \cdot (\pi A e)]/[2 \cdot C_f\,(S_{wett}/S_{wing})]\}^{0.5}$$

$$AR = (b^2/S_{wing}) \text{ where } b = \text{wing span}$$

$$(L/D)_{max} = \{[C_f\,(S_{wett}/S_{wing}) \cdot (\pi e) \cdot (b^2/S_{wing})]/[2 \cdot C_f\,(S_{wett}/S_{wing})]\}^{0.5}$$

$$= (\pi/2)^{0.5} \cdot (C_f/e)^{-0.5} \cdot [b/(S_{wett})^{0.5}]$$

As these aircraft will be operating at similar values of $(C_f/e)^{0.5}$ and approximately the same relationship to $(L/D)_{max}$, the cruise L/D will be a function of $[b/(S_{wett})^{0.5}]$ or $[AR/(S_{wett}/S_{wing})]$.

The cruise lift/drag ratio based on estimates from the aircraft geometry plotted against an aspect ratio parameter, is shown in Fig. 11.4.

Typical values of (S_{wett}/S_{wing}) for these aircraft are now required. S_{wett} is a function of the fuselage size (hence passenger capacity), and also the wing area (hence landing field length). S_{wing} is a function of landing field length. This, of course, assumes that these aircraft have similar maximum lift coefficients in the landing configuration. On the above basis some correlation between (S_{wett}/S_{wing}) and [passenger capacity/(landing field length)^2] might be expected. This is shown in Fig. 11.5.

With our chosen aspect ratio an assessment of the cruise lift/drag ratio can now be made. The aircraft wing loading and thrust loading can now be chosen. The actual engine thrust and wing size will depend on the aircraft take-off mass.

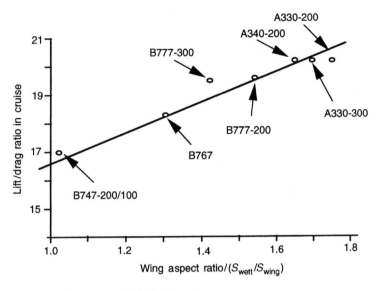

Fig. 11.4 Lift drag ratio for current wide-bodied aircraft.

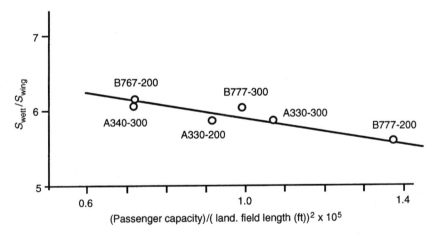

Fig. 11.5 Wetted area ratio for current wide-bodied aircraft.

Take-off mass

An approximate assessment of this mass is required (as described in Chapter 7).

$$M_{TO} = M_E + M_{PAY} + M_F$$
$$= M_{PAY}/[1 - (M_E/M_{TO}) - (M_F/M_{TO})]$$

Empty mass ratio

The ratio of operating empty mass to *MTOM* can be obtained from current aircraft as shown on Figs 11.6 and 11.7 for two class and three class seating respectively.

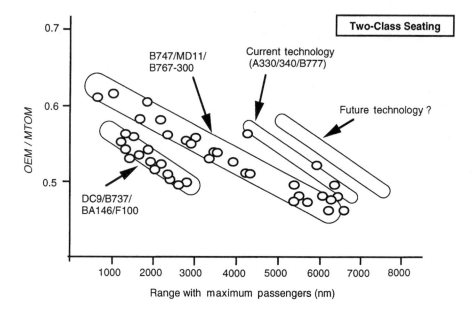

Fig. 11.6 Empty weight ratio for range (2 class seating).

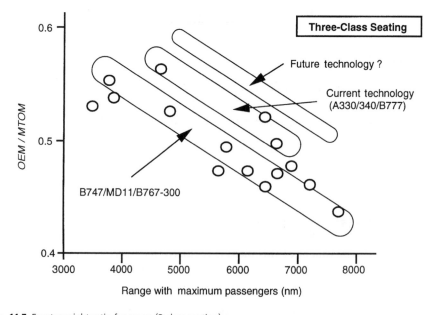

Fig. 11.7 Empty weight ratio for range (3 class seating).

This 'empty mass' ratio will lie between approximately 0.5 and 0.6 depending upon the aircraft type. It should be borne in mind that if technology improvements in the structure (e.g. composite materials) are introduced the ratio may reduce. In general the aircraft structure represents about 25–30% of the take-off mass. This data will give a first estimate of M_E/M_{TO}.

Fuel mass

Aircraft fuel mass has to allow for the design flight profile plus the specified reserves. At this stage it is impossible to calculate each stage of the flight plan accurately. However, an estimate of the cruise L/D could be made (Fig. 11.4) and could be used to check the cruise fuel requirement.

An assessment of the principal engine characteristic will enable an estimate of the specific fuel consumption to be obtained. A statistical chart such as Fig. 11.8 enables the design range to be converted to a still-air range. This makes it possible to assess all the fuel (including the reserve fuel) as cruise fuel.

The Breguet Range is given by:

$$R = (L/D) \cdot (V/SFC) \log e\,(M_1/M_2)$$

where: SFC = engine cruise characteristics taken from flat part of cruise loop
M_1 = take-off mass
R = still-air range in nm (from design specified range and Fig. 11.8)
$M_2 = M_1 - M_F$

The fuel load (M_F) is given by:

$$M_F = M_1 - M_2$$

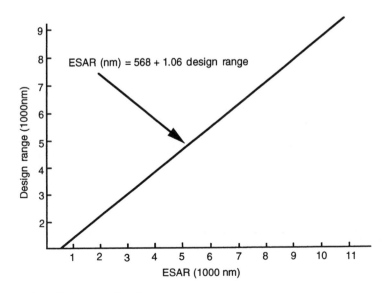

Fig. 11.8 Equivalent still air range estimation.

Therefore:

$$M_F/MTOM = \frac{M_F}{M_1} = 1 - \frac{M_2}{M_1} = 1 - \left(\frac{M_1}{M_2}\right)^{-1}$$

Estimates are now available for M_E/M_{TO} and $M_F/MTOM$. These can be inserted in the mass equation to derive an initial take-off mass.

Summary of initial estimates

An approximate value for the take-off mass, wing area and engine size has now been determined. The next stage, in which all the design criteria implicit in the requirements can be taken into account, can now be undertaken to give the initial estimate of the aircraft design.

The method employed has been devised to show whether any of the initial requirements are dominant and the effect of relaxing such requirements. It is based on the premise that the passenger capacity, cruise speed and the design range are fixed. The passenger capacity and cruise speed are determined by the market requirements and the design range by the geography of the area in which the particular type of aircraft under consideration will be operating. The weight and drag methodology of Chapters 7 and 8 along with the performance methods of Chapter 10 are required. If these methods are programmed on spreadsheets, variations of wing area, engine size and take-off mass can be easily assessed.

Preliminary estimating procedures

The initial estimates can now be used as a guide to select three values for wing area, and at each wing area select three values for take-off mass and three values for engine size. This will give nine aircraft designs at each wing area. The drag and operating weight empty can be determined for each combination of wing area, engine size and take-off mass. The range can then be estimated for each of the combinations. The appropriate combination of engine size and take-off thrust at each wing area to give the required design range can now be found. This method is shown diagrammatically on Fig. 11.9 and the derivation of the take-off mass to give the design range on Fig. 11.10.

The analysis above will enable Fig. 11.11 to be constructed. This shows a range of aircraft with differing engine sizes, wing areas and the take-off weight at which the design range is met.

The aircraft which meet the performance requirements have to be deduced. The various performance requirements have now to be superimposed on Fig. 11.11 as shown on Fig. 11.12.

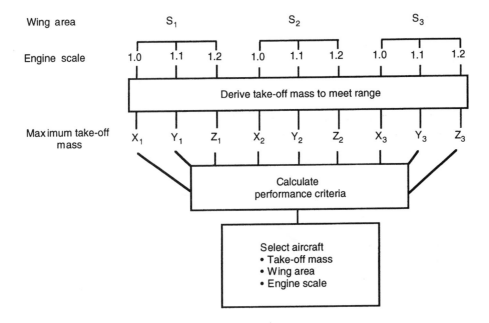

Fig. 11.9 Parametric study plan.

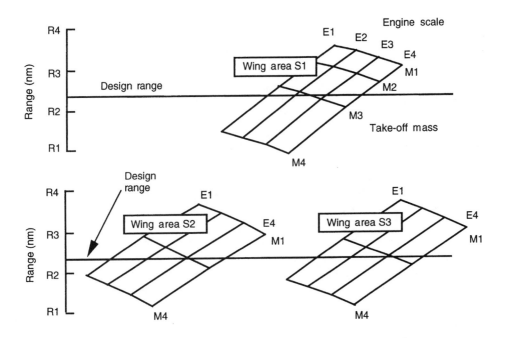

Fig. 11.10 Parametric study plots.

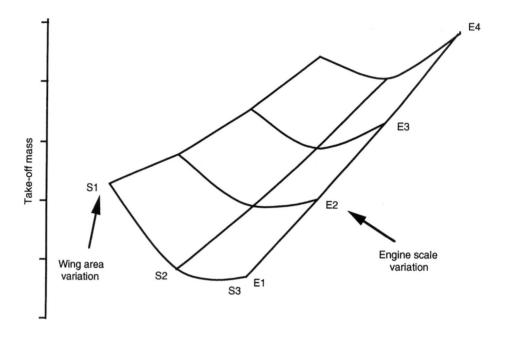

Fig. 11.11 *MTOM* vs engine size and wing area.

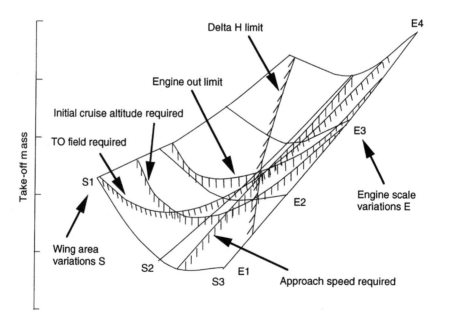

Fig. 11.12 *MTOM* carpet plot.

Example   289

Example

The following example illustrates the foregoing method.

Engine data

The engine data used is shown in Figs 11.13 to 11.19.

These charts show installed engine data (i.e. including the effects of air and power off-take, and intake pressure recovery). The data is in a non-dimensional form referred to a sea-level static take-off thrust (F_n^*) of 81 237 lb (361 kN).

Aircraft specification

Seats:	
in a 2-class configuration	305
in a 3-class configuration	265
Range with 305 seats	6500 nm
Take-off field length (SL, ISA + 15°C)	8000 ft
Approach speed	EAS 145 kt
Cruise speed	M0.85
Initial cruise altitude capability	35 000 ft
Minimum margin above optimum cruise altitude	2000 ft
Engine-out ceiling	16 000 ft
Landing field length	5750 ft
Wing characteristics:	
Sweepback at quarter chord	33°
wing thickness/chord at root	0.13
wing thickness/chord at tip	0.09
aspect ratio	10.0

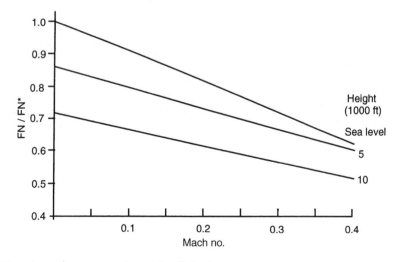

Fig. 11.13 Engine performance: maximum take-off thrust.

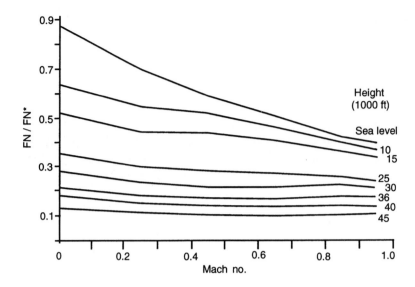

Fig. 11.14 Engine performance: maximum continuous thrust.

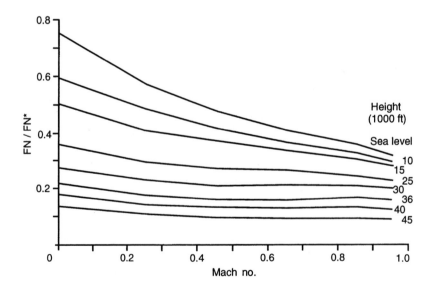

Fig. 11.15 Engine performance: maximum climb thrust.

Example   291

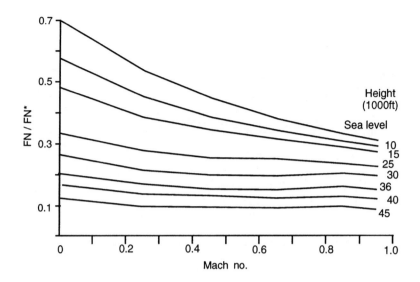

Fig. 11.16 Engine performance: maximum cruise thrust.

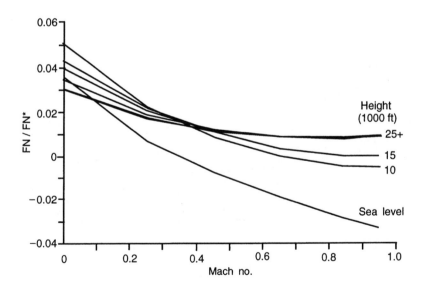

Fig. 11.17 Engine performance: descent/idle thrust.

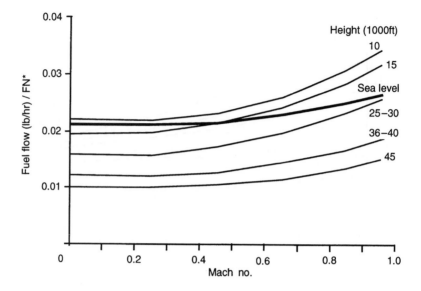

Fig. 11.18 Engine performance: descent/idle fuel flow.

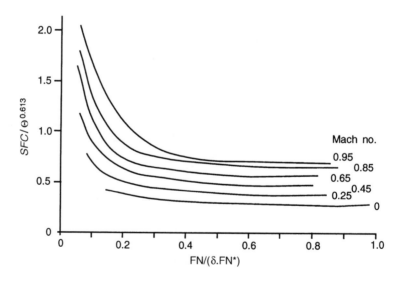

Fig. 11.19 Engine performance: non-dimensional SFC loops.

Example   293

Initial estimate of take-off mass

For this class of aircraft at the design range with some allowance for new technology it would be expected that the $(M_{OE}/MTOM)$ would be around 0.53 [from Fig. 11.6]. Now we know that:

$$MTOM = M_{pay} + M_{OE} + M_F$$

which can be rearranged to give:

$$MTOM = M_{pay}[1 - (M_{OE}/MTOM) - (M_F/MTOM)]^{-1}$$

Hence:

$$M_F/MTOM = 1 - (M_1/M_2)^{-1}$$

where: M_1 = initial aircraft mass
M_2 = final aircraft mass

From Fig. 11.8, the equivalent still-air range for a 6500 nm design range would be 7470 nm. Equivalent still-air range assumes that all fuel is used at cruise. Now, passenger capacity (3 class)/(landing field length)$^2 = 0.8 \times 10^{-5}$.
From Fig. 11.5 $(S_{wett}/S_{wing}) = 6.1$, and $(AR/6.1) = 1.639$ and $(L/D)_{cruise} = 20$.
A cruise Mach number (Mn) of 0.85 gives a true air speed at 35 000 ft of 490 kt. If we look at the engine data (Fig. 11.19), at M0.85:

$$SFC/\theta\ 0.615 = 0.7\ SFC = 0.59$$

$$(L/D) \cdot (V/SFC) = 20 \cdot 490/0.59 = 16\,610$$

Note, the parameter above is referred to as the range parameter.
Substituting the above values into the range equation:

$$R = (L/D) \cdot (V/SFC)\log e\,(M_1/M_2)$$

gives:

$$7470 = 16\,610\log e\,(M_1/M_2)$$

This equation gives:

$$M_F/MTOM = 0.362$$

A payload of 305 passengers at 95 kg each gives 28 975 kg or 63 879 lb. Hence take-off mass can be calculated as:

$$MTOM = 63\,879/(1 - 0.53 - 0.362)$$

$$= 591\,470\,lb\ (268\,240\,kg)$$

Wing area is most likely to be determined by the approach speed at maximum landing mass (M_{land}). On this type of aircraft $(M_{land}/MTOM) = 0.72$. $C_{L_{max}}$ (for approach speed $= 1.3\ V_{stall}) = 2.3$.

$$C_L\ \text{approach} = 2.3/1.3^2 = 1.36$$

$$W = C_L \cdot q \cdot S$$

Thus wing loading is:

$$W/S = C_L \cdot q$$

q at $145 \, \text{kt} = 71.4 \, \text{lb/sq ft}$. Hence at landing:

$$W/S = 1.36 \times 71.4$$
$$= 97.1 \, \text{lb/sq ft}$$

and at take-off:

$$W/S = 97.1/0.72$$
$$= 134.8 \, \text{lb/sq ft}$$

Hence wing area is given by:

$$S = 591\,470/134.8$$
$$= 4388 \, \text{sq ft}$$
$$= 398 \, \text{m}^2$$

This calculation gives us the starting point for take-off weight and wing area. What about thrust? For this aircraft, top of climb may size the engine. At this condition, $Mn = 0.85$, $(L/D) = 20$, altitude $= 35\,000 \, \text{ft}$.

For operational reasons we need a climb rate of 300 ft/min.

Weight at the top of climb is:

$$0.98 \times MTOW = 0.98 \times 591\,470$$
$$= 580\,000 \, \text{lb}$$

Thrust required for zero rate of climb is:

$$580\,000 \, (L/D) = 580\,000/20$$
$$= 29\,000 \, \text{lb}$$

F_n/W for a 300 ft/min climb is given by:

$$F_n/W = (V_C/V)$$
$$= 300/(490 \cdot 101.3)$$

Note that 101.3 is the conversion of knots to ft/min. Hence

$$F_n = 3505 \, \text{lb}$$

Hence total thrust required is:

$$29\,000 + 3505 = 32\,505 \, \text{lb}$$

At 35 000 ft altitude the engine thrust lapse rate is 0.2. There are two engines on the aircraft. Therefore each engine will need 81 260 lb (361 kN) of sea-level static thrust.

We now have some idea of the likely size of the aircraft and its engine.

Example   295

MTOM	591 500 lb (268 249 kg)
Wing area	4400 sq ft (398 m²)
Engine thrust	81 300 lb/eng (361.7 kN)
[sea level static (SLS)]	

From the charts, the baseline engine has a thrust (SLS) of 81237 lb (361 kN).

The above estimates enable a selection of wing area and engine scale to be made to start our design procedure.

Design procedure

For the purposes of this example the following design points were chosen:

wing area (sq ft)	4000		5000		6000
engine scale	0.9	1.0	1.1	1.2	

Wing   The wing geometry was based on:

aspect ratio $= 10.0$
sweepback at $1/4$ chord $= 33°$
t/c at root $= 0.13$
t/c at tip $= 0.09$

Fuselage   This was based on a survey of current aircraft projects. A fuselage with two aisles was chosen with the following dimensions:

fuselage diameter $= 20.33$ ft
fuselage length $= 206$ ft

Preliminary design procedure – results

The drag and weight information can now be derived as shown (in Imperial units) in Fig. 11.20 and 11.21 respectively.

The (profile + wave) drag data is shown at a lift coefficient of 0.5; in practice the drag will need to be adjusted for other lift coefficients. In addition the lift-dependent drag needs to be added and an Oswald efficiency factor of 0.86 has been assumed in this example. This drag and weight data (shown in Imperial units) together with the engine data can now be combined to produce Fig. 11.22. At our design range of 6500 nm Figure 11.23 can be derived from Fig. 11.22.

This gives the various combinations of engine scale, wing area and take-off weight which meet the range criteria. The performance criteria must now be satisfied. Figs 11.24 to 11.28 show how the performance criteria vary across the various combinations of take-off weight, engine scale and wing area.

The selected (or specified) minimum performance criteria as shown in each of the previous plots can now be superimposed upon Fig. 11.23 to produce Fig. 11.29.

For this example the design point (in terms of engine scale and wing area) has been selected to give minimum take-off weight to meet the combined specification. At this stage in the design process selecting minimum take-off weight represents a

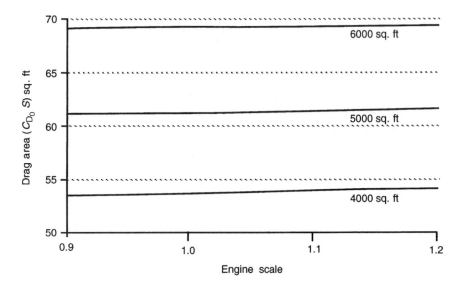

Fig. 11.20 Profile + wave drag against engine scale.

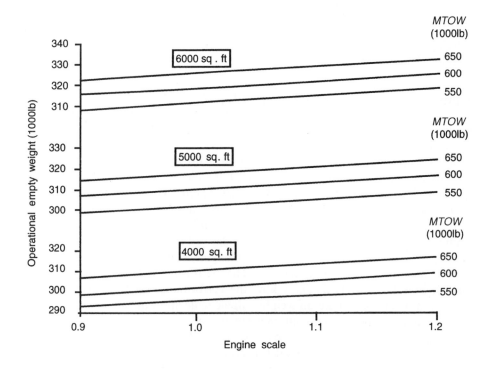

Fig. 11.21 Operational empty weight versus engine scale.

Example 297

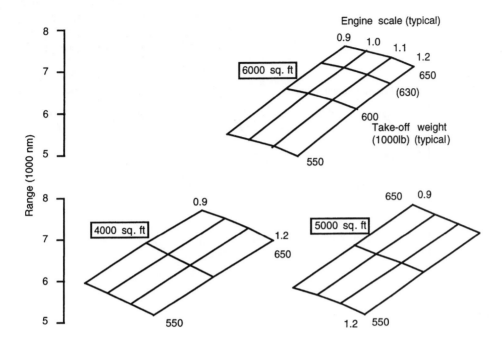

Fig. 11.22 Parametric plots.

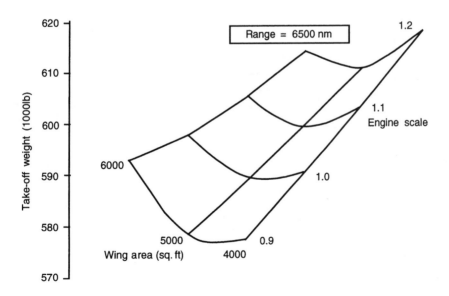

Fig. 11.23 *MTOW* carpet plot.

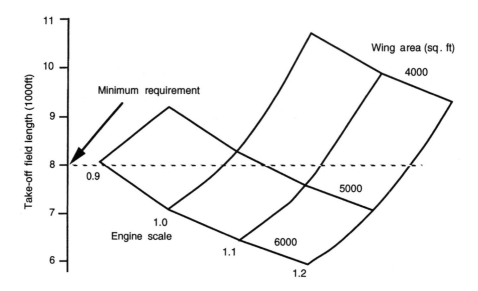

Fig. 11.24 Take-off performance carpet plot.

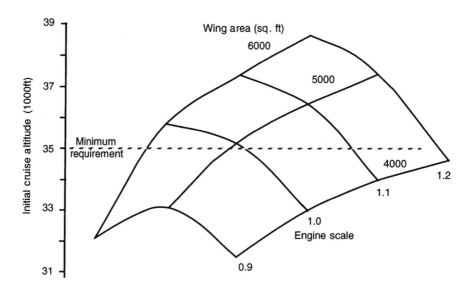

Fig. 11.25 Initial cruise altitude carpet plot.

Example 299

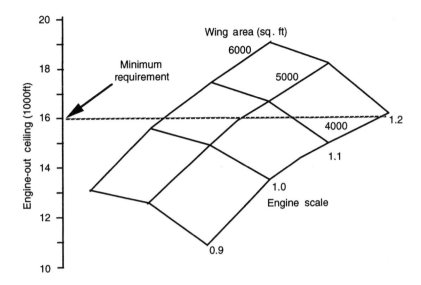

Fig. 11.26 Engine-out ceiling carpet plot.

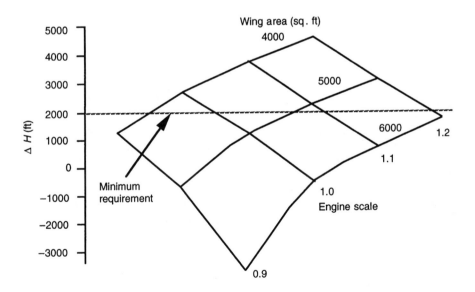

Fig. 11.27 Altitude margin (ΔH) carpet plot.

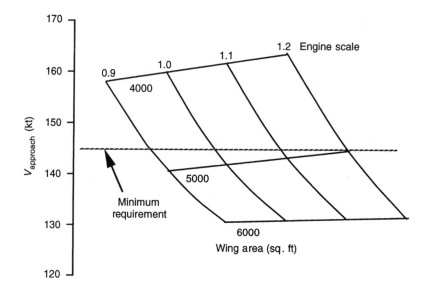

Fig. 11.28 Approach speed (kt) carpet plot.

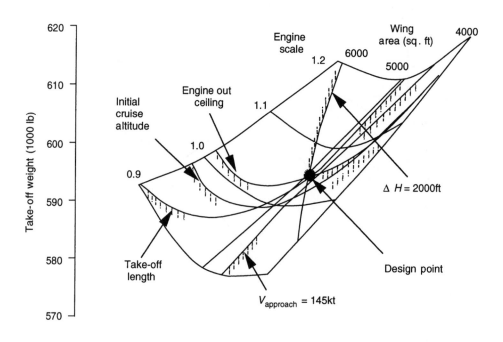

Fig. 11.29 Take-off weight carpet plot showing design constraints.

Example   301

Table 11.1 Comparison between initial approximation and new design values

	Preliminary design	Approximation
Take-off mass (lb)	595 250	591 500
Wing area (sq ft)	4950	4400
Engine thrust (lb)	87 580	81 300

reasonable overall design criterion. However, other criteria can be used such as minimum fuel burn or minimum direct operating cost if or when these are thought to be more appropriate.

The preliminary design can now be compared with the original approximation (Table 11.1).

It is easy to examine changes to the performance or specification by reading data off the charts at the new values for the requirements and superimposing on the design chart.

Initially the choice of wing aspect ratio was based on current aircraft with a similar specification. The only way to establish the validity of this original assumption is to repeat the analysis over a range of aspect ratios.

So far engine size and wing area have been considered to be variables; however, the aircraft designer is often presented with a fixed choice of engine. In our example the engine scale is unity. At this engine size the engine-out ceiling requirement is out of line with the other requirements. This would suggest that the designer approach the engine company for an increase in maximum continuous power to bring the requirements more into line. Alternatively the designer could approach those responsible for setting the aircraft specification, asking for a concession on the engine-out ceiling height.

In the case of a fixed sized engine a design chart similar to Fig. 11.29 could be produced using wing aspect ratio and wing area as the two variables.

Take-off mass has been used as the optimising criterion on the basis that there is a direct relationship between $MTOM$ and aircraft price/operating cost. Using the methods described in Chapter 12 and the aircraft design case studies, aircraft price or operating cost could be used instead of $MTOM$.

12

Aircraft cost estimations

Objectives

Aircraft design decisions have significant influence on the first cost and operating expenses of the aircraft. It is therefore important to understand the cost implications of aircraft manufacture and operation and to take these into account when deciding the aircraft configuration and performance.

Consideration of cost aspects is especially significant in the preliminary design phase of aircraft projects as fundamental decisions are taken which will be influential in the overall cost of the project. Such decisions affect the cost of manufacturing and equipping the basic aircraft and the subsequent cost of operating it over the route structure of an airline. It is therefore essential to understand the cost estimation methods to be used by the customer when comparing competitive aircraft, in order to make sensible design choices.

This chapter introduces the methods by which aircraft operating costs are estimated. These methods are used in the preliminary project phase to allow comparisons to be made between different aircraft configurations and to assess the best choice of values for all the aircraft parameters.

Indirect costs (those not directly related to the aircraft parameters, for example those associated with marketing and sales expenses) are only briefly covered, whereas direct costs are described in enough detail to allow estimating methods to be incorporated into the aircraft design process. The principal cost functions are described and typical values given.

The chapter concludes with a specimen calculation and some reference data to be used in student project work.

After completing this chapter you should understand the main characteristics of cost estimation and be able to predict direct operating cost (DOC) on projected aircraft designs.

Introduction

The main financial criterion on which the aircraft design should be judged is the return on investment (ROI) to the company. The difficulty of using this parameter

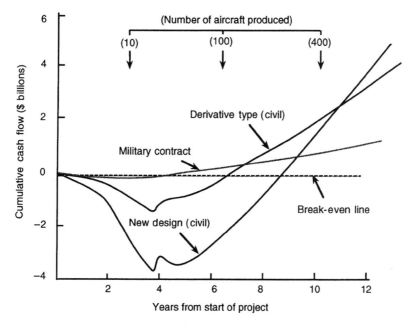

Fig. 12.1 Project cash flow 150 seat aircraft ($1995) (source AIAA-86-2667).

lies in the inherent variability in the nature of aircraft manufacture and operations and the associated substantial initial investment (i.e. prior to the inflow of revenue from sales of aircraft). The delay in recovering expenditure will result in negative cost flow in the early years of production. During this time customers (airlines) make only a small payment to the manufacturers, to reserve a position in the aircraft production schedule; this is called an option. From a project design study of a 150 seat regional aircraft conducted by McDonnell–Douglas (MD) some years ago, the cash flow history was forecast. This is shown in Fig. 12.1 together with cash flow histories for a developed (stretched) existing civil aircraft and a military project.

The 'return' as measured in the ROI parameter will depend on the terms and conditions of the investment loans and not solely on the aircraft technical and operational performance. Nevertheless, the overall financial balance sheet for the project must provide sufficient confidence to investors for the project to be externally financed. The degree of confidence held by the investors will be conditioned by many factors. Such factors will include the technical analysis of the aircraft performance in comparison with competing designs and the expectation of the total world market for the aircraft. The design team will be expected to provide such analysis.

A major difficulty in assessing the financial success of a project is the long timescales involved in aircraft design and development. The factors which influence financial viability are often considered over shorter time cycles than those associated with typical aircraft developments. Within this environment it is not surprising that funding is sometimes difficult to find for new projects. Such reluctance leads to national/governmental involvement in aircraft projects. This further complicates a pure financial analysis as political issues may be have to be considered.

Although aircraft manufacturers are in business to exploit advances in technology the financial restrictions mentioned above make them somewhat conservative in the development of new designs. If a revolutionary innovation ran into technical/operational difficulties it could ruin the company. A lot of technical effort needs to be put into building confidence in new methods before they are accepted for production.

The difficulties of using ROI as the design criterion have led to the adoption of life-cycle cost analysis. This involves the summation of all the cost elements associated with purchasing, operating and supporting the aircraft throughout the operational life. Although this parameter can be used for both military and civil projects, the components in the total cost equation differ for each sector. For civil aircraft a lot of data is available from past aircraft and operations, on which to base such analysis.

Each airline has developed its own methods for estimating operating expenses related to their particular type of operation, flight patterns, aircraft fleet and accounting procedures. The aircraft manufacturer will have to subject the projected aircraft, in competition with others, to the particular cost method of potential customer airlines. Since these methods vary widely between different operators it is necessary for the design team to use a standardised method for cost analysis, especially in the early stages of the design. Such methods are used to provide guidance in the choice of values for the fundamental operational and design parameters.

It is difficult to rationalise the design of the aircraft to different cost methods so a choice has to be made. Whichever method is chosen it can be used only to show the relative cost variation between different designs. The method will not predict actual costs as these vary so widely over different operational practices.

There are several different standardised methods available but they all trace their origin to the 'Standard Method of Estimating Comparative Direct Operating Costs of Turbine Powered Transport Airplanes' (Air Transport Association of America, Dec. 1967). Most standard cost methods only estimate the direct operating costs of the aircraft. The total cost of owning and operating an aircraft is the sum of indirect operating costs (IOC) and DOC. Although both costs may be influenced by the type of aircraft under consideration (e.g. fleet mix) it is common practice to consider the two cost components separately.

In an inflationary economic climate, values for costs are highly time-dependent; therefore some effort must be made to secure current prices for the various elements that make up the total operating cost. Alternatively, old prices must be 'factored' to account for changes since publication. This factoring requires the use of an inflation index. Traditionally such an index is difficult to determine but as the cost method used in aircraft design is employed only to estimate relative costs the exact evaluation of the index may be less significant than for absolute cost predictions. Cost values are however 'date sensitive' and any published data must show the year to which costs refer. American airlines are compelled to submit cost information to the government and this is collated and published annually in

'Aircraft Operating Costs and Performance Report' CAB Report (this data is used in *Aviation Week* quarterly/annual reports). These reports are useful sources of cost updating information. Although in project studies only relative costs are considered care must be taken to ensure that the influence of individual variables is truly represented in the cost equation.

Indirect operating cost (IOC)

IOC estimation methods deal with those costs not directly attributable to a particular aircraft type or the flying costs of a particular operation. It may include some or all of the following expenses:

- facility purchase cost and facility depreciation
- facility leasing cost
- facility maintenance cost
- ground equipment depreciation
- ground equipment maintenance costs
- maintenance overheads
- headquarters overheads
- administration and technical services
- advertising, promotion and sales expenses
- public relations cost expenses
- booking, ticket sales and commission
- customer services
- training

In order to classify the above parameters for particular airlines the American Civil Aeronautics Board (CAB) requires American airlines to report indirect costs associated within the following headings: aircraft and traffic servicing, promotion and sales, passenger services, general and administrative overhead, ground property and equipment maintenance and depreciation expenses.

Indirect operating costs obviously vary over a wide range depending on the type of operation and activity of the airline. Standard methods of estimating these costs are available (e.g. 'Boeing Operating Cost Ground Rules') and data are published on airline actual expenses (e.g. the USA CAB annually publishes statistics and data in journals such as the *Flight International* and *Aviation Week* annual reviews). Although aircraft design may have significant influence on indirect costs, for example by requiring new maintenance facilities and the introduction of new skills for advances in technology, it is difficult to quantify the exact cost of the inter-relationship. Airline management and operational aspects are predominant factors in the indirect costs and these are outside the control of the aircraft designer. It is therefore usual to ignore the effects of indirect cost on the selection of aircraft design parameters. As the variation in direct operating costs of competing aircraft and airlines narrows, the influence of IOC will become much more significant and aircraft sales teams will use reductions in the indirect costs to show their product to advantage over competitors (e.g. through savings in fleet commonality).

Indirect operating costs are not insignificant; they can account for between 15 and 50% of the total operating expenses (i.e. up to the same cost as the direct operational expenses).

Direct operating costs (DOC)

Under the DOC category all the costs associated with flying and direct maintenance must be considered. Figure 12.2 shows a typical breakdown of these costs.

Some standardised DOC methods do not include all the factors shown in Fig. 12.2. The cost components may be considered under four broad headings:

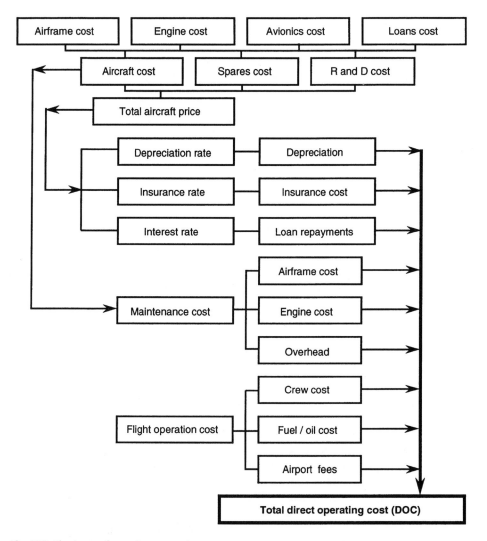

Fig. 12.2 Direct operating cost components.

- standing charges
- flight costs
- maintenance costs
- cost parameters

A description of each of these headings is given below.

Standing charges

These are the proportion of the costs that are not directly linked to the aircraft flight but may be regarded as the 'overhead' on the flight. Such costs, in order of importance, include:

1. depreciation of the capital investment
2. interest charges on capital employed
3. aircraft insurance

Depreciation and interest charges are sometimes quoted as an item and referred to as 'ownership' costs. Items 2 and 3 are ignored in some cost methods as they are small components compared to the cost of depreciation; however, these items will be discussed first.

Insurance cost
This is directly related to the risks involved and the potential for claims following loss. The airworthiness authorities oversee the observance of the safety standard; therefore the risk of accident is well established. For the insurance companies the associated technical risk is relatively easy to estimate as it is associated directly with the probability of failure of the total aircraft system. Above this baseline risk is the possibility of losing aircraft due to non-technical occurrence (e.g. terrorism) and the subsequent potential for personal claims. Such risks are difficult to determine in advance due to the sometimes transient nature of the problem. Insurance companies will therefore vary their fee in relation to the nature of the operation (e.g. geographical areas of flights) and the level of airline security. The annual premiums for aircraft insurance vary between 1 and 3% of the aircraft price. If insurance is to be included in the cost method a value of 1.5% is considered typical.

Interest charges
Interest charges are impossible to quantify in a general analysis as the banks and government agencies will charge various fees to different customers. Such charges will be dependent on the world economic climate, local exchange rates, the credit standing of the purchaser and the export encouragement given by the national government of the airline or manufacturer. Further complications may arise due to 'off-set' agreements made between the two trading partners (manufacturer and airline). For all these reasons, and because most of these factors are outside the influence of the aircraft manufacturer, many cost estimation methods ignore this cost component but it is necessary to include the interest costs in any business plan. Current national base interest rates should be determined and used for such analysis.

Depreciation

Depreciation will be dependent on many factors including the capital involved, the airline purchasing policies, the accounting practices of the financial loan companies, the competition for capital and the overall world economic conditions at the time the aircraft is purchased. As the aircraft is always maintained to a fully airworthy condition throughout its life, it will have an identifiable value when sold (known as the residual value). As with any capital item, the residual value will reduce as the aircraft ages. The depreciation period will be dependent on the accountancy policy of the airline and the expected development of the routes for which the aircraft is bought to service. Typically, an aircraft may be considered to be at the end of its useful life after 15–20 years with zero residual value. With civil aircraft design methods reaching a mature technology state the useful life of aircraft is progressively extending with 20–30 years likely to become more common in the next decade. The choice of depreciation period and the estimation of residual value is made by the purchasing airline (* e.g. 12 years depreciation to 15% residual value).

The main parameter in the evaluation of depreciation is the total price of the aircraft. The initial aircraft price used in the evaluation of depreciation will include an allowance for the capital required to provide for spares holding (typically 10–15% of the aircraft initial price). Therefore

$$\text{aircraft initial price} = \text{factory cost of the aircraft} + \text{spares cost}$$

The percentage of the first cost of the aircraft to be depreciated per year can be determined as:

$$[(\text{initial price} - \text{residual value})/\text{initial price}]/(\text{depreciation period}/100)$$

For example in the case above * the equation will give:

$$[(P - 0.15P)/P]/[12/100] = 7.08\%$$

where P = aircraft manufacturer's price (including airframe, engines, systems and online equipment).

The estimation of aircraft price is complicated by many non-engineering factors (e.g. market conditions, competition, politics, off-set trade agreements, international collaboration on manufacture, etc.). The president of one of the largest aircraft manufacturers once said, 'The price of an airplane . . . has little direct relationship to the design and production costs of the vehicle.' Nevertheless the price of an aircraft must be sufficiently high to allow a reasonable return on investment for the manufacturer and sufficiently low to allow an adequate return on the purchase investment made by the operator.

A cynical view of this process is to state that the initial price of the aircraft (and engine) is set at what the customer will pay and not related to the technical manufacturing factors. This may not be exactly true but it is known that the initial sales prices of a new aircraft type are always set low to enable the manufacturer to gain market penetration. Also, when the aircraft type is old the manufacturer will quote low prices against competing new aircraft, in the knowledge that the development costs for his aircraft have been recovered. The airline will order airframe and engines by separate contract and will bargain for the best price for

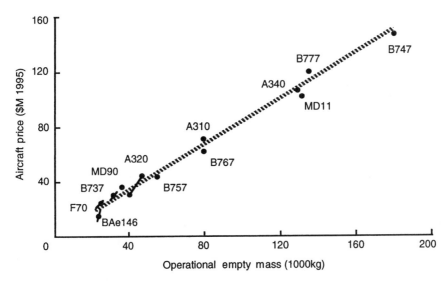

Fig. 12.3 Aircraft purchase price against OEM (1995) (data source Avmark).

each from different suppliers. The price of aircraft (airframe plus engines) is quoted in annual surveys in the aeronautical press. These data identify the current market price for the aircraft and not the cost of manufacturing the aircraft. However, the data can be used to determine aircraft price in the project design phase, as shown in Fig. 12.3.

With the complexity and uncertainty of factors associated with the market price for aircraft it is surprising that there is not more variability in the price of aircraft per pound of aircraft empty mass (OEM). Such a crude plot should not be used for particular aircraft analysis as the range of aircraft size covered is too extensive (100–500 passengers). A more detailed analysis should be made around the size of aircraft under consideration (e.g. if only aircraft less than 200 000 lb were analysed on the data above a different average-line would have been drawn). There is a strong argument in favour of using aircraft parameters other than OEM for determining aircraft price (e.g. maximum take-off mass, aircraft speed, number of passengers, etc.). Each design study should consider which parameter is most appropriate.

There are several data sources available from which aircraft price information can be obtained (e.g. The *Avmark Aviation Economist* database of jet airliner values). These should be used to build a database for the particular design. Care must be taken when using price data to normalise the values to account for inflation and the devaluation of currencies. In some cost evaluation methods there is an option to allow for leasing of aircraft to be substituted for purchasing.

There are several methods available for determining aircraft manufacturing costs from the configuration and system details (e.g. J Burns, SAWE Paper No 2228, 1994). The total cost can be considered as the sum of the design costs (which is a function of the technical complexity of the design), the overhead cost of development (which is a non-recurring cost independent of the number of aircraft produced) and the production cost (which is directly related to the number of aircraft produced, the complexity of the design and the number of co-operating manufacturing companies). The design and overhead costs occur in the early years of the design and manufacture cycle. In the early years these costs will be a large component of the sales cost, but as the design matures and sales increase the design and development cost will be repaid and eventually form a relatively small element in the sale price. For a constant production volume, the manufacturing cost will start low (because some of the early production cost will be regarded as development) but quickly stabilise to a constant value. The capital requirements for each cost will depend on the type of aircraft, the complexity of the design and the method of manufacture. A new aircraft design will involve more design and development capital than a derivative (e.g. stretched) aircraft. The MD study for a 150 seat regional jet mentioned earlier showed the estimated total programme cost for a new and derived design as shown in Fig. 12.4.

The overhead cost may be regarded as a small part of the total cost only if sufficient numbers of aircraft can be produced (and sold). Dividing the total cost associated with the design, development and manufacturing cost, by the number produced gives the unit cost (aircraft break-even price) as shown in Fig. 12.5.

This analysis determines the average cost of producing each aircraft, evaluated at the end of the productive life of the design. The problem for the manufacturers is that they do not know at the start of the project how many aircraft will be produced so the average price study is not used as a basis for fixing the initial aircraft price. Setting a constant price for a new aircraft sale ($57M and $50M on the above figure) allows a break-even production quantity to be predicted (i.e. about 250 and 350 aircraft respectively). One aircraft manufacturing company president commented that the break-even production value for an aircraft type was always about 100 more than your current sales!

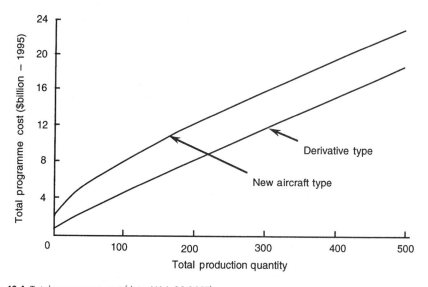

Fig. 12.4 Total programme cost (data AIAA-86-2667).

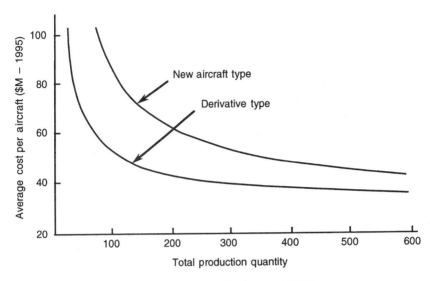

Fig. 12.5 Effect of production quantity on aircraft price (data AIAA-86-2667).

The break-even values are used to draw the programme cash flow diagrams shown in Fig. 12.1. The discontinuity in the curves in the third year arises from the receipt of deposits for future sales options. A customer will be asked to pay a small deposit to secure an option position (date) on the production line. About two years ahead of delivery the manufacturer will ask for progressive stage payments up to about one third of the sale price. These cash inputs help with the manufacturers' cash flow and show goodwill from the customer.

Note, in the above analysis the manufacturer's cash flow break-even point lies between six and nine years from the start of the project. The variability is caused by the uncertainty in the number of aircraft produced. A stabilised production volume will only arise after about six years from the start of the production process.

The differences between new and derived designs in the above analysis illustrates the variability in the assessment of aircraft production cost. Although constant sales price was assumed in the analysis, experience shows this is not the normal market practice.

As mentioned earlier, the average price will be discounted at the start and towards the end of the production run. The only rule that seems to apply to the price of aircraft is that it is set to suit the market and not as a calculation from the design and production cost. However, on average, the manufacturer needs to make a profit from the sales to stay in business and the airline must make a profit from the purchase.

In some studies it is desirable to include the engine cost as a separate item from the aircraft price. In this way the optimum aircraft design (from an aircraft DOC standpoint) can be determined. Engine price is largely dependent on the take-off thrust with a small overhead to account for non-thrust dependent cost components. In Chapter 9 the following relationship was introduced for the cost of the

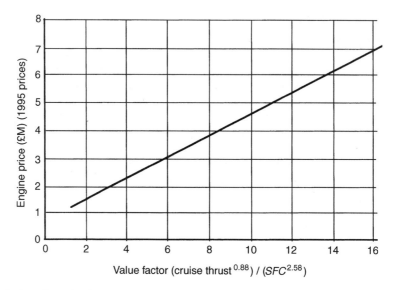

Fig. 12.6 Engine price estimation.

engine: The value factor based on 1995 market prices below is used in the trade-off graph (Fig. 12.6) which is reproduced here.

$$\text{value factor} = \frac{(\text{cruise thrust})^{0.88}}{(SFC)^{2.58}}$$

Both thrust and SFC are at the maximum cruise rating at $M = 0.8$ at $35\,000\,\text{ft}$.

Leasing

For accountancy purposes, sometimes related to local taxation policies, some airlines lease their aircraft from a third party in preference to outright purchase. The cost method reflects this option by eliminating the standing charge element and then calling the DOC value a 'Cash DOC'.

Maintenance costs

Prediction of maintenance costs is complicated by the lack of definition for items to be included under this heading. Setting up a maintenance facility is an expensive outlay for the airline. Some such facilities run as a separate business. The capital cost of buildings, the administration costs and the cost of special equipment may be regarded as an indirect cost on the total maintenance operation and included in the IOC evaluation. This suits the aircraft manufacturer as the evaluation of DOC would be proportionately reduced. The attribution of the maintenance overhead burden forms the biggest variability in the different standard methods for estimating DOC. Some airlines contract-out their aircraft and engine maintenance to other airlines or specialised maintenance companies. In these cases the total charge for maintenance will be automatically set against DOC as the cost will be directly attributable to a specific aircraft.

Maintenance charges include labour and material costs associated with routine inspections, servicing and overhaul (for airframe, engines, avionics, systems, accessories, etc.). There will also be some non-revenue flying involved and this will be charged to the maintenance account.

Generalised estimation of the total costs involved has always presented difficulties due to the variability of maintenance tasks for different aircraft and the variations in the types of operation. All the standard DOC methods include procedures for estimating maintenance costs but care must be taken when adapting these standardised methods to particular designs.

It is common practice to divide the maintenance tasks into airframe and engine components giving:

total maintenance cost = cost of airframe maintenance + cost of engine

maintenance + maintenance burden (overhead cost)

The airframe and engine maintenance costs are further sub-divided into labour and material components, for example:

airframe maintenance cost = cost of labour + cost of materials

Furthermore, each component cost is considered to be the sum of a cost per flight (flight overhead) and the cost per flying hour (recurrent charge). For example the cost of airframe labour maintenance is defined as:

airframe labour cost = (labour cost per stage + labour cost per flying hour

× stage time) × labour rate

The total maintenance burden is often based on a standing charge to cover the overhead cost of providing the maintenance service plus an hourly cost, as shown below:

maintenance burden = standing cost + hourly charge × block time/stage

Hence the components of the maintenance cost are:

1. airframe maintenance labour cost per stage
2. airframe maintenance labour cost per flying hour
3. airframe maintenance material cost per stage
4. airframe maintenance cost per flying hour
5. engine maintenance labour cost per stage
6. engine maintenance labour cost per flying hour
7. engine maintenance material cost per stage
8. engine maintenance material cost per flying hour
9. maintenance burden standing charge
10. maintenance burden hourly (flying) charge

Items 5–8 are sometimes reduced to a simple maintenance cost per flying hour obtained from the engine supplier.

Flight costs

This cost element comprises all the costs which are directly associated with the flight. The following items are summed to give the total flying costs per hour:

1. crew cost
2. fuel and oil usage
3. landing and navigational charges

Crew costs

These include the salaries for the flight and cabin staff. Crew productivity has increased over the last ten years with the increased acceptance of two-pilot operation. The number of crew is dictated by airworthiness standards and labour union agreements. Typically there will be two flight crew for smaller aircraft travelling shorter stages, three for heavier aircraft and for long-range flights it is sometimes necessary to have more than one set of flight crew on the flight. The number of cabin staff is associated with the number of passengers (30–50 passengers per cabin attendant is typical). Annual utilisation of crew varies depending on staff contracts. Eight hundred hours a year is typical for a medium size regional jet aircraft operation. Wage rates differ between airlines and between aircraft type so it is difficult to generally assess crew costs. The following relationship can be used:

crew costs = (annual cost of flight crew member × number of flight crew

+ annual cost of cabin staff member × number of cabin staff)

× (flight block hour)/(crew utilisation per year)

Note, flight and cabin staff utilisation is likely to be different, cabin staff having a higher utilisation.

Crew cost will include overheads for enforced stop-overs on long-range schedules. These expenses are sometimes treated as a general operating cost and therefore considered as indirect costs.

Fuel and oil

The cost of fuel and oil is relatively easy to estimate providing the price of fuel can be accurately predicted. Next to depreciation, fuel cost represents the most significant cost parameter in the design. A historical survey of fuel price over a 20-year period shows the difficulty of making this prediction. Figure 12.7 shows how the price of fuel has varied between 50 and 350% of the baseline value in only a 20-year period (due mainly to the imposition of cartel trading by the oil suppliers in the late 1970s). With such variability it is difficult to confidently predict fuel price in the future.

Cost parameters

All the preceding costs are calculated on an hourly (flight) basis. They are all summed to produce the direct operating cost of the aircraft per flight hour. The

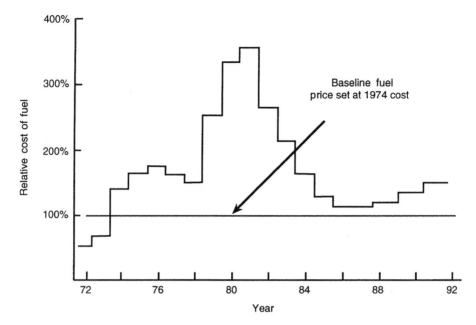

Fig. 12.7 Fuel price variability (data source AIAA-86-2667).

cost of flying a particular route (known as the **stage cost**) is found by multiplying the hourly DOC by the block time (hours). The stage cost can be divided by the block distance to show the **mile cost**. The mile cost can be divided by the maximum number of seats flown on the stage (note not the number of seats occupied as this will vary with airline schedules) to provide the **seat mile cost**.

Some manufacturers quote all the above cost parameters on a **cash** basis, by which they mean that the aircraft standing charge is not included in the total cost. Their reasoning for doing this (apart from making the DOC figure much lower) is that it is becoming common practice for airlines to lease their aircraft on an annual basis and therefore this cost becomes a different part of the company balance sheet. Be careful when using quoted DOC figures from manufacturers or the aeronautical press to understand if the **total** or **cash** method of calculating the values has been used.

Design influence

Aircraft project designers are seen to influence costs directly by the basic configuration of the aircraft (system complexity, aircraft size, engine size, etc.) and the selected performance (cruise speed, range, etc.). All these aspects will have a substantial input to the cost model through the standing charges, the fuel used and the maintenance required. The designers also influence cost indirectly through airline economics (market size, ticket price, aircraft performance, passenger appeal). These indirect factors feed into the cost analysis through revenue potential, the demand for the aircraft type, market development and ultimately to commonality and type derivations. It is important for the designers to recognise these influences in the early stage of the aircraft project so that the design can meet the market potential and thereby maximise success of the project.

It will be necessary to conduct several parametric trade-off studies to fully understand the various competing aspects of the design. It is unusual for the design team to consider the aircraft as a 'point-design' aimed exclusively at one market sector. The initial design will be slightly compromised to allow for future stretch (increased payload and/or range) to extend the market for the aircraft type.

Example

To illustrate how the direct costs are calculated consider an aircraft with the following details:

number of seats	300
range at maximum payload	7200 nm
cruise speed	M0.825
(at a representative altitude this equates to a speed of 243 m/s)	
aircraft maximum take-off mass	243 200 kg
engine take-off thrust (two engines)	370 kN (each engine)
cruise SFC	0.55
fuel consumption	5500 kg/hr (each engine)
aircraft utilisation	4200 hr/year
engine maintenance	$190/hr/engine
airframe maintenance	$660/hr (labour) and $218/hr (materials)

Estimating the standing charges

We shall first estimate the 'standing charge' element in the DOC calculation. From a graph similar to Fig. 12.2, drawn for aircraft of similar size (weight) to the proposed design, we can determine the aircraft price. Using the $MTOM$ quoted above and assuming an empty weight fraction of 58% (typical of this size aircraft) we estimate the empty weight and then the aircraft price:

$$OEM = 0.58 \times MTOM = 0.58 \times 243\,200 = 141\,100 \text{ kg}$$

From the cost graph (Fig. 12.3) at this weight:

$$\text{total aircraft price} = \$M118$$

Using the quoted engine SFC and derating the take-off thrust to the cruise condition allows an estimate of the engine price to be made using the value function shown in Chapter 9:

$$\text{value function} = (\text{cruise thrust})^{0.88}(SFC)^{-2.58}$$

The graph of current engine prices gives a value of $M9.8 per engine. There are two engines; therefore cost of engines $= \$M19.6$, making the cost of airframe $= \$M98.4$ and giving:

$$\text{total aircraft factory price} = \$M118$$

Assuming the cost of aircraft spares is 10% of the airframe price and the cost of engine spares is 30% of the engine price we can determine the total spares cost:

$$\text{spares cost} = (0.1 \times 98.4) + (0.3 \times 19.6) = \$M15.7$$

Adding the spares cost to the total aircraft factory price gives:

$$\text{total investment cost} = 118 + 15.7 = \$M133.7$$

Assuming that this cost is depreciated to 10% over a 16-year operational life gives an annual cost of:

$$\text{depreciation costs/year} = 0.9 \times 133.7/16 = \$M7.52$$

Assuming interest on investment cost is set at 5.4% per year, the annual cost will be increased by:

$$\text{interest/year} = 0.054 \times 133.7 = \$M7.22$$

Assume insurance is charged at 0.5% of aircraft cost:

$$\text{insurance/year} = 0.005 \times 118 = \$M0.59$$

Hence the total annual standing charge is:

$$\text{total standing charge/year} = 7.52 + 7.22 + 0.59 = \$M15.33$$

Dividing this by the quoted aircraft utilisation per year of 4200 hours gives:

$$\text{standing charge/flying hour} = \$3650$$

Estimating the flying costs

Assume two flight crew at $360 per hour and nine cabin crew at $90 per hour are used:

$$\text{crew costs/hr} = (2 \times 360) + (9 \times 90) = \$1530$$

Assume landing fees are charged at $6 per ton (of aircraft MTO):

$$\text{landing charge} = 0.006 \times 243\,200 = \$1459$$

Assume navigational charges of $5640 per flight for this type of aircraft.
Assume ground handling charges of $11 per passenger per flight.
For a 300-passenger aircraft this gives:

$$\text{ground handling} = 11 \times 300 = \$3300$$

Therefore airport charges are:

$$\text{total airport charge} = 1459 + 5640 + 3300$$
$$= \$10\,399 \text{ per flight}$$

In order to relate this cost to the aircraft flying hours it is necessary to determine the block time of the flight. For the cruise phase of 7200 nm and with the aircraft travelling at M0.825 (a cruise speed of 473 kt) the time taken is:

$$\text{cruise time} = 7200/473 = 15.22\,\text{hr}$$

To this must be added the time for climb and descent and the time on the ground. As part of the 7200 nm stage distance will be covered in the climb and descent phases and the aircraft speed will be reasonably high in these parts of the flight profile, assume 10 minutes lost time. Assume 20 minutes for start-up, taxi-out and take-off, eight minutes for hold prior to landing and five minutes for landing and taxi to stop:

$$\text{total extra time on flight} = 10 + 20 + 8 + 5 = 43\,\text{min}\,(= 0.72\,\text{hr})$$

Hence:

$$\text{block time} = 15.22 + 0.72 = 15.94\,\text{hr}$$

The above airport charges can now be related to flying hours:

$$= 10\,399/15.94 = \$669/\text{hr}$$

The fuel consumption is quoted as 5500 kg/hr per engine at cruise. During the non-cruise phases the fuel consumption would be higher than this but on this long-range design it will be sufficiently accurate to ignore these increases.

Converting the fuel consumption to US gallons using the conversion factors and fuel density quoted in Data E, gives:

typical jet fuel density $= 800\,\text{kg/m}^3$
volume of fuel used $= 5500/800$ (per engine) $= 6.875\,\text{m}^3 = 6875$ litres
(conversion: 1 US gallon $= 3.785$ litres)
volume of fuel $= 1816$ US gallons
With two engines, the aircraft will burn $(2 \times 18\,116) = 3632$ gallons/hr.
The cost of fuel is assumed to be 70c/US gallon. Therefore:

$$\text{Fuel cost} = 3632 \times 0.7 = \$2542/\text{hr}$$

Maintenance costs

These are relatively difficult to estimate in the initial project stage so values are used that are representative of the aircraft and engine types:

engine (labour + materials) $190/hr/engine
airframe (labour) $660/hr
airframe (materials) $218/hr

$$\text{total maintenance cost} = (2 \times 190) + 660 + 218 = \$1258/\text{hr}$$

Total direct operating cost

(a) Total DOC per flying hour. This is the sum of all the above component costs:

standing charge	= $3192	(35% total)
crew cost	= $1530	(17%)
airport charge	= $669	(7%)
fuel cost	= $2542	(27%)
maintenance cost	= $1258	(14%)
total cost	**$9191**	

$$\text{Total stage cost} = 9191 \times 15.94 = \$146\,500$$
$$\text{Mile costs} = 146\,500/7200 = \$20.35$$
$$\text{Seat mile cost} = 20.35/300 = 6.78c$$

(b) Total cash direct operating cost/hour. For those operators that lease their aircraft the 'standing charge' estimated above may not form part of the direct operating cost analysis as the leasing contract may be accounted as an annual charge unrelated to the aircraft operation. In such cases the DOC calculation is termed the 'Cash DOC' and will have values as shown below for our specimen aircraft:

crew cost	$1530	(26% total)
airport charge	$669	(11)
fuel cost	$2542	(42)
maintenance cost	$1258	(21)
total cost	**$5999**	

$$\text{Total stage cost} = 5999 \times 15.94 = \$95\,624$$
$$\text{Mile cost} = \$95\,624/7200 = \$13.28$$
$$\text{Seat mile cost} = \$13.28/300 = 4.43c$$

Note how the significance of crew and fuel cost increases in the cash DOC compared to the normal DOC. In this type of analysis the fuel cost becomes nearly half the total operating cost. Manufacturers will want to quote this figure if their aircraft is more aerodynamically efficient than their competitors, particularly if their aircraft is newer and more expensive to buy than the competition!

Reference data

Typical values for parameters used in modern direct cost estimation models are given below.

1. Flight profile (at ISA conditions)

1. Start-up and taxi-out = 20 minutes (10 minutes for short-haul operations).
2. Take-off and climb up to 1500 ft*.
3. Climb from 1500 ft* to initial cruise altitude.
4. Cruise at specified speed to include any stepped climb manoeuvre+.
5. Descent to 1500 ft*.
6. Hold (eight minutes) at 1500 ft.
7. Landing and taxi-in = five minutes
   (* note maximum speed below 10 000 ft is 250 kt in US airspace)
   (+ only possible with minimum rate of climb of 500 ft/min or more).

2. Reserves

1. Diversion = 200 nm (after hold).

2. Hold = 30 minutes at 1500 ft at minimum drag speed (clean).
3. Contingency = 5% of stage fuel.

3. Payload
Either (a) fuselage volume limit or (b) zero fuel weight limit.

4. Block time
Flight time plus 25 minutes long-haul, 15 minutes short-haul).

5. Utilisation

(a) Long-range (6500 nm is typical) = 4800 hours/year.
(b) Medium/short-range = 3750 hours/year.

6. Investment
The sum of:

(a) aircraft price (A) = manufacturer's price (M) + buyer's equipment (B) + modification costs* + capitalised interest on stage payments**
    where: * = 6% of $M + B$, ** = 2.5% of M
(b) airframe spares = 10% of (A less engine cost)
(c) engine spares = 30% of total installed engine cost

7. Depreciation
(a) Useful life = 14–20 years for new designs.
(b) 10% residual value.

8. Interest
5% of total investment (approximately 11% on 100% financing).

9. Insurance (values vary between 0.35 and 0.85)
0.5% of aircraft price (A).

10. Crew costs (1989 cost value, ref. AEA)
(a) Flight = $710 long-haul (493 short-haul)/flying hours (minimum two crew).
(b) Cabin = $90/operating hours (1 per 35 passengers).

11. Landing fees (1989 cost value, ref. AEA)
$6 per metric ton.

12. Navigation changes (1989 cost value, ref. AEA)
(Stage length/5) $\times (MTOM/50)^{0.5}$

where: stage length is in km and $MTOM$ is in metric tonnes.

13. Ground handling charges (1989 cost value, ref. AEA)
110 $\times$ payload (metric tonnes).

14. Fuel price

Use current price (e.g. $0.22/kg, i.e. between 65 and 80c per US gal).

15. Maintenance

Maintenance costs are dependent on the way aircraft are used (e.g. route structure), the maintenance practices adopted by the airline and the age of the equipment (airframe and engines). Evaluation of the costs involves a detailed knowledge of factors which are unlikely to be available in the early design stages. Standardised methods for calculating maintenance costs are published (e.g. AEA, Boeing) and these should be used as soon as sufficient data is available (e.g. typical operating profiles). Until this time values from aircraft of a similar size and performance can be used or the simplified formulae below.

(a) Airframe direct cost (US$/block hour) (1994):

$$C_{AM} = 175 + 4.1\, M_{OET}$$

where M_{OET} = aircraft operational empty mass in metric tonnes.

Typical values range from 300 for small regional jets to over 1000 for the B747–400.

(b) Engine direct cost per engine (US$/block hour) (1994):

$$C_{EM} = 0.29\, T$$

where T = engine thrust (kN).

The engine maintenance cost is a function of several engine parameters. The above calculation is appropriate to modern medium bypass (typically five) engines.

(c) Total aircraft maintenance cost is

$$C_{AM} + C_{EM} \cdot N_E$$

where N_E = number of engines.

The maintenance costs above include all costs associated with the facility (e.g. overhead/burden).

Some comments on the above data

Absolute values for costs are time-dependent due to the effects of inflation in world economies. The figures above should be increased in line with inflation indices (typically between 3 and 6% per annum).

The data above is given as guidance to be used when better information is not available. It is appropriate to aircraft of conventional layout and materials. It does not include costs associated with specialised equipment such as thrust reversers, nacelle noise alleviation structures, freight doors and cargo equipment, extra passenger services and sophisticated aircraft instruments and avionics. Cost estimation is regarded as a 'black art'. Each airline and manufacturer will have developed methods and parameters appropriate to their own operations. In preliminary aircraft design it is necessary to show the trade-offs that are possible in the assumptions above. This will allow significant variations from the standard values to be assessed and allowances made to the aircraft specification if appropriate.

Miscellaneous definitions for airline DOC calculations

1. Utilisation

Aircraft utilisation is used for the calculation of depreciation and maintenance costs. It is defined as:

> Revenue hours is that time associated with the block time calculation below and does not include training, positioning for schedule, or any other non-revenue flying. Utilisation depends on the type of flight operation, time-round time, stand-down time, maintenance time, etc. Utilisation will obviously vary from one airline to another and for difference aircraft types. Increasing utilisation directly reduces costs and reflects an increased efficiency; therefore much effort has been put into raising utilisation.

Disregarding airline operational aspects the most significant factor affecting utilisation is block time. Long duration schedules without the non-revenue time stops associated with shorter flights show highest utilisation (as in the previous example calculation).

2. Block time (T_b) (hours)

This includes the total time spent from starting engines to engines 'off'. It includes the following components:

$$T_b = T_{gm} + T_{cl} + T_d + T_{cr} + T_{am}$$

where: T_{gm} = ground manoeuvre time including one minute for take-off
$\qquad$ = 0.25 hr
$\qquad T_{cl}$ = time to climb including acceleration from take-off speed to climb speed
$\qquad T_d$ = time to descend including deceleration to manual approach speeds
$\qquad T_{am}$ = time for air manoeuvres (no credit for distance) = 0.1 hr
$\qquad T_{cr}$ = time at cruise altitude (including traffic allowance)

This is evaluated by:

$$T_{cr} = [D + (K_a + 20) - (D_c + D_d)]/V_{cr}$$

where: D $\quad$ = trip distance (i.e. stage length) CAB (statute rules)
$\qquad D_c$ $\:$ = distance to climb (statute miles) including distance to accelerate from TO speed to climb speed
$\qquad D_d$ $\:$ = distance to descend (statute miles) including distance to decelerate to normal approach speed
$\qquad V_{cr}$ = average true airspeed in cruise (mph)
$\qquad K_a$ = airway distance increment
$\qquad\quad$ = $(7 + 0.015\,D)$ for $D < 1400$ statute miles
$\qquad\quad$ = $0.02\,D$ for $D > 1400$ statute miles

These values should be estimated within the following conditions:

- climb and descent rates shall be such that 300 ft/min cabin pressurisation rate is not exceeded;
- in transition from cruise to descent the cabin floor angle shall not change by more than 4° nose down;
- the true airspeed used should be the average speed attained during the cruising portion of the flight including the effects of step climbs, if used;
- zero wind speed and standard temperature shall be used for all performances.

3. Block speed (V_b)
This is defined as:

$$V_b = D/T_b$$

4. Flight time (T_f) hours
This is defined as:

$$T_f = T_b - T_{gm}$$

5. Unit costs
It is common practice to quote aircraft costs in ways other than the aircraft hourly cost. The following costs are often used.

1. *Cost per aircraft mile* (usually quoted in cents/mile or pence/mile)

$$C_{am} = C_{ah}/V_b \times 100$$

2. *Cost per short-ton mile*

$$C_{st} = C_{am}/(\text{payload in short tons})$$

(Note: 1 short-ton = 2000 lb)

3. *Cost per seat mile* (or alternatively defined as cost per passenger mile)

$$C_{sm} = (C_{st} \times \text{passenger unit weight})/2000$$

(Passenger unit weight can be assumed to be 205 lb, the 2000 converts the short ton unit to pounds.)

Note cost estimates 2. and 3. assume 100% load factor. If lower load factors are to be assumed it is necessary to make more detailed modifications to the earlier cost analysis (e.g. fuel costs) to account for reduced aircraft weight.

6. Weights
To establish a constant method of cost estimation it is necessary to precisely define various weight terms.

1. *Payload*   Within the limitations of the volumetric payload capacity and the structural payload capacity, the payload is equal to:

aircraft take-off weight − (operating weight + weight of fuel)

2. *Volumetric payload capacity* is equal to:

(no. of passenger seats × weight of the passenger plus food) × (gross volume of all freight and baggage holds × density of freight and baggage)

3. *Structural payload capacity* Two limits to payload may frequently be imposed by structural strength considerations:

(a) maximum zero fuel weight
(b) maximum landing weight

In the latter case the take-off weight must be such that the following total does not exceed the maximum landing weight:

operating weight + payload + reserve fuel and oil

The reserve fuel = (fuel carried + ground burn fuel consumed).

4. *Operating weight* The operating weight is the weight of the aircraft fully equipped and ready for operation, including flight and cabin crew and their baggage, but less fuel, oil, and payload. It includes expendable items such as anti-icing fluid, humidification, drinking, washing and galley water, and passenger amenities associated with the role of the aircraft.
If the fuel required for anti-icing, cabin heating, auxiliary power units, etc. is drawn from the main tanks or if their operation affects engine consumption, a suitable allowance must be added to the fuel carried.

5. *Weight of crew members* The weights of male crew members and of stewardesses shall be taken as 165 lb (75 kg) and 143 lb (65 kg) respectively plus baggage allowances of 44 lb (20 kg) each for services involving overnight stops away from base. It is to be assumed that the weight of the crew's food, when carried, is covered within the total allowance for passengers' food.

6. *Passenger weight* The inclusive weight of a passenger, baggage and food, termed the 'passenger-unit weight', is to be taken as 205 lb (93 kg). This figure, which is specified for general purpose calculations of passenger mile cost, is a weighted mean of loads realised in operation and is based on medium-range services; it is made up of 156 lb passenger weight, 44 lb baggage weight and 5 lb food.

Table 12.1

Service category	Baggage	Amenities, including food		
Short service (local)	33 lb (15 kg)	4 lb (1.8 kg) for all seats		
		Economy	Tourist	1st class
Continental and feeder-line (duration less than 6 hours)	44 lb (20 kg)	7 lb (3.2 kg)	12 lb (5.5 kg)	25 lb (10.4 kg)
Intercontinental services (duration greater than 6 hours)	66 lb (30 kg)	9 lb (4.1 kg)	16 lb (7.3 kg)	32 lb (14.6 kg)

7. *Density of freight and baggage* The density of freight and baggage is to be taken as 10 lb/cu ft (160 kg/m³) on the gross volume of the freight and baggage holds. For use in more specific calculations the values in Table 12.1 may be regarded as fairly typical.

8. *Unit weights of fuel and oil* The following unit weights of one Imperial gallon (note, 1 US gallon = 0.8 Imperial gallon) of fuel and oil are to be employed for the cost methods.

Turbine Fuel	8 lb	(3.63 kg)
Lubricating Oil	8.9 lb	(4.04 kg)
Methanol-water	9.1 lb	(4.13 kg)

13

Parametric studies

Objectives

The previous chapters in this book have described the design process as an iterative method conducted within a well defined design area, bounded by a number of rigid constraints and satisfying a fixed set of operational requirements. This process is satisfactory to determine the initial layout but the resulting design is highly conditional on the parameters defining the aircraft specification. It is important that the designers understand the significance of the choice of the values used for the design specification and their influence on the aircraft configuration. Methods are required that allow the designers to gain this understanding.

It is unusual for aircraft manufacturers to concentrate on one particular configuration. Even when the initial set of requirements seems rigid, the manufacturers will be expecting to extend the design into new markets. For example, most new civil aircraft designs will eventually form part of a family of aircraft with different payload/range specifications. This will involve stretching (and in some cases shortening) the original design. Furthermore, the aircraft manufacturer will want to offer the design with alternative engines so that a competitive engine market is maintained. Also in some cases the civil aircraft may be converted into military transports or used for maritime patrol and reconnaissance.

For all the reasons mentioned above the aircraft configuration selected in the project stage may not be exactly matched to the initial specification. The degree of compromise on the initial design is a difficult decision area for the designers. The initial configuration will not be successful if it is made too inefficient compared to its competitors, but without the potential for later development the future success of the project would be jeopardised.

The method by which the designers select the initial 'baseline' design involves the use of parametric studies. This chapter describes how the aircraft initial design layout is used as the focus for further parametric studies. After a brief general introduction to the methods used, an example is presented showing how different types of parametric study are used in a case study. This example may be used to guide student project work and to allow a comprehensive understanding of the design area before finalising the aircraft configuration.

Further examples of parametric studies are included in the case studies following Chapter 15.

Introduction

Parametric studies are normally conducted around a known design. This is called the 'baseline aircraft'. The studies are used to determine the sensitivity of the design parameters at this design point. For those parameters shown to be sensitive to change, the designers will need to make a careful selection of the value to be used on the proposed design or to make adjustment to the layout to reduce the dependency on the parameter. Conversely, for those parameters that are seen to be uncritical to the effectiveness of the layout the designers may have more flexibility in the selection of the value to be used.

For most aircraft designs there are many parameters that could be investigated in such studies. To keep the multi-variable design problem manageable it is preferable to consider only one or two variables in any specific study. For example in a parametric study of wing layout, the baseline layout may be held constant except for changes to the aircraft wing area and aspect ratio. In another case, the suitability of different types of engine may be studied with everything else on the configuration kept constant. Further parametric studies may be conducted to investigate the introduction of a new technology (e.g. by showing the influence of new materials and manufacturing methods on aircraft structure weight). The list of possible studies is long and it is up to the design team to use experience to select the types of study and the parameters to vary. The team must select the type of study on the basis of those aspects that are most influential to the particular design.

Nine-point studies

For a parametric investigation involving two variables the classical nine-point study can be used. In a particular aircraft design study, variable x could be 'wing loading', variable y the 'thrust loading' and the objective function could be 'aircraft maximum take-off weight'. At each of the nine points an aircraft will be fully designed to the appropriate values x and y. When all nine designs have been determined a carpet plot can be drawn to show the variation of the designs against aircraft weight as shown in Fig. 13.1.

Note, that for each design point all aircraft parameters (geometry, mass, performance, cost, etc.) will have been evaluated. Similar carpets could be plotted for any of the overall design parameters. The 'baseline design point' is often chosen as the centre point in the nine-point study to show positive and negative changes.

In association with the x and y variables a third parameter could be varied. Variable z could be the 'number of passengers'. Such three-variable studies produce a series of carpet plots as shown in Fig. 13.2.

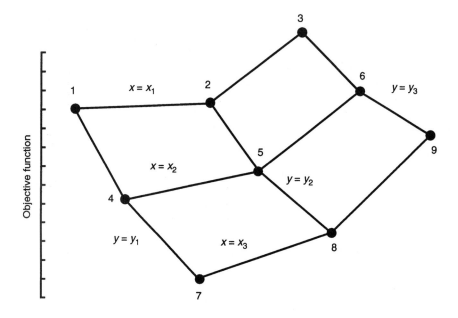

Fig. 13.1 Classical nine-point study.

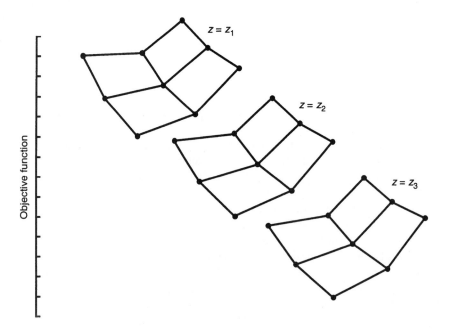

Fig. 13.2 Repeated nine-point studies.

Example 329

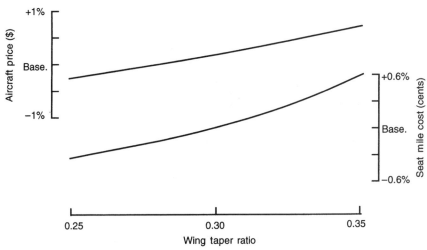

Fig. 13.3 Single-parameter studies (fixed engine size).

Single-variable studies

Single-parameter studies are sometimes used to show the sensitivity of the design to a particular variable (e.g. direct operating cost, DOC, and aircraft price vs taper ratio). In such studies all other parameters are held constant (e.g. wing area and engine size). The change in the variable will have secondary effects on some of the operating characteristics (e.g. climb rate, range) and care must be taken that the resulting aircraft in the study are feasible (i.e. meet all the aircraft specifications and constraints). Figure 13.3 shows a typical single-variable study, in this case an investigation into wing taper ratio.

Example

To illustrate the use of parametric studies in aircraft project design this example considers the design of a small (50-seat) regional jet. The basic specification calls for a 48-seat aircraft flying over a 1000 nm single stage length with normal reserves and operating out of a 5300 ft field at + 25° ISA.

The details of the baseline design are listed below and the aircraft configuration is shown in Fig. 13.4.

Overall length	24.5 m (80.4 ft)
Overall height	7.9 m (25.9 ft)
Wing span	22.6 m (74.2 ft)
Wing area	55 m² (591.6 ft²)
Aspect ratio	9.31
MTOM	18 730 kg (41 300 lb)
Payload	48 passengers @ 100 kg each (or 5700 kg cargo)
OEM	11 730 kg (25 860 lb)
Design range	1000 statute miles at full payload
Take-off field length (ISA-SL)	1550 m (5100 ft)

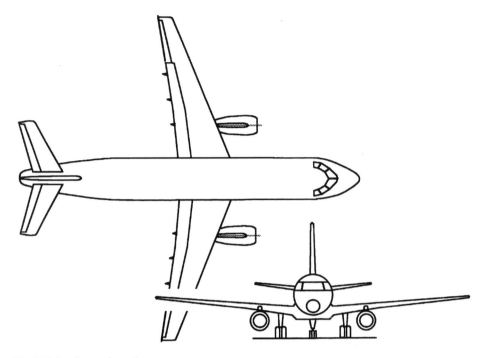

Fig. 13.4 Baseline configuration.

The first parametric study to be conducted concerns the sensitivity of the choice of wing aspect ratio. Point design studies were conducted at a range of input values of aspect ratio from 8.5 to 10.5. This involved a complete redesign of the baseline aircraft to meet the required specification shown above. At each of the selected aspect ratio values the aircraft weight, wing area and DOC changed from the baseline design. Various aircraft parameters are plotted in Fig. 13.5. The combined effects of structure weight and fuel usage resulting from the change in aspect ratio is seen in each parameter. The influence of span loading on climb performance and thereby the WAT (weight, altitude, temperature) constraint is clearly shown at low values of wing aspect ratio. The study shows the optimum choice of aspect ratio to be 9.22 for minimum DOC.

Varying aircraft geometrical parameters, as in the study above, is relatively simple as these are direct inputs to the estimating equations used in the design process. Altering a principal operational parameter (e.g. landing field length, LFL) involves much more iteration in the design calculations. Nevertheless, it is important to show the sensitivity of such parameters on aircraft geometry as shown in the next study (see Fig. 13.6).

As all the details of the aircraft are evaluated at each value of the parameter (LFL above) it is possible to plot other sensitivities (e.g. empty weight, fuel usage, etc.). When all the operating parameters have been analysed it may be necessary to review the original specification.

More fundamental changes to the aircraft specification may be linked to future design options. For example, it is likely that the aircraft engine will be developed

Example   331

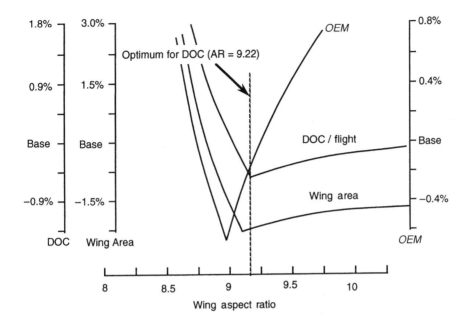

Fig. 13.5 Aspect ratio study (fixed engine size).

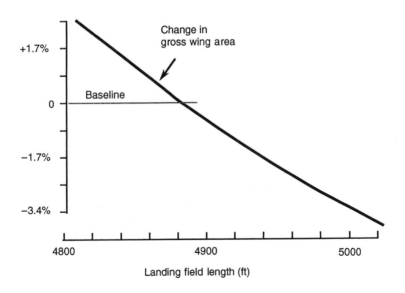

Fig. 13.6 Landing field length study.

during the design life of the aircraft, providing more thrust. To study the influence of engine changes further sensitivity studies can be conducted using the engine scaling relationships described in Chapter 9. In the current study the baseline engine is assumed to be developed into more powerful versions. The engine scale is the ratio of the developed thrust to the original baseline aircraft value. Figure 13.7 shows the effect on aircraft size, performance and costs which results from the adoption of more powerful engines on the original specification.

The curves in Fig. 13.7 show the conflicting influences between different parameters. For example, aircraft price will increase due to the more expensive and larger aircraft and engines but due to the increased thrust available the cruise speed will increase thereby reducing the stage time and raising the possibility of higher utilisation per year. Such studies may dictate an initial specification which is slightly less optimum for the early designs but has the capacity to handle future developments more efficiently.

This type of study may be better analysed by a conventional two-parameter

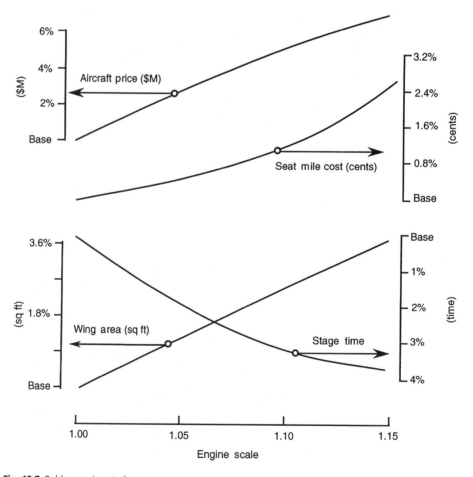

Fig. 13.7 Rubber engine study.

Example   333

study. The stretch potential of the original aircraft (increasing capacity to 56 seats) with a larger engine is shown in Figs 13.8 and 13.9.

The aircraft variables chosen in the stretch studies illustrated in the figures were the stage and field lengths. The top-most point on the study (i.e. maximum stage, minimum field) was shown to be impractical with the selected engine. A small (8%)

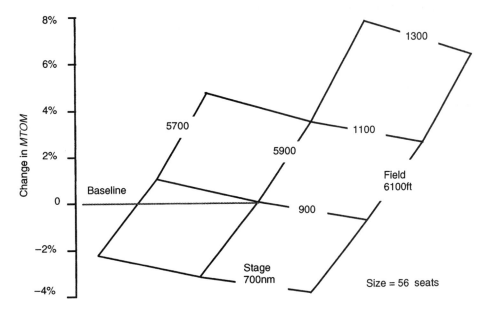

Fig. 13.8 Aircraft stretch study (maximum take-off mass).

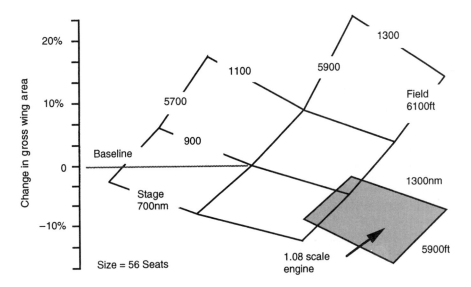

Fig. 13.9 Aircraft stretch study (gross wing area).

thrust increase was investigated and the result of the top four conditions is shown by the shaded segment in Fig. 13.9. The relative position of the missing segment in the main graph and the shaded segment indicates the sensitivity of the design to engine thrust improvement.

Finally, the regional-jet design study was extended to consider engine designs. Such studies assume that the engine size is exactly matched to the aircraft requirements (known as 'rubber' engines). Although totally impractical in reality as engines are only available at specific sizes, these studies show the best combination of aircraft and engine integration. Such studies show the compromise that must be accommodated by the design in using an existing engine sized differently from the optimum. Studies for the regional jet project involved three parameters (number of passengers, range, required field length). The results are shown in Fig. 13.10.

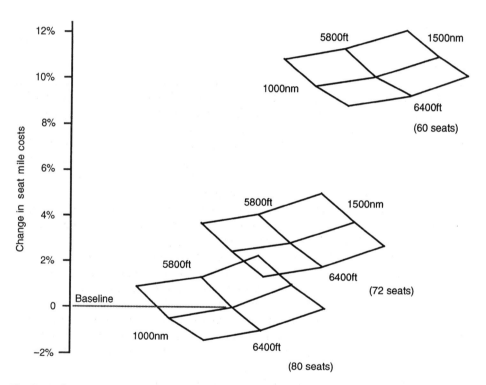

Fig. 13.10 Three-parameter study.

14

Aircraft type specification

Objectives

At the end of the project phase the design team will know a lot of details of
the aircraft. It is important to record these in a report that other people can
understand. The conventional format for this report is called the Aircraft Type
Specification. This includes all the detailed information that has been derived
for the baseline design. The report is a complete description of the aircraft that
the design team expect to be produced. The Type Specification therefore forms
a focus for the conclusion of the project design phase.

For academic studies, the Type Specification acts as a useful planning sheet
for the various detail studies that need to be conducted. The various headings
are used as an aide-memoire to the student. They also act as a final report
framework covering all the areas of the design which must be considered.
Obviously, for some students' work the level of detail in some of the specialist
areas mentioned in the Type Specification may be too fine to be considered
(this must not be used as an excuse for avoiding work!). One of the main uses
of this section of the book, for student project work, is in the organisation
and planning of the project. The main headings can be used to identify the
various tasks to be undertaken. They can be used to allocate and plan the
progress of work throughout the course.

This section can be used to guide the students in their preparation of the
final project report. For student work the individual parts of the type record
are not equal; their relative importance will depend on the nature of the study.
For example, for some students' coursework the 'performance' section may
be the main area of specialisation and for others the system descriptions will
be the central issues.

Finalisation of the Type Specification often forms the conclusion to the
work of the initial project design group. This chapter therefore concludes
the first part of the book and leads to the design studies in the next part.

Type record

From the start of the conceptual phase and all the way through the subsequent design processes, details of the aircraft configuration are continuously refined. As time allows, more information about the layout is sought to enable better estimates of the aircraft parameters to be produced. This will involve progressively considering more details of the aircraft structure and systems. For example, at the start of the process the design knowledge about the tail will only consist of a guessed tail arm and a crude estimate of area. At the end of the project phase the detail design of the tail, fin, rear fuselage, the control surfaces, control movements and forces, and all the aircraft systems will be known. To record such detail the Aircraft Type Specification is created and used to summarise the current state of the aircraft design. It is the responsibility of the design team leader to ensure that the information and data in the report are accurate and can be guaranteed by the company. It is therefore a document that is continually evolving as design decisions are made right up to the time that the design is frozen.

The Type Specification document is a complete description of the manufacturer's liability for the design. It is a statement of what will be provided by the manufacturers for the sale price. As such, it will eventually form part of the contract of sale and serve as the guarantee from the manufacturer. With this background the document is treated seriously in the design organisation and will not contain any speculative statements, potentially misleading information or inaccurate data. The report will consist of textural descriptions, drawings, numeric data, graphs and charts. The type of information presented in the report will obviously vary with the type of aircraft but the following sections are representative.

1. Introduction
2. General design requirements
3. Geometric characteristics
4. Aerodynamic and structural criteria
5. Weight and balance
6. Performance
7. Airframe
8. Landing gear
9. Power plant (and systems)
10. Fuel system
11. Hydraulic systems
12. Electrical systems
13. Avionics
14. Instruments and communications
15. Flight controls and flight deck
16. Passenger cabin layouts
17. Environmental control systems
18. Safety systems and emergency exits
19. Servicing
20. Airport compatibility
21. Exceptions to regulations

A brief description a each of these sections is given below.

1. Introduction

The introduction will define the status of the report, specify commercial confidentiality, give a brief description of the aircraft type (a simplified three-view general arrangement drawing of the aircraft may be presented), specify the main functional items (e.g. number and position of crew, description of passenger accommodation, freight/cargo specification) and provide contractual sales statements and guarantees.

2. General design requirements

This section lists the airworthiness and operational regulations and other documents to which the aircraft is designed (e.g. JAR-25, JAR-E, etc., FAR part 25, part 33, part 34, part 36, etc.). It will define the certification procedures adopted and the Type Approval to be claimed. Specialised requirements (e.g. interchangeability, cross-wind performance, ice protection) will be included. The exterior and internal finish standards for the 'white-tail' aircraft will be described. (The term 'white tail' is used to describe an aircraft built without an airline contract, i.e. awaiting a sale and therefore without an airline livery.)

3. Geometric characteristics

All the principal geometric features of the aircraft will be listed together with a fully itemised general arrangement (GA) drawing of the aircraft. Typical data will include:

1. overall dimensions (overall length, overall span, overall height, clearances);
2. wing geometry [gross area, span, mean chord, aspect ratio, dihedral, sweepback angles (leading edge, quarter chord position, trailing edge), datum planes, flap areas, flap angles, aerofoil description (including thickness)];
3. tailplane and elevator (gross area, span, quarter chord position, aspect ratio, dihedral, aerofoil, setting angle, sweepback, control movements);
4. fin and rudder [gross area, height (span), quarter chord position, aspect ratio, aerofoil, sweepback at quarter chord, rudder movements];
5. control surfaces, if not specified in 2, 3, 4 above (aileron area aft of hinge, total aileron area, aileron span, aileron movements, similar parameters for elevator of rudder);
6. fuselage (external) (overall length, maximum width, maximum depth, door height above ground);
7. fuselage (internal) (flight deck geometry, passenger cabin length, cross-sectional shape, typical seating arrangements including seat pitch, aisle width, aisle height, baggage hold volumes, overhead stowage volume, doors and window geometry);
8. landing gear (track, wheelbase, tyre sizes, bogie descriptions);
9. power plant (engine description, overall length, width, height, thrust line offset to fuselage datum);
10. fuel tanks (description, volumes and capacities).

The geometry section will include a three-view GA of the aircraft suitably dimensioned together with appropriate component geometry drawings (e.g. passenger cabin section and plan, etc.).

4. Aerodynamic and structural criteria

In this section all the parameters that affect the aerodynamics and structural analysis are listed. The following are typical sub-sections:

1. flight profiles
2. flight envelopes
3. gust envelopes (fatigue spectrum)
4. landing factors (vertical velocity of descent, runway load classification number)
5. operating speeds (stall speeds, flap speeds)
6. flap settings
7. design weights (take-off mass, maximum landing mass, zero fuel mass, empty mass, payload)
8. floor loadings
9. pressurisation profiles
10. towing
11. jacking and ground equipment provision

The data and information in the above list may be illustrated by graphs, charts and drawings showing the detail criteria.

5. Weight and balance

A detailed weight statement of the aircraft showing the component weight (similar to the weight summary described in Chapter 7) will be included in this section. Specification of the weights for payload, operational items and fixed equipment will be given. Centre of gravity limits for all operational conditions will be shown (usually by a weight–balance diagram). Weight guarantees will be quoted.

6. Performance

Documentation for the aircraft will include a Performance Manual in which the aircraft performance will be defined in detail. This will contain aircraft aerodynamic data [drag polars at low speed and various flap settings, high speed (cruise) drag polars, lift curves, asymmetric drags and thrust (power) required curves]; engine data (at various air speeds, altitude, temperatures, thrust ratings, fuel usage); field performance [take-off and landing distances at various aircraft weights, airfield altitude and temperatures, flap settings, and poor weather (slush) performance]; and manoeuvring performance (climb, descent, cross-wind performance, etc.).

Some of the data will be 'certified' (meaning that it has been tested during the flying for the Certificate of Airworthiness) and some of the data will be used for flight operations and planning purposes. The Type Specification manual will contain a summary of the performance manual. Specific guarantees on aircraft performance form part of the contractual documentation between the airframe and engine manufacturers and the operator. The data included in this section will include a general description of the performance methods (i.e. a specification for contractual purposes), take-off and landing charts (at various aircraft weights and ISA conditions), payload–range charts (with assumptions made for non-cruise segments, e.g. ground running), ferry range, and speed definitions.

7. Airframe

This section provides an introduction to the airframe design and construction details. Statements will be made on the general design philosophy, materials used, corrosion protection, fatigue philosophy, damage tolerance and fabrication, maintainability philosophy. Each of the major aircraft structural framework components will be described. The list may contain:

1. wing general description
2. centre section
3. outer panels
4. flaps and ailerons
5. other wing features (e.g. lift dumpers, tip design)
6. fuselage general description
7. nose section (including cockpit)
8. forward fuselage
9. centre fuselage (including wing joint)
10. aft fuselage
11. tail section (including empennage attachment)
12. empennage general description
13. horizontal stabiliser and elevator
14. vertical stabiliser and rudder
15. engine installation
16. nacelles and pylons

Each of these descriptions will be illustrated by detail drawings of the component (basic geometry and structural members) showing the layout and specifying the structural materials.

8. Landing gear

Following a general description of the undercarriage each component will be described. The list may include:

1. main landing gear
2. nose landing gear
3. landing gear doors
4. retraction and deployment (locks)
5. brakes
6. nose wheel steering

Each of the above descriptions will be illustrated by drawings showing the design and operation of the undercarriage and its systems.

9. Power plant (and systems)

For each of the engine options available, the engine, external drives and accessories, engine controls, engine systems, and engine mounting are described:

1. general description
2. engine description
3. engine mountings
4. engine controls
5. engine systems (starting, fuel, hydraulic, pneumatic, electric)
6. fire protection system
7. engine lubrication, drains and vents

Various drawings will accompany these descriptions showing pipework and block diagrams of the system.

10. Fuel system

The description of the fuel system will include the tankage, engine equipment, quantity measurement, refuelling and dumping, ventilation, controls and instrumentation. Several diagrams will be used to illustrate these descriptions.

11. Hydraulic systems (see below)

12. Electrical systems (see below)

13. Avionics (see below)

14. Instruments and communications

Sections 11, 12, 13, 14 follow a similar pattern of presentation to that of the fuel system. After a general description of the system, each of the main component parts is described. Block diagrams and drawings are used to amplify the descriptions.

15. Flight controls and flight deck

The aircraft flight control system is becoming a significant feature of new designs as more computer integration is incorporated. The general design philosophy will be presented followed by detail descriptions of the ailerons, elevator and rudder, primary flight controls and trim systems. The flight control section may also include a description of flap/airbrakes/lift dumper operations. As appropriate, diagrams and drawings will be used to illustrate the various system descriptions. A description of the crew station (cockpit) will give details of the provision of controls, instrumentation, services and seating in the area. The pilot's view from the seating position will be shown (by diagram) and compared to design requirements.

16. Passenger cabin layouts

The fuselage cabin section accommodates passengers and freight holds. This section describes the facilities under the following headings:

1. passenger cabin
2. freight holds
3. access/escape/emergency systems
4. cabin crew equipment
5. services

Several drawings and diagrams illustrate these topics.

The passenger cabin layout will be described to show the fixed services (galleys, toilets, overhead storage, seat rails and windows). Interior trim and furnishings will be described but these will be the subject of discussions with specific operators. The layout and volumetric sizes of the baggage/freight compartments will be shown together with the loading doors.

Access (doors) to the various areas and services will be shown and escape routes (emergency hatches, etc.) described. Other emergency equipment (aisle lighting, oxygen, escape-chutes, etc.) will be described. All cabin crew equipment (stewards station, galley, kit storage, reserve crew station, etc.) will be described.

Finally a section will be devoted to the service facilities and aircraft turn-round. Location of ground service vehicles around the aircraft will be shown (by diagram).

17. Environmental control systems

This section includes all the systems associated with fuselage pressurisation, auxiliary power unit (APU), ventilation, air conditioning and ice protection (e.g. wing leading edge, windscreen, engine intake, etc.).

18. Safety systems and emergency exits

These systems will include:

1. fire detection and extinguishing systems
2. oxygen
3. escape
4. bird strike
5. electrical (e.g. emergency lighting)
6. warning devices and instrumentation

Descriptions of these systems with associated block diagrams will be presented.

19. Servicing

All aspects of aircraft ground servicing, maintenance and reliability will be described under this heading:

1. aircraft turn-round
2. towing and steering
3. locks and ground restraint
4. engine change
5. auxiliary power unit (APU)
6. jacking, hoists and slings
7. ground support requirements
8. test equipment and standard couplings
9. maintenance procedures
10. reliability

20. Airport compatibility

For civil aircraft it is essential to provide data on the aircraft characteristics with regard to airport facilities. A separate manual 'Airplane Characteristics – Airport Planning' will be produced by the manufacturers to fully detail the aircraft capabilities. For the Type Specification all that is needed is an introduction to the following topics:

1. aircraft geometry
   - clearances
   - door sizes and locations
   - cargo compartments
   - sensors
2. ground manoeuvring
   - turning radii
   - cockpit visibility
   - runway paths
3. terminal services
   - turn-round
   - service connections and requirements
   - towing
4. engine noise
5. wing vortex data
6. pavement data
   - landing gear geometry
   - landing gear loadings
   - pavement requirements (load classification number, Federal Aviation Administration methods)

21. Exceptions to regulations

This is a straightforward section listing the contractual issues on which the aircraft is designed and providing the legal framework forming the contract of sale. Appendices will list equipment and suppliers.

Illustrations

The Type Specification contains several engineering drawings, schematic diagrams, system block diagrams, graphs, charts and diagrams. The list below is not exclusive but provides a guide to the type of supporting illustrations to the text of particular sections.

Aircraft three-view GA
Internal plan view (e.g. cabin arrangement)
Flight envelopes
Fatigue spectrum
Undercarriage utilisation spectrum
Weight and centre of gravity diagram
Floor loading
Cockpit view diagram
Centre section fuselage joint
Tailplane structure
Main undercarriage GA (including installation)
Nosewheel steering
Engine power take-offs
Starter installation
Electrical system
Antenna locations
Hydraulic system
Flying control systems
Cabin pressurisation schedule
Central warning panel
Ejection seat installation
Communication system
Access panels
Visibility from flight deck

Runway paths
Internal side view (fuselage package)
Aircraft geometry
Mission profiles
Undercarriage vertical velocity spectrum
Runway loadings
Fuselage structural framework
Fuselage cross-section
Wing structural framework
Flap details
Fin structure
Nose undercarriage GA (including installation)
Engine installation
Engine control system
Fuel system and tankage
External lighting
Avionics
Pneumatic systems
Environmental control system
Instrumentation
Oxygen system
Ground escape system
APU arrangement
Ground service
Turning radii

Conclusion

Clearly, the detail contained in the above descriptions refers to the aircraft data at the end of the project design phase and just prior to release of the design into the detail design phase. Prior to this time the level of detail will be appropriate to the degree of analysis conducted and the time available to refine the configuration. Knowledge of the requirements of the Type Specification document serves as a goal throughout the project design. The document should be started early in the project analysis phase and continuously updated as aircraft detail design unfolds. In a large organisation in which several separate departments are working on the project, the Type Specification acts as a common description of the aircraft. In such circumstances changes to the document must be strictly controlled by an Amendment Committee (or Management Review Body). The significance and thereby the status of the Type Record cannot be overestimated as it forms the legal and organisational framework for the design of the aircraft.

15

Introduction to spreadsheet methods

Objectives

Aircraft design is an iterative process, and therefore very labour intensive. The designer is constantly striving to meet all the design specifications and airworthiness requirements, whilst offering the most cost-effective aircraft to the customer. In the late 1940s, a number of engineers began to address this problem by attempting to automate the design process, using a mixture of mathematical and semi-empirical relationships. However, the calculations still needed to be carried out by hand. With the introduction of computers, designers realised that many of the labour intensive tasks could be achieved using computer programs but early programming languages were very specialised and difficult to use. More recently, the power of desktop computers has enabled calculations to be performed using spreadsheet programs. This section describes the use of spreadsheet programs for student projects in aircraft design. This chapter is followed by several case studies using spreadsheet methods.

Introduction

Computer programs used for aircraft design work have traditionally only been available for use on mainframe type computers. In general they have been written in high-level, procedural languages, such as ADA and FORTRAN. This is fine for industrial usage, but such systems require the user to be fluent in the appropriate computer language and know about the architecture of the program.

The power and flexibility of modern desktop computers has increased dramatically in the past few years. However, there is still a gap between the predicted benefits of utilising computers and the actual benefits derived. If a computer program or system is difficult to use, then prospective users will resort to their traditional design methods, i.e. doing calculations by hand or batching work to large computers.

To exploit the capability of personal computers to their full extent, a more user-orientated approach is required when creating software. This must involve higher learning rates, low error rates and increased usability of output data. What is required is not a traditional high-level (procedural) program, but a program which has many of the required functions within its own structure and only requires a user to input the desired analytical method. Spreadsheets[1] appear to fit this description with their ability to carry out all but the most complex functions required.

This chapter details the methodology used to produce a spreadsheet suitable for student project design work. The prime requirements for such programs are a modular design which will allow a variety of point estimates to be made and the ability to perform parametric studies. The analysis modules necessary for initial project design work are described and the analytical methods used are briefly described. In the following chapters spreadsheet programs are then applied to several example aircraft projects and the results are described.

Spreadsheet layout

The layout chosen for an aircraft design spreadsheet will determine the overall usability. Considering the need for modularity and the ability to perform parametric studies the layout shown in Table 15.1 is recommended.

Column A contains a descriptive label of the numerical value, either input or calculated. Column B contains the variable name for the descriptive label. This will either be a standard symbol such as S for gross wing area or an abbreviation of the descriptive label. Finally, column C contains either numerical data (input) or a formula to estimate the value from other data in the spreadsheet.

Using this layout, column C may be copied and pasted several times to the right of itself. Each of the pasted columns is a separate aircraft design and key input parameter, such as wing area, may be varied in each column. Thus a parametric study has been set up with, for example, wing area as an input parameter. The spreadsheet will then calculate the output values for each column and key output variables, for example maximum take-off weight, may be plotted against wing area.

Alternatively, for large spreadsheets, models may be generated to alter specific parameters in a specific manner. As a particular parameter is varied, the required output parameters may be passed to another worksheet or workbook for use later.

Table 15.1 Spreadsheet layout

Column A	Column B	Column C
Descriptive Label	Variable Name	Numerical Value/Formula

Modularity

It is important that a modular layout is developed. This will make the spreadsheet simpler to use, and allow modules to be easily changed for different design studies. Typical modules required for initial project design are:

1. input data
2. atmospheric conditions
3. geometric calculations
4. mass model
5. cruise aerodynamic calculations
6. take-off performance
7. landing performance
8. second segment climb performance
9. centre of gravity model
10. cost model
11. direct operating cost (DOC) model

The above list gives an indication of the modules required to successfully produce a satisfactory model from the initial specification. Each of the modules will be discussed in turn.

1. Input data

Input data consists of the basic information used to define the shape, required performance and capabilities of the design. Much of this information is generated as the initial estimates phase, whilst other parameters will be used directly from the design specification, e.g. payload, range, cruise altitude, etc. The data is used by the analytical modules to determine the weight, performance, cost, etc. of a design. Clearly the more detailed the analytical methods used, the more detailed the input data must be. A balance has to be struck between the time available to obtain the required input data and the accuracy of the data output. In most practical cases between 100 and 150 input values are required.

2. Atmospheric conditions

Central to any aircraft performance analysis is the need to know the atmospheric conditions at a particular altitude. The model will be based around the international standard atmosphere (ISA) characteristics for the troposphere. However, many transport aircraft now fly above the tropopause (i.e. higher than 11 km) and thus, in general, it is also necessary to model the characteristics of lower stratosphere.

While ISA is normally adopted, the ability to increase/decrease the temperature (Δtemp) to suit non-standard conditions is desired. The module must be used at least twice [for the airfield conditions (altitude, Δtemp) and

	A	B
1	**Input data:**	
2	**Take-off:**	
3	Altitude (m)	0
4	Δtemp (K)	0
5	**Cruise:**	
6	Initial cruise altitude (m)	1000
7	Δtemp (K)	0
8	**Constants:**	
9	S/L ISA temperature (K)	288.15
10	Pressure at S/L	101 325
11	Viscosity at 273.15 K (kgms^{-1})	0.00001714
12	Lapse rate (K/m)	0.0065
13	Pressure at 11 000 m (Pa)	=((B9-B12*11000)/B9)^(9.81/(287*B12))*B10
14	Density at 11 000 m (kgm^{-3})	=B13/(287*(B9-B12*11000))
15	**Calculations:**	
16	**Take-off:**	
17	Temperature (K)	=(288.15 + B4)-(B9*B3)
18	Pressure (Pa)	=(B17/B9)^(9.81/(287*B12))*B10
19	Density at altitude (kgm^{-3})	=B18/(287*B17)
20	Absolute viscosity (kgms^{-1})	=(B17/273.15)^(3/4)*B11
21	Kinematic viscosity (m^{4}s^{-1})	=B20/B19
22	Speed of sound (ms^{-1})	=SQRT(1.4*287.3*B17)
23	**Cruise:**	
24	Temperature (K)	=IF(B6>11000,(B9-B12*11000),(B9-B12*B6)) + B7
25	Pressure (Pa)	=IF(B6>11000,EXP(9.81/(287*B24)*(11000-B6))*B13,(B24/B9)^(9.81/(287*L))*P0)
26	Density (kgms^{-3})	=IF(B6>11000,EXP(9.81/(287*B24)*(11000-B6))*B14,(B25/(287*B24)))
27	Absolute viscosity (kgms^{-1})	=(B24/273.15)^(3/4)*B11
28	Kinematic viscosity (m^{4}s^{-1})	=(B27/B26)
29	Speed of sound (ms^{-1})	=SQRT(1.4*287*B24)

Fig. 15.1 Atmospheric conditions module.

also the cruise condition (altitude, Δtemp)]. Other mission segments may require further atmospheric conditions to be determined. A spreadsheet module suitable for calculations in both the troposphere and stratosphere is shown in Fig. 15.1. Note, in this example variable names have not been used as the module is reasonably compact.

3. Geometric calculations

The geometric calculations module determines the detailed geometry from the basic data input. These calculations include determination of wetted areas, mean aerodynamic chords, etc. Methodology described earlier in the book [Figs 7.19 and 8.2(b)] may be used, together with standard geometrical methods. Output may be checked directly using the built-in graphing capabilities of spreadsheet packages. Figure 15.2 shows the determination of equivalent wing planform for a cranked trailing edge wing, along with the mean aerodynamic chord (MAC) for both the actual and equivalent wing planforms.

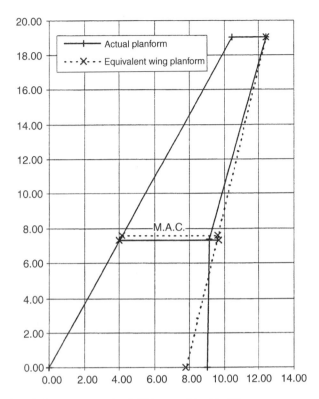

Fig. 15.2 Presentation of wing geometry using built-in spreadsheet facilities.

4. Mass module

The mass module determines the overall mass breakdown for the aircraft. This module is generally iterative as many specific mass items are dependent on the maximum take-off mass of the aircraft. An initial estimate of the maximum take-off mass is provided and the user can iterate this value manually to arrive at a final answer. Convergence will normally be completed after approximately ten attempts. Alternatively, the solver facilities available in many modern spreadsheet programs may be used to rapidly determine the maximum take-off mass. For particularly complex spreadsheet models the built-in solvers are sometimes unable to determine a solution. In such cases, macros may be generated to automate the iteration. An example macro for this task using Visual Basic (Microsoft Excel) is shown below:

```
Sub Macro1()
Sheets("Sheet1").Select
10   If (Abs(Range("C12"))>0.01) Then
        Range("C11").Select
        Selection.Copy
        Range("C10").Select
        Selection.PasteSpecial Paste:=xlValues, _
            Operation:=xlNone, _
```

```
        SkipBlanks: = False, Transpose: = False
    GoTo 10
  End If
End Sub
```

The macro assumes that the worksheet is called 'Sheet1' and that the computed mass is in cell $C11$. The initial estimate is defined in cell $C10$. The difference between the two values is computed in cell $C12$. If the absolute difference between the initial estimate and the computed mass is greater than 0.01 then the initial estimate is replaced with the computed value. Although very crude it converges rapidly to a solution. Sheet names and cell references can easily be altered to suit the particular layout of your model.

5. Cruise aerodynamic module

The primary output from the cruise module is the aircraft cruise lift/drag ratio. This is used to compute the fuel required for the mission under analysis and hence determine the maximum take-off mass for the mission. To do this, the module computes the drag characteristics for the cruise flight condition using geometrical information. In it simplest form, only a single cruise altitude will be analysed. However, more advanced models may include stepped profiles with a number of cruise altitudes to more closely reflect operating practice. It is convenient to check the engine power setting at this stage to make sure that the installed thrust is sufficient to meet the initial cruise requirements (normally 300 ft/min climb rate).

To enable a drag prediction to be made input data for this module include the cruise conditions, geometric calculations and initial mass estimate. Engine data is also required to convert the thrust required at the initial cruise altitude to a static sea-level thrust.

6. Take-off module

The take-off performance module determines the take-off field performance of the aircraft. As well as computing the standard take-off field length it is often necessary to compute the balanced field length. Here, the distance is computed for both the 'accelerate–go' and the 'accelerate–stop' phases. For the accelerate–stop phase, braking is handled in a similar manner to the landing phase (see Chapter 10). The balanced field length is then determined by altering the decision speed, V_1, until both the 'accelerate–go' and the 'accelerate–stop' distances are equal. Again this iteration may be handled within spreadsheets using the built-in solver or specialist macros.

Geometric data, aerodynamic characteristics and take-off mass are used to determine aircraft drag during the take-off run. Installed engine thrust is then used to determine the acceleration during the take-off run. If engine data is available to determine the thrust decay with aircraft speed the accuracy of the results is improved.

7. Landing module

The landing performance module determines the landing field performance of the aircraft. As with take-off, aircraft drag is calculated using mass estimates and geometric data. Depending upon the accuracy required, the landing phase may be broken into a small number of segments or an incremental procedure used. The latter allows the effect of brake application, reverse thrust, speedbrakes and lift dumpers to be analysed but at considerable increase in computing time and complexity.

8. Second segment climb module

The second-segment climb performance module determines the climb gradient of the aircraft with one engine inoperative (OEI). As discussed in Chapter 10, a civil transport aircraft must meet specific climb gradients depending on the number of engines (as defined in the airworthiness requirements). The drag model in this case must take proper account of the increased drag from the wind-milling effect of the inoperative engine and also the increased drag due to control deflections associated with asymmetric powered flight.

9. Centre of gravity module

The centre of gravity module is essential for determination of the correct location of the wing to balance the aircraft. Moments are taken about the nose for each component using the masses calculated in the mass module. The centre of gravity is then calculated by summing the moments and masses to determine the aircraft centre of gravity distance from the nose. This may then be expressed as a percentage of the wing mean aerodynamic chord (MAC). If desired this module can be expanded to consider variations in passenger loading together with the centre of gravity of the aircraft in the operating empty condition.

10. Aircraft cost and DOC module

The aircraft cost module determines the 'price' the airline will need to buy the aircraft. This may be estimated using a simple cost per kilogram relationship or by a detailed model using formulae to predict development and production costs. The airframe and engine size are the primary input variables for the aircraft cost module.

The direct operating cost module determines the operating costs as perceived by the airline. This is often a useful parameter for comparing designs. It may also be used to optimise a given design. Input variables include the fuel used, crew salaries, maintenance charges and the purchase cost of the aircraft.

Summary

Together with second-segment climb performance, the field and cruise performance modules form the specific flight cases that 'size' the engine for a given design. An optimum design will require the same 'size' engine to meet all three requirements. In practice this may not be possible as often one of the three requirements may be dominant and therefore set the critical requirement. Twin-engined aircraft are often found to be second-segment climb critical, while four-engined aircraft are likely to be climb critical (i.e. top of climb performance). The field case may be critical for any aircraft if the allowable runway is short.

Spreadsheet format

Unfortunately the number of modules described produces quite a large spreadsheet. If the all the modules are arranged on a single worksheet using the layout suggested the spreadsheet will be very long (>2000 rows). It soon becomes tedious scrolling up and down past each module. However, modern spreadsheets are three-dimensional in nature. A traditional spreadsheet contains a series of rows and columns. Three-dimensional spreadsheets continue this tradition but contain pages within a single file or workbook. Data may be passed between the different worksheets just as it can be passed between cells on the same page. It is then sensible to place individual modules on separate sheets. This provides a strong structure to a spreadsheet model and reduces the length of each page resulting in less scrolling and better usability. All sheets except the input data sheet may be locked to prevent changes by other users. This is a useful feature for student project work in a group environment where one member may be responsible for specific model development.

Data handling

Whilst many of the aircraft design methods are analytical in nature and thus relatively straightforward to enter into the spreadsheet model, experimental data, such as engine performance data, are sometimes more difficult to enter. Where the data are one-dimensional (i.e. dependent on one variable) curve fit and regression techniques may be used to generate functions for the data that can then be entered directly into the spreadsheet. Care must be taken that such functions match the original data throughout the range of parameters anticipated, whilst being numerically efficient, i.e. high-order polynomials should be avoided. If at certain points the functions are no longer valid, appropriate error handling must be included to limit input parameter variation or else additional functions must be included to control certain parameters for extreme ranges of input values.

In many cases data are two-dimensional, i.e. a parameter is dependent on two variables. Linear interpolation may be used to determine values from these data. Care must be taken with linear interpolation as this may upset some types of optimisation routines due to the sudden change of gradient between successive cells of data. Alternatively, polynomial interpolation, used frequently for procedural programs, may be applied to spreadsheet methods. A user-defined function is used to read in the tabular data from a specified worksheet and then determine the required value for the input parameters provided. A method adapted from Fortran[2] and suitable for use with Microsoft Excel is shown below:

```
Function Polin2(x1, x2, Sheet, Offset, N_columns, N_rows)
Dim m, n, NMAX, MMAX As Integer
Dim dy, y, x1a(10), x2a(10), ns, w, den, ya(10, 10), c(10),
d(10) As Double
Dim j, k, Length, Column_offset, Row_offset As Integer
Dim ymtmp(10), yntmp(10) As Double
Dim X_Location, Y_Location, Location As String
    ' Determine Off-set
Length = Len(Offset)
Column_offset = Asc(Left(Offset, 1))
Row_offset = Right(Offset, (Length - 1))
    ' Set-up x1a(m)
For i = 1 To N_columns
    X_Location = Chr(Column_offset + i) + CStr(Row_offset)
     x1a(i) = Sheets(Sheet).Range(X_Location).Value
Next i
    ' Set-up x2a(n)
For i = 1 To N_rows
    Y_Location = Chr(Column_offset) + CStr(i + Row_offset)
    x2a(i) = Sheets(Sheet).Range(Y_Location).Value
    Next i
    ' Set-up ya(N_columns, N_rows)
For i = 1 To N_columns
    For j = 1 To N_rows
        Location = Chr(Column_offset + i) + CStr(j +
Row_offset)
        ya(i, j) = Sheets(Sheet).Range(Location)
    Next j
Next i
For j = 1 To N_columns
    For k = 1 To N_rows
        yntmp(k) = ya(j, k)
    Next k
    ns = 1
    dif = Abs(x2 - x2a(1))
    For i = 1 To N_rows
        dift = Abs(x2 - x2a(i))
        If (dift) < (dif) Then
        ns = i
        dif = dift
    End If
```

```
            c(i) = yntmp(i)
            d(i) = yntmp(i)
            Next i
              ymtmp(j) = yntmp(ns)
            ns = ns - 1
            For m = 1 To (N_rows - 1)
                   For i = 1 To (N_rows - m)
                          ho = x2a(i) - x2
                          hp = x2a(i+m) - x2
                          w = c(i+1) - d(i)
                          den = ho - hp
                        If (den) = 0 Then
                                   End
                        End If
                        den = w/den
                        d(i) = hp * den
                        c(i) = ho * den
                   Next i
                   If (2 * ns) < (N_rows - m) Then
                          dy = c(ns+1)
                   Else
                          dy = d(ns)
                          ns = ns - 1
                   End If
                   ymtmp(j) = ymtmp(j) + dy
            Next m
      Next j
      ns = 1
      dif = Abs(x1 - x1a(1))
      For i = 1 To N_columns
            dift = Abs(x1 - x1a(i))
              If (dift) < (dif) Then
                    ns = i
                    dif = dift
            End If
              c(i) = ymtmp(i)
            d(i) = ymtmp(i)
      Next i
      y = ymtmp(ns)
      ns = ns - 1
      For m = 1 To (N_columns - 1)
            For i = 1 To (N_columns - m)
                   ho = x1a(i) - x1
                   hp = x1a(i+m) - x1
                   w = c(i+1) - d(i)
                   den = ho - hp
                   If (den) = 0 Then
```

```
            End
        End If
        den = w/den
            d(i) = hp * den
            c(i) = ho * den
    Next i
    If (2 * ns) < (N_columns - m) Then
            dy = c(ns + 1)
    Else
            dy = d(ns)
            ns = ns - 1
    End If
    y = y + dy
Next m
Polin2 = y
End Function
```

Example

The function 'Polin2' requires the table location, number of rows, columns, and the two input variables to be provided, as shown in Fig. 15.3.

In this the data represent the maximum climb thrust available as a fraction of maximum static thrust for a given altitude and Mach number. Since the formula that calls the function includes the sheet name (in this case called 'Engine Data') the function may be applied any number of times to different tables on different worksheets. Thus look-up tables for engine data and, say, aerodynamics may be separated from the analytical modules.

	A	B	C	D	E	F
1				Mach No.		
2	Alt (ft)	0	0.2	0.4	0.6	0.8
3	0	0.854	0.700	0.592	0.513	0.45
4	10000	0.688	0.579	0.496	0.446	0.400
5	15000	0.608	0.517	0.446	0.401	0.371
6	25000	0.450	0.388	0.346	0.325	0.308
7	30000	0.383	0.333	0.300	0.283	0.275
8	35000	0.325	0.283	0.258	0.246	0.242
9	40000	0.254	0.221	0.203	0.192	0.192
10						
11						
12	Estimate	0.25				
	0	= Polin2(B12,A13,"Engine","A2"				
		,5,7)				

Fig. 15.3 Function Polin2.

Conclusions

The chapter has discussed how spreadsheets may be used to analyse aircraft designs using all the methods developed previously for procedural programs. The use of spreadsheets enables aircraft models to be developed more rapidly, with less difficulty and yet convey the unique problems associated with aircraft design, performance and optimisation to engineering students.

References

1. Jenkinson, L. R. and Rhodes, D. P. (1995). 'Application of Spreadsheet Analysis Programs to University Projects in Aircraft Design', 1st AIAA Aircraft Engineering, Technology and Operations Congress, Los Angeles, CA., September 19–21, 1995.
2. Press, W. H., Teukolsky, S. A., Vetterling, W. T. and Flannery, B. P. (1992). 'Numerical recipes in FORTRAN', Second edition, Cambridge Press, 1992.

16

Advanced regional jet

Summary

This study was undertaken using spreadsheets based on the design methods presented in the first part of the book. The initial specification called for an advanced regional airliner with a capacity of 70–100 passengers. The design brief emphasised the importance of fuel cost to the design as it was felt that in future a 'carbon tax' may be levied on aviation fuel. Initial research identified that many of the current 100-seat regional airliners are simply 'down-sized' variants of larger aircraft. By basing the proposed family around a basic capacity of 70 passengers, it was felt that a more efficient design may be produced.

The introduction to the project briefly describes how the full specification was defined using data on current regional aircraft and current airline operating practices. Initial estimates of the size and shape of the aircraft are then made in order to determine the necessary inputs to the spreadsheet model. The model is then used to optimise the baseline design. The minimising objective function was the aircraft direct operating cost (DOC). The chapter concludes with data for the optimised design, together with a three-view general arrangement drawing.

Introduction

Many 70–100-seat regional aircraft are now more than 20 years old providing a strong market for the introduction of a new regional aircraft. Advances in technology should provide improvements over types such as the Douglas DC9 and Boeing 737–200. It is interesting to note that many aircraft in this category have been 'down-sized' from larger models. Such down-sized aircraft (e.g. Fokker 70, Boeing 737–500/600) are not necessarily efficient designs. Although the Fokker 100 was a new design in 1988, it was in essence a development of the Fokker F.28 and production has now ceased with the demise of Fokker. The Avro RJ series, formally known as the British Aerospace BAe 146, has sold slowly, partly due to its four-engined configuration and perceived increased complexity.

Table 16.1 Current 70–100 seat regional aircraft

Aircraft type	Nominal capacity (seats)	Design range (nm)
Avro RJ 70	70	1952
Avro RJ 85	85	1782
Avro RJ100	100	1671
Avro RJ115	100	1794
Boeing 737–200	115	1900
Boeing 737–500	108	1700
Fokker 70	70	1080
Fokker 100	107	1290
McD DC9–10	85	1311
McD DC9–30	105	1725

More recently, McDonnell–Douglas launched the MD95, a re-engined DC9–30 with a new cockpit and updated systems. This aircraft, now re-branded the Boeing 717, still utilises the DC9–30 wing and so does not employ the latest advances in technology introduced over the past decades.

In view of existing aircraft in this market segment it was decided that the baseline model should have a capacity for 70 passengers with the potential to produce a family of larger aircraft based around the initial model. Emphasis has been placed on fuel efficiency from the outset as there is a risk of fuel price increases in the future via environmental taxes (referred to as a 'carbon tax').

From the initial brief given for the project, the aircraft capacity is fixed at 70 passengers. This leaves several key parameters that must be determined in order to complete the specification of the aircraft. These are:

- design range
- cruise speed
- field length
- initial cruise altitude

The design range of the aircraft will affect the total weight and cost of the aircraft as well as the operational flexibility for the airline operator. There are several methods by which the design range may be chosen. In the first instance, a direct comparison with current regional aircraft can be made. Table 16.1 provides the basic characteristics of several regional jet transport aircraft (taken from Data A).

The data shows that the design range increases with aircraft size. However, these data give no indication of how these aircraft are currently operated. Current operating practices may be analysed by investigating current airline operating schedules. The routes flown by current regional aircraft can be identified and the ranges flown determined by estimating the great circle distance between airport pairs. This is often a lengthy exercise. Example stage lengths obtained using this approach are shown in Table 16.2.

For this study the number of routes and the distances flown by regional aircraft have been analysed from European and US airline schedules and Data C. For a given aircraft type it was found that approximately 90% of these routes were less

Table 16.2 Typical stage lengths flown by regional aircraft

Route	Aircraft type	Nominal stage length (nm)
LGW-EDI	Avro RJ	309
LHR-BSL	Avro RJ	389
FRA-BUH	BAC 1–11	780
PHL-SYR	Douglas DC9	198
YYZ-YWG	Douglas DC9	806
LHR-LYS	Fokker F.28	406
CVG-PIT	Fokker 100	222
LGW-GLW	Fokker 100	321
MPH-CLT	Fokker 100	443
STL-BWI	Fokker 100	639
ORY-STO	Fokker 100	849

than half that of its design range. It is likely, however, that many operators choose to fly two sectors or more before refuelling. This will tend to reduce turn-round times and enable aircraft to be re-fuelled at the airline's main base.

After analysing the data, a design range of 1250 nm was decided for the initial aircraft model. Later, derivatives may be developed with longer ranges. Cruise speed for the design was determined by considering existing regional aircraft. Maximum operating speed (V_{mo}) (EAS) and Mach number are listed in Table 16.3 (taken from Data A). The normal operating Mach number is approximately 0.05 below the maximum value (Mn_{mo}). Thus from the table cruise Mach numbers are seen to be between M0.68 and M0.79.

Over the relatively short stages flown, the higher speeds offer little advantage in time saving, yet increase the thrust required and hence the total fuel burnt. Thus it was decided that a cruise speed of M0.70 be adopted. This relatively slow speed has the advantage of reducing the required wing sweep required which consequently will improve the take-off/landing performance of the aircraft. High-speed cruise altitudes are also listed in Table 16.3 (taken from Data A). The cruise altitude varies between 25 000 and 29 000 ft. Higher cruise altitudes are more desirable to improve lift/drag ratio and reduce engine specific fuel consumption. The difficulty with higher altitudes for short-haul flights is the time taken to reach

Table 16.3 70–100 seat regional aircraft cruise performance

Aircraft type	Cruise altitude	V_{mo} (EAS kt)/Mn_{mo}
Avro RJ 70	29 000	300/M0.73
Avro RJ 85	29 000	300/M0.73
Avro RJ100	29 000	305/M0.73
Avro RJ115	29 000	305/M0.73
Boeing 737–200	25 000	350/M0.84
Boeing 737–500	26 000	340/M0.82
Fokker 70	26 000	320/M0.77
Fokker 100	26 000	320/M0.77
McD DC9–10	25 000	350/M0.84
McD DC9–30	26 000	350/M0.84

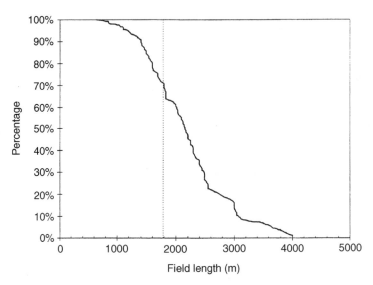

Fig. 16.1 Distribution of field lengths at major European airports.

the cruise altitude. If this time could be reduced, higher cruise altitudes may be more feasible. On this basis an initial cruise altitude of 35 000 ft was set with a time to climb to cruise altitude of less than 24 minutes.

The field length is an important requirement for the design. If the field length requirement is too short the design will incur significant cost penalties. Too great a length, however, and the aircraft will be unable to operate from certain airfields reducing its market potential. Using field length information from Data C, cumulative percentage vs field length was plotted in Fig. 16.1 for UK, French and German airports. The 50th percentile corresponds to a field length of around 2100 m (6890 ft). Thus designing for this field would provide access to only half of the airports. Reducing the field length to 1800 m (5900 ft) increases the available airports to 70%. Reducing the field length further produces a small gain in the number of airports but at an increase in aircraft cost.

This approach is quite broad-based and thus it is useful to narrow down the number of airports to those identified during the stage length exercise. Comparison should also be made with field lengths required by existing aircraft which are shown in Table 16.4.

In this case analysis of current regional aircraft field lengths suggests that a relatively relaxed field length of 1800 m (5900 ft) might provide economic benefits without unduly penalising the operational flexibility of the design.

The complete specification can now be summarised:

- capacity 70 passengers
- design range 1250 nm
- cruise speed M0.70
- field length 1800 m
- initial cruise altitude 35 000 ft
- time to climb to cruise altitude < 25 minutes

Table 16.4 Regional aircraft field performance

Aircraft type	Take off field length (m)	Landing field length (m)
Avro RJ 70	1440	1170
Avro RJ 85	1646	1192
Avro RJ100	1829	1265
Avro RJ115	1829	1265
Boeing 737–200	1829	1350
Boeing 737–500	1832	1360
Fokker 70	1296	1210
Fokker 100	1856	1321
McD DC9–10	1615	1411
McD DC9–30	1777	1317

Now that the aircraft specification has been completed, the conceptual design phase can begin. This may be broken into three areas:

1. initial estimates
2. geometric sizing
3. optimisation of baseline design

Analysis of existing aircraft

Before the new design can be sized it is useful to analyse existing aircraft of a similar specification to determine key parameters such as operating empty mass (MOE) ratio and lift/drag ratio. These parameters are then used together with the payload/range specification to determine an estimate for the aircraft maximum takeoff mass ($MTOM$). From this information the wing and tail surfaces may be sized. The performance estimates which lead to the definition of powerplant size can then be made.

A key indicator of a design's structural efficiency is the empty weight ratio (i.e. the operating empty weight divided by the maximum take-off weight). In general, the empty weight ratio decreases with increasing take-off weight, mainly due to the fact that larger aircraft fly over greater ranges and hence have a higher fuel weight Operating empty weight ratios are listed for civil jet transport aircraft in Data A. Inspection of the data shows that the values vary from just over 0.60 for short-haul aircraft to around 0.45 for large, long-range aircraft. Values for the empty weight ratio for the regional aircraft mentioned previously are plotted in Fig. 16.2. Lines are shown connecting developed or stretched versions of the basic aircraft. It is clear from the data that the basic model has the highest operating empty weight ratio and that this reduces as the aircraft is stretched. Conversely 'shrinking' (e.g. Fokker 70) increases empty weight ratio.

This effect is important when considering the stretch potential of the aircraft. To increase commonality it is desirable to maintain the same wing and tail surfaces across the family. As a result of this philosophy the initial model is penalised with higher than necessary empty weight ratio.

Lift/drag ratio information is more difficult to estimate at this stage and is not

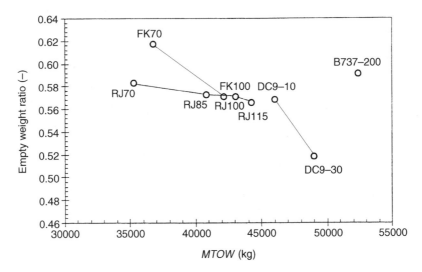

Fig. 16.2 Empty weight ratios of current regional aircraft.

widely reported as manufacturers regard this data as commercially sensitive. It is, however, possible to make crude estimates of lift/drag ratio using published data and the Breguet Range equation. Normally this equation is used to determine the range flown or fuel mass required for new designs. For existing aircraft information on cruise speed, fuel mass, payload and engine specific fuel consumption (SFC) are available and this enables lift/drag ratio to be estimated.

For example using data for the Avro RJ70 from Data A and data for the Textron AL502R from Data B the following values can be found:

design range	1952 nm
engine cruise SFC	0.7 lb/hr/lb
cruise speed	432 kt
typical passengers	70
$MTOM$	40 823 kg
MOE	23 360 kg

Assuming a standard passenger mass of 75 kg and baggage of 20 kg, the payload is:

$$M_P = 70 \times 95 = 6650 \, \text{kg}$$

The fuel mass for this payload is then:

$$M_F = 40\,823 - 23\,360 - 6650 = 10\,813 \, \text{kg}$$

The equivalent still air range ($ESAR$), as shown in Fig. 11.8, is computed as:

$$ESAR = 1.063 \times 1952 + 568 = 2641 \, \text{nm}$$

The estimated lift/drag (L/D) ratio is then:

$$L/D = ESAR \times (c/V) \times \ln(W_1/W_e)$$

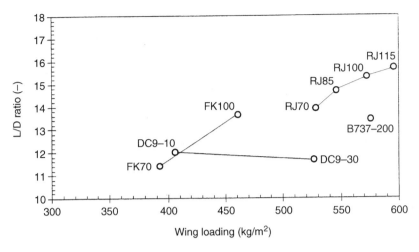

Fig. 16.3 Estimated *L/D* ratios for current regional aircraft.

where: $c = $ engine cruise SFC
$V = $ cruise speed
$W_1 = MTOM \simeq$ aircraft mass at start of cruise
$W_2 = MTOM - M_F = $ aircraft mass at end of cruise

Hence:

$$L/D = 2641 \times \frac{0.7}{432} \times \ln \frac{40\,823}{40\,823 - 10\,813} = 13.91$$

Lift/drag ratios have been estimated for other regional aircraft using Data A for aircraft data and Data B for engine *SFC* data. These are plotted in Fig. 16.3 against aircraft wing loading. High wing loadings tend to increase aircraft lift/drag ratio but a smaller wing will increase take-off and approach speeds and thus have a detrimental effect on field performance. The RJ series have significantly higher lift/drag ratios compared to other aircraft. This may be partly due to the high-wing configuration and its associated drag and lift benefits.

Initial estimates

Once the specification is complete and the preliminary analysis of similar aircraft has been conducted, it is possible to estimate the maximum take-off mass of the design and proceed to define the aircraft geometry. Based on the empty weight ratio analysis, a reasonable value for our aircraft is 0.60. From the lift/drag ratio estimates an initial value of 12.5 seems appropriate for a low-wing configuration. These figures are now used with the specification to estimate the fuel mass and then the overall aircraft mass.

The Breguet Range equation is used to estimate the fuel mass ratio for the design. The cruise speed is determined by first calculating the speed of sound for the required altitude. This may be done using ISA atmospheric properties, or by reference to Data D. At 35 000 ft, the speed of sound is $296.5\,\text{ms}^{-1}$ giving a cruise of speed of $207.6\,\text{ms}^{-1}$ (404 knots). The ESAR may be estimated using the expression presented in Chapter 11:

$$\text{ESAR} = 1.063 \times 1250 + 568 = 1897\,\text{nm}$$

Engine data for the BPR 3.0 engine listed in Chapter 9 (Fig. 9.18) is suitable for this study and gives an SFC of 0.635 lb/hr/lb at 35 000 ft and M0.70. A comparison with Data B shows this to be close to the **RR-BMW BR715**, a modern regional aircraft powerplant. Using this data, the fuel weight ratio is estimated for the design using:

$$\frac{M_F}{M_{TO}} = 1 - e^{\left[\frac{-Rc}{v_D^L}\right]}$$

$$= 1 - e^{\left[\frac{-1897 \times 0.635}{404 \times 12.5}\right]}$$

$$= 0.212$$

where R = equivalent still air range (ESAR)

Assuming a standard passenger weight of 75 kg and 20 kg of baggage per person, the design payload is 6650 kg. The maximum take-off weight can be estimated using:

$$M_{TO} = \frac{M_P}{1 - \dfrac{M_F}{M_{TO}} - \dfrac{M_{OWE}}{M_{TO}}}$$

$$= \frac{6650}{1 - 0.212 - 0.60}$$

$$= 35\,413\,\text{kg}$$

This compares well with the Fokker 70 aircraft with a maximum take-off weight of 36 800 kg. The spreadsheet method allows us to quickly illustrate the sensitivity of the design to payload/range changes as shown in Table 16.5.

Once the maximum take-off weight has been calculated, the wing area and engine thrust can be estimated. Again, little information is available at this point, so a comparison with existing regional aircraft is made. Data A shows that for similar aircraft the maximum wing loading (M_{TO}/S) varies between 400 and $577\,\text{kg/m}^2$. Based on these data, a wing area of $75\,\text{m}^2$ was chosen. The corresponding wing loading is $472.1\,\text{kg/m}^2$.

Table 16.5 Effect of payload/range on overall aircraft mass

	1000 nm	1250 nm	1500 nm	1750 nm
50 Pax	21 990	25 077	29 018	34 220
70 Pax	30 786	35 107	40 625	47 908
90 Pax	39 582	45 138	52 232	61 596
110 Pax	48 377	55 169	63 839	75 284

For most civil aircraft, the engine will be sized around either the take-off field length, second-segment climb gradient or top of climb requirements. At this stage, any of the three cases can be used to size the engine. However, the simplest option is often to estimate the thrust required at the top of climb. This calculation requires an estimate for the aerodynamic drag polar which in general form is:

$$C_D = C_{D_0} + kC_L^2$$

where: C_{D_0} = profile drag coefficient
k = induced drag coefficient (see below)
C_L = lift coefficient

The profile drag coefficient, C_{D_0} typically varies between 0.013 and 0.024 for civil transport aircraft. Lower values are associated with large, long-range aircraft with relatively small wings and high-wing loadings. Higher values tend to be associated with short-haul aircraft with relatively large wings and low-wing loadings. Based on this knowledge a value of $C_{D_0} = 0.022$ will be assumed for this exercise.

The induced drag coefficient, k, may be expressed as:

$$k = \frac{1}{\pi(AR)e}$$

where: AR = wing aspect ratio
e = Oswald's efficiency factor

The Oswald's efficiency factor typically varies between 0.75 and 0.85 for civil transport aircraft. Higher values are associated with more modern aircraft with advanced wing sections. Lower values relate to first and second generation civil jet transport aircraft. We will assume an aspect ratio, AR of 9.0 for the initial estimates. It would be possible to conduct a parametric optimisation later in the design process to determine the 'best' value for aspect ratio.

Using the method presented in Chapter 8 with an aspect ratio of 9.0 and a nominal taper ratio of 0.25 gives:

$$C_1 = 1.01$$

$$C_2 = 1.02$$

The lift-dependent drag coefficient can now be computed as:

$$\frac{dC_D}{dC_L^2} = \left(\frac{1.01 \times 1.02}{\pi \times 9.0} + 0.0004 + 0.35 \times 0.022\right) = 0.044$$

This corresponds to an Oswald's efficiency value of 0.798.

The drag polar for the clean configuration is thus:

$$C_D = 0.022 + 0.044C_L^2$$

The lift coefficient at the initial cruise altitude, assuming 2% fuel burnt during climb, is:

$$C_L = \frac{0.98 \times 35\,413 \times 9.81}{\frac{1}{2} \times 0.38 \times 207.6^2 \times 75} = 0.555$$

The total drag coefficient is given by:

$$C_{D_0} = 0.022 + 0.044 \times 0.555^2 = 0.0356$$

The drag is then calculated as:

$$D = \frac{1}{2} \times \rho \times V^2 S C_D$$

$$= \frac{1}{2} \times 0.38 \times 207.6^2 \times 75 \times 0.0356$$

$$= 21\,798\ V$$

At the initial cruise altitude a climb rate of 300 ft/min (1.524 ms^{-1}) is normally required for civil airline operations. The climb angle associated with this rate is:

$$\gamma = \frac{V_C}{V}$$

$$= \frac{1.524}{207.6} = 0.007\,341\ \text{rads}$$

The thrust required, T, is then calculated using:

$$\gamma = \frac{T - D}{W}$$

$$0.007\,341 = \frac{T - D}{0.98 \times 35\,413 \times 9.81}$$

$$T = 24\,348\ \text{N}$$

The thrust must now be converted to a sea-level static value using the BPR 3.0 engine data from Chapter 9 [Figs 9.18(a–f)]. The thrust available at 35 000 ft and Mach 0.7 will be somewhat less than the maximum value at sea-level static conditions. This reduction in thrust with increasing aircraft speed and altitude is normally termed thrust lapse. Engine lapse rate is defined as the ratio of thrust at a given aircraft speed and altitude to that at sea-level static. For this engine the lapse rate is approximately 0.25. Therefore the equivalent sea-level static thrust is 97 391 N. Assuming two engines, each engine must provide 48 696 N (10 945 lb) of thrust at sea-level static thrust setting. The corresponding thrust/weight ratio is 0.280 which is slightly lower than that of the Fokker 70 at 0.341.

Once initial estimates have been made for the aircraft maximum take-off weight, wing area and thrust required, the geometry can be defined for the baseline aircraft.

Configuration

Before detailed sizing of the aircraft is initiated, the aircraft configuration must be determined. It is useful to consider a wide range of configurations and then assess each for its strengths and weaknesses. It is likely that no single configuration

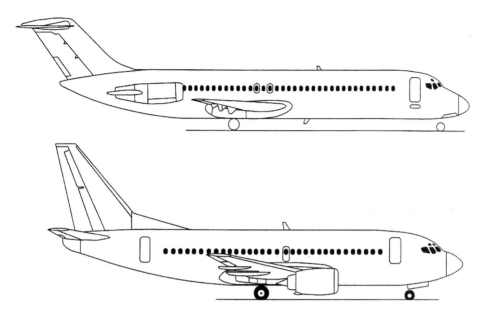

Fig. 16.4 Alternative configurations for baseline design.

is a clear favourite. In this case two configurations will be assessed through the preliminary design phase before a final decision is made. The two configurations: are (a) wing-mounted engines and (b) rear-fuselage-mounted engines. The former configuration is used for a number of larger aircraft but the smallest example is the Boeing 737. In contrast rear-fuselage-mounted engines have been used on a number of short-haul regional aircraft including the Douglas DC9, Fokker F.28 and F.100, Caravelle and BAC1-11. For comparison the Boeing 737 and Douglas DC9 configurations are shown in Fig. 16.4. Analysis conducted during the project suggested that the wing-mounted engines configuration is slightly lighter but suffers from inferior take-off performance due to the trailing edge flap discontinuity. Concerns over foreign object ingestion and the ability to use larger powerplants on stretch variants with wing-mounted engines suggested that the rear-fuselage option was a better choice. This also provides benefits such as a quieter cabin environment.

Fuselage

The fuselage design is centered around the choice of fuselage cross-section. This defines the seating layout, length of the aircraft and the ability to stretch the aircraft. Seating layouts for regional aircraft vary between four, five and six abreast. Too few seats across and the aircraft will have a high finesse ratio possibly incurring a weight penalty. The aircraft will also be difficult to stretch as the fuselage finesse ratio will become even higher and the cabin interior seem too 'tube' like. Too many seats abreast and the basic model will have a low finesse ratio

Table 16.6 70–100 seat regional aircraft fuselage cross-sections

Aircraft type	Seats abreast	Fuselage diameter (m)
Avro RJ	5/6	3.56
Boeing 737–500	6	3.73
Fokker 70/100	5	3.30
McD DC9–10/30	5	3.60

possibly incurring penalties from the small tail arm. The aircraft can, however, be stretched easily. The fuselage cross-section characteristics of several regional aircraft are shown in Table 16.6.

For the largest regional jets six abreast seating is standard with a fuselage diameter of around 3.70 m. The Avro RJ series has a diameter of 3.56 m somewhere between the typical five or six abreast fuselage diameter. This provides a generous five abreast layout, but a rather cramped six abreast layout. Noting that a slightly larger diameter would provide a more generous five abreast cabin, similar to that found on wide-body aircraft with the alternative for six abreast seating, a decision was made to use a 3.76 m fuselage diameter. Cross-sections for five and six abreast seating are shown in Fig. 16.5.

Once the cross-section has been chosen the next step is to generate the seating layout and fuselage planform. For short-haul aircraft it is common to offer only two seating classes (business and economy). For charter operations a single class is typical. For this aircraft a generous 34-inch (0.85 m) single-class seat pitch will be assumed. A two-class variant would give business class at 38-inch (0.95 m)-pitch and economy at 31-inch (0.775 m)-pitch.

Reference must be made to the airworthiness requirements specifying the number, size and location of emergency exits and operational requirements defining the

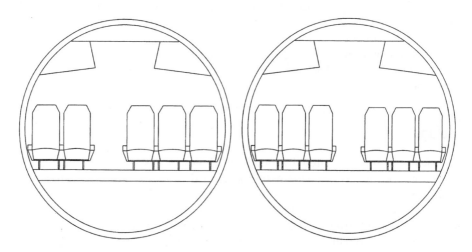

Fig. 16.5 5 and 6 abreast cross-section.

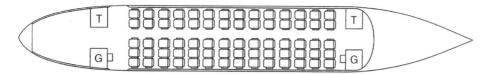

Fig. 16.6 5 abreast seating layout.

number/size of toilets and galleys. Information on both of these subjects is provided in Chapter 5. For the initial 70-seat model the basic features are:

- 34-inch (0.864 m) seat pitch
- two toilets and two galleys
- one Type I exit and two Type III exits per side (exited limited seating 110)

Using this information and allowing for toilet and galley areas, the cabin seating plan may be constructed. Forward of the cabin, space is provided for the cockpit. Behind the cabin space must be made available for the fuselage-mounted engines and the tail surfaces. Around these requirements the fuselage must provide an aerodynamic shape. The final layout is shown in Fig. 16.6.

Wing

From the initial estimates two wing parameters have been defined for the wing: gross area and aspect ratio. Before the external wing shape can finalised, the wing taper ratio and quarter chord sweep angle are required. A lower taper ratio will tend to reduce wing mass and increase induced drag (Oswald's efficiency decreases). It is also important to check that there is adequate chord near the tip for the placement of control surfaces. An initial value may be determined from existing aircraft. Values are seen to vary between 0.196 and 0.356. A sensible value is 0.25 which may be adjusted by optimisation studies later in the design process.

The quarter chord sweep angle must be chosen in conjunction with the wing thickness/chord ratio. Typically wing thickness/chord ratio is expressed for both the wing root and tip. A rough estimate of the mean chord from Chapter 6 is:

$$\left(\frac{t}{c}\right)_m = \frac{(t/c)_r + 3 \cdot (t/c)_t}{4}$$

An initial estimate for the root and tip thickness/chord ratios is 0.15 and 0.12. Using the above equation the average thickness/chord ratio is 0.1275.

For a cruise Mach number of 0.70, Fig. 6.6 shows that a quarter chord sweep angle of 17.5° is appropriate.

Horizontal tail

At this stage a detailed analysis cannot be undertaken so use is made of tail/wing area ratios (S_H/S) and tail-volume coefficient $(S_H L_H/S\bar{c})$ from similar aircraft (Table 16.7).

Table 16.7 Regional aircraft horizontal tail coefficients

Aircraft type	S_H/S	$S_H L_H/S\bar{c}$
Avro RJ 70	0.202	0.707
Avro RJ 85	0.202	0.805
Avro RJ100	0.202	0.876
Fokker 70	0.232	0.880
Fokker 100	0.232	1.176
McD DC9–10	0.295	1.176
McD DC9–30	0.275	1.190

The value of S_H/S varies between 0.202 and 0.344. Older aircraft such as the Douglas DC9 tend to have higher values. The horizontal tail-volume coefficient varies between 0.707 and 1.432. This parameter is dependent on the tail arm and hence the tail-volume coefficient increases as an aircraft is stretched. Thus the baseline design is the most critical in terms of horizontal tail sizing.

At this stage the horizontal tail arm is not known, so the initial estimate will be based on area ratio. For the baseline design a value of 0.20 is chosen giving a horizontal tail gross area of $15.0\,\text{m}^2$. The remaining generated parameters (e.g. AR, (t/c), Λ) are, like the wing, chosen from comparison with similar aircraft.

Vertical tail

Vertical tail information (S_V/S for the area ratio and $S_V L_V/Sb$ for the volume coefficient) from Data A is shown for T-tail regional aircraft in Table 16.8. The vertical tail area ratio for the Fokker 70/100 and Douglas DC9 is between 0.13 and 0.17. In contrast the Avro RJ series have significantly higher values at 0.20. This is due to the different location of the engines on the Avro RJ series.

Normally the vertical tail area is defined by the engine-out case where sufficient area is required to provide adequate directional control. Since the engines are wing-mounted on the Avro RJ series, the engine-out turning moment is increased requiring a larger vertical tail. The difference in configuration is also evident in vertical tail volume coefficients where typical values are around 0.05–0.07.

Again, the vertical tail arm is not known at this stage so the vertical tail is sized

Table 16.8 Regional aircraft vertical tail coefficients

Aircraft type	S_V/S	$S_V L_V/Sb$
Avro RJ 70	0.201	0.098
Avro RJ 85	0.201	0.110
Avro RJ100	0.201	0.117
Fokker 70	0.132	0.053
Fokker 100	0.132	0.064
McD DC9–10	0.172	0.074
McD DC9–30	0.161	0.074

assuming an area ratio of 0.16. This gives a vertical tail area of $12\,\text{m}^2$. The remaining geometrical parameters are chosen from similar aircraft.

Geometry summary

The baseline aircraft geometry is now complete and summarised below.

1. Fuselage parameters:
   - diameter   3.76 m
   - overall length   28.5 m
   - nose-section length   9.0 m
   - centre-section length 15.0 m
   - tail-section length   8.0 m

2. Wing parameters:
   - gross area $75.0\,\text{m}^2$
   - aspect ratio   9.0
   - taper ratio   0.25
   - root (t/c) 0.15
   - tip (t/c) 0.12
   - quarter chord sweep $17.5°$

3. Horizontal tail parameters:
   - gross area $15.0\,\text{m}^2$
   - aspect ratio   4.9
   - taper ratio   0.45
   - root (t/c) 0.12
   - tip (t/c) 0.10
   - quarter chord sweep $30°$

4. Vertical tail parameters:
   - gross area $12.0\,\text{m}^2$
   - aspect ratio   1.1
   - taper ratio   0.7
   - root (t/c) 0.13
   - tip (t/c) 0.11
   - quarter chord sweep $40°$

5. Nacelle:
   - length   3.8 m
   - maximum diameter   1.3 m

Aerodynamics

Estimation of profile drag

The airframe geometry can now be used with the methods presented in Chapter 8 to estimate the profile drag coefficient. The data shown in Table 16.9 correspond to

Table 16.9 Baseline aircraft profile drag estimate

Component	Wetted area (m²)	C_f	F	C_{Dc}
Wing	124.1	2.68×10^{-3}	1.449	64.3×10^{-4}
Horizontal tail	29.3	2.90×10^{-3}	1.412	19.2×10^{-4}
Vertical tail	24.7	2.62×10^{-3}	1.376	14.2×10^{-4}
Fuselage	285.9	1.92×10^{-3}	1.103	80.7×10^{-4}
Nacelle	15.2	2.56×10^{-3}	1.25	15.1×10^{-4}
Secondary items				19.2×10^{-4}
Total				0.0212

a cruise altitude of 35 000 ft and M0.70. The total profile drag coefficient is 0.0201, slightly lower than the value used for the initial estimates. The largest contribution is from the fuselage (42%), followed by the wing (36%).

Estimation of induced drag coefficient

The induced drag coefficient was estimated earlier using the method provided in Chapter 8 assuming current technology level. The value of 0.044 corresponds to an Oswald's efficiency of 0.80.

Estimation of maximum lift coefficients

The method presented in Chapter 8 can be used to estimate the maximum lift coefficient for landing. A sectional two-dimensional lift coefficient of 1.7 is assumed for the clean wing. This gives an actual clean wing lift coefficient of:

$$C_{L_{max}} = 0.9 \times 1.7 \times \cos 17.5 = 1.46$$

For the baseline design it is assumed that the aircraft has double-slotted fowler trailing edge flaps and full-span leading edge slats. Analysis of similar aircraft suggests that 0.75 would be a reasonable value for the ratio of flap span to wing span. Using this value with a knowledge of the fuselage width and wing geometry the flapped wing area is found to be 48.26 m². The lift increment due to trailing edge flaps is computed as follows.

From Chapter 8 for double-slotted fowler flaps, maximum deflection 30°:

$$\Delta C_{L_{max}} = 1.6$$

$$\frac{c'}{c} = 1.2475$$

$$\Lambda_{HL} = 6.6°$$

$$\Delta C_{L_{max}} = 1.6 \times 1.2475 \times \frac{48.26}{75} \times \cos 6.6 = 1.28$$

For leading edge slats:

$$\Delta C_{L_{max}} = 0.4$$

$$\frac{c'}{c} = 1.05$$

$$\Lambda_{HL} = 20.9°$$

$$\Delta C_{L_{max}} = 0.4 \times 1.05 \times \frac{56.96}{75} \times \cos 20.9 = 0.30$$

The maximum lift coefficient in the landing configuration is thus 3.0. For the take-off phase a lower flap angle is used, around 10–15°. For the initial analysis a lift coefficient of 2.24 will be assumed. The choice of value will, however, affect the take-off field length and climb gradient with one engine inoperative and thus may be changed later in the design process.

Mass estimate

A spreadsheet model has been developed using the methods presented in earlier chapters to rapidly assess the design. The structure of modules is similar to that described in Chapter 15. Geometrical and operational data is used to determine aerodynamic and mass characteristics. Aerodynamic output is used to estimate fuel mass and, in turn, the maximum take-off mass ($MTOM$). The process is iterative, the previous estimate for $MTOM$ being used to initiate the calculations but, as shown in Table 16.10, a substantial reduction is predicted.

The operating empty weight ratio is seen to be 0.596, very close to the value assumed for the initial estimates. In contrast the cruise lift/drag ratio is 14.43 which partly accounts for the lower $MTOM$.

Performance

Field performance

Take-off and landing performance has been estimated using the methods described in Chapter 10. The average thrust available during the ground roll is determined from engine lapse rate data for the bypass ratio 3.0 engine presented in Chapter 9 (Fig. 9.18). The components and total distances forming the take-off field length are shown in Table 16.11.

The landing performance is shown in Table 16.12. The largest portion of the landing distance is that due to braking. The factored landing distance increases the total distance further to account for pilot variability.

Second segment climb

The second-segment climb gradient is discussed in detail in Chapter 9. Since this aircraft is of twin-engined configuration with a high take-off lift coefficient it is

Table 16.10 Baseline aircraft mass statement

Mass breakdown	
Wing group (kg)	3070
Fuselage group (kg)	3730
Tail group (kg)	391
Undercarriage group (kg)	1114
Surface controls group (kg)	549
Nacelle group (kg)	259
Airframe mass (kg)	8140
Propulsion group (kg)	4503
Hydraulic and electrical systems group (kg)	646
Cabin and safety systems (kg)	468
Instrumentation (kg)	243
Systems (kg)	1357
Furnishings (kg)	2220
Aircraft empty mass (kg)	16 220
Operational items group (kg)	333
Flight crew (kg)	186
Cabin crew (kg)	186
Operational empty mass (kg)	16 925
Payload (kg)	7000
Zero fuel mass (kg)	23 925
Fuel mass (kg)	4486
Maximum take-off mass (kg)	28 411

Table 16.11 Take-off field length

Take-off segment	Distance (m)
Ground roll	894.1
Ground distance from rotation to screen height	197.3
Take-off distance	1092
Factored take-off distance	1255

Table 16.12 Landing field length

Landing segment	Distance (m)
Ground distance from screen height to flare	163
Ground distance covered during flare	82
Free roll	103
Ground roll during braking	455
Landing distance	804
Factored landing distance	1339

likely that the climb gradient with one engine inoperative may be critical in defining the aircraft thrust requirements. Using the method described in Chapter 8 the aircraft drag polar for the take-off configuration with undercarriage up is:

$$C_{D_0} = 0.042 + 0.044C_L^2$$

To this must be added the drag of the windmilling engine and that due to control surface deflection during flight with asymmetric power. A first estimate of engine windmill drag may be obtained from:

$$\Delta C_D = \frac{0.3 \times A_f}{S}$$

where A_f is the fan cross-sectional area and S is the wing reference area. For our regional jet, $A_f = 1.4\,\text{m}^2$. This results in a drag increment of 0.0056. Trim associated with asymmetric flight may be assumed initially to be 5% of the aircraft profile drag. The overall profile drag coefficient is now 0.0497.

Again using data from Chapter 9 the engine lapse rate at the reference climb speed is 0.821. The static thrust of the remaining engine is 44.34 kN. This gives a net thrust of 36.74 kN. At the reference climb speed of 121.4 kt, the lift coefficient is 1.56 producing a total drag coefficient of 0.1568 and a drag force of 28.09 kN. The climb gradient is then estimated using:

$$\gamma = \frac{T - D}{Mg}$$

$$= \frac{36\,742 - 28\,086}{28\,411 \times 9.81} = 0.031$$

The 3.1% climb gradient is significantly higher than the 2.4% requirement for twin-engined aircraft. This is surprising but partly due to the high value for wing aspect ratio. This reduces induced drag which is a significant proportion of the second-segment climb drag. Also the specified high cruise altitude dictated a relatively high thrust/weight ratio. This may be reviewed later in the study, see below.

Optimisation

Three parameters were individually investigated for optimisation allowing the engine size to vary:

- wing area
- aspect ratio
- minimum DOC

Since both the take-off, landing and second-segment climb requirements were comfortably met, the wing area was analysed to investigate whether the top of climb thrust requirements could be reduced, allowing engine size also to be reduced. For this exercise the spreadsheet analysis program was used to compute the overall design characteristics for several different wing areas. The results are shown in Fig. 16.7.

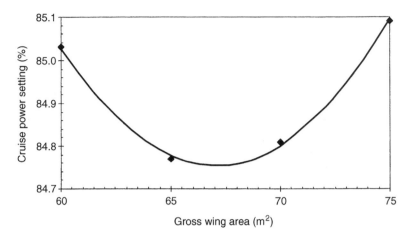

Fig. 16.7 Optimum wing area.

The minimum cruise power setting is seen to occur for a wing area of approximately $67\,\text{m}^2$, compared with the baseline figure of $75\,\text{m}^2$. However, since there is a requirement for growth from the basic design it is desirable to 'over-wing' the initial aircraft. On this basis a wing area of $70\,\text{m}^2$ was chosen in consideration of the small penalty in thrust that will be incurred.

The wing aspect ratio and direct operating cost (DOC) are closely related. Increasing aspect ratio tends to reduce cruise drag, increase cruise lift/drag ratio and reduce the amount of fuel required. This will reduce DOC per aircraft seat mile. However, if aspect ratio is increased further, the increased structural mass of the wing will offset the fuel savings and the DOC per ASM will increase. Determining the optimum value is complex as results are dependent on assumptions used in the DOC calculation, particularly the cost of fuel. This is further complicated by the fact that the aircraft will not typically enter service until around five years after the conceptual design phase and then may stay in production for 20 years. The analysis must consider the sensitivity of results to changes in fuel price in the future. In this scenario fuel prices may increase with introduction of a 'carbon tax', so the baseline fuel price was set at a typical present day value of $0.50 per US gallon. To assess the sensitivity of this assumption DOC per ASM was then estimated as a function of aspect ratio for fuel prices of $0.50, $1.00, $1.50 and $2.00 per US gallon as shown in Fig. 16.8.

For present fuel prices the aspect ratio for minimum DOC is approximately 8.0 (similar to the value used on current regional transport aircraft). Increasing the fuel price to $1.00 per US gallon increases the optimum aspect ratio to 11.00. Increasing the fuel price further increases the optimum aspect ratio to around 12. The highest value ever used on a swept wing jet transport aircraft is 10.0 on the Airbus A330/340 series. Values around 12.0 create structural and aeroelasticity problems that may not be solved with current materials (aluminium alloys). In view of this a value of 10.5 is chosen to reflect the probable increase in fuel prices relative to current values.

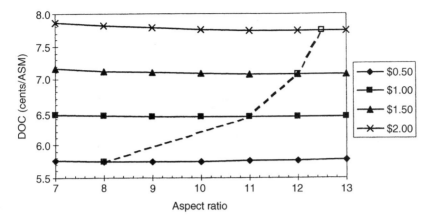

Fig. 16.8 Effect of fuel price and aspect ratio on DOC.

Final design

The final design is shown in Fig. 16.9. The large fuselage cross-section is apparent giving the aircraft a 'stubby' appearance.

Data for the final design are shown in Table 16.13.

The final design exhibits an operating empty weight ratio of 0.599, the slight increase compared with the baseline design being attributable to the higher wing aspect ratio of the final design. Lift/drag ratio is increased to 14.76 with a cruise altitude of 35 000 ft.

Time to climb to cruise altitude

The time required to climb to the initial cruise altitude of 35 000 ft was calculated using a simplified approach. Since ambient conditions, true airspeed and engine thrust vary with altitude, the climb phase was broken down into a series of segments. The conditions at the mid-point of each segment were then used to compute the climb gradient and the time to climb through the segment. The total time is then obtained by summing the time taken for each segment. The results from the spreadsheet are shown in Table 16.14.

The results show that the final design just achieves the specification to reach the initial cruise in under 25 minutes. In view of this an alternative cruise option was analysed with a cruise altitude of 31 000 ft. This reduces the time to climb to 18.5 minutes, but the cruise lift/drag ratio decreases to 13.62, reducing the design range by 107 nm to 1143 nm.

Alternative payloads

The spacious fuselage allows 84 passengers to be accommodated by changing to a six abreast seating layout. If take-off mass remains unchanged the fuel mass is reduced and the range decreases substantially to 578 nm.

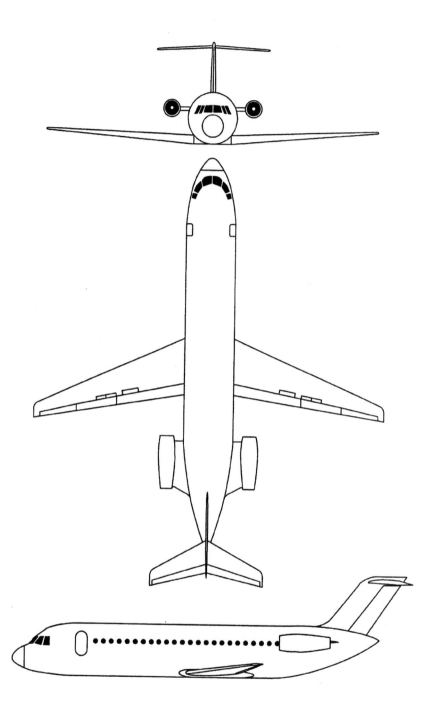

Fig. 16.9 Final design 3 view general arrangement.

Table 16.13 Final design characteristics

FUSELAGE	
Diameter (m)	3.76
Length (m)	28.5

WING	
Area (m^2)	70.0
Span (m)	25.1
Aspect ratio	10.5
$C_{L\,max}$ (take-off)	2.24
$C_{L\,max}$ (landing)	3.0

MASS (kg)	
Operational empty mass	17 023
Fuel mass	4395
Maximum TO mass	28 418

ENGINE	
Static thrust (kN/engine)	44.3

PERFORMANCE	
Cruise Mach number	0.70
Initial cruise altitude (ft)	35 000
Cruise L/D	14.76
FAR take-off field length (m)	1750
FAR landing field length (m)	1404
Design range (nm)	1250

ECONOMICS (1992 Prices)	
Aircraft price ($M)	19.58
Aircraft direct operating cost (cents)	6.42

Table 16.14 Time to climb to cruise altitude

Segment	Average rate of climb (m/min)	Time taken (min)
0–7000 ft	797.28	2.68
7000–15 000 ft	741.2	3.29
15 000–23 000 ft	491.2	4.97
23 000–31 000 ft	324.0	7.53
31 000–35 000 ft	216.0	5.64
Total time (min)		24.1

Table 16.15 Gross-weight options

	84 pax/1250 nm	96 pax/1250 nm
Operational empty mass (kg)	17 740	18 552
Maximum take-off mass (kg)	30 762	32 781
Cruise L/D ratio	15.25	15.59
Factored take-off distance (m)	1561	1680
Landing distance (m)	1484	1552
Second segment climb gradient	0.042	0.034
DOC per ASM (cents)	5.44	4.84

If the seat pitch is also reduced to 29 in (0.725 m) with a six abreast layout the capacity is further increased to 96. Calculations show that the theoretical design range is just about 11 nm. This is clearly unrealistic and so aircraft maximum take-off mass must be increased. This is done for different payloads.

Gross-weight options

The spreadsheet model was used to compute requirements for an aircraft with 1250 nm range with 84 and 96 passengers. The results are shown in Table 16.15.

Maximum take-off mass increases by approximately two and four tonnes respectively compared with the 70-seat model. Wing area remains unchanged thus increasing wing loading and hence cruise lift/drag ratio. This, combined with the increased passenger capacity, reduces the DOC per ASM for each model. With 84 passengers, the DOC per ASM is 15% below the baseline aircraft. For 96 passengers, DOC per ASM is 24.6% below the baseline aircraft. This clearly illustrates the advantages of the larger aircraft and correspondingly the penalty cost of increasing passenger space on the initial variant with five abreast seating if the market can sustain larger capacity aircraft.

Summary

This chapter described the design of an advanced regional airliner from the development of a specification, through to prediction of initial estimates and computation of the baseline design. Simple optimisation studies on the baseline design have been presented together with an investigation of growth variants.

17

Military transport aircraft

Summary

The choice of a military aircraft design study in a book devoted to civil aircraft design may seem strange but it has been included to illustrate how the methods described in the book can be applied, with care, to other types of aircraft. The military transport aircraft provides a good example of how the payload and operational requirements sometimes dictate the overall configuration of the aircraft.

This study was undertaken in 1993 by two undergraduate students. At about the same time, the McDonnell–Douglas C-17 was undergoing trials with the USAF as a supplemental airlifter to the ageing fleet of Lockheed C-5 Galaxy airlifters. The C-17 was designed to transport out-size loads direct to front line battle zones using semi-prepared airstrips around 3000 ft (1500 m) long. This requirement led to several compromises being made during the design in order to attain the required performance. These compromises were mainly associated with the restricted capacity that could be carried (both weight and volume) and the reduced range. In contrast, the C-5 Galaxy requires a 10 000 ft (3000 m) paved runway when operating with a full payload.

This study illustrates the process of setting the initial design specification with particular reference to the operational requirements. The design process is followed through from the initial estimates to detailed trade-off studies.

Introduction

The Gulf War in 1991 highlighted the need for heavy airlifters to transport artillery, vehicles, helicopters and supplies to areas quickly and efficiently. Such was the shortfall in airlifter capacity that civil reserve aircraft were called upon to assist the USAF fleet of Galaxy and Starlifter aircraft. These aircraft were generally freighter versions of large civil transport aircraft such as the Boeing 747 and Douglas DC-10. These aircraft exhibited good payload/range characteristics, but the height of the main deck meant that only palletised cargo could be accommodated. In response to this deficiency, the USAF has since requested that additional military airlifters be procured. With the rising costs of the C-17 program, a restart of C-5 Galaxy production was even proposed. The Galaxy was originally designed in the late 1960s and combines dated aerodynamic technology with second generation turbofan engines. Since this aircraft was designed the Soviet Union has developed two military airlifters which substantially outperform the ageing Galaxy aircraft.

In response to the need for additional airlifters, this study represents the conceptual design of a new military airlifter. It is anticipated that the introduction of super-critical wing technology and advanced third generation turbofan engines developed for civil transport applications will significantly improve performance and reduce costs compared with the existing C-5 Galaxy and C-17 Globemaster III aircraft.

Specification

The specification for a military airlifter may appear to be similar to that of a civil transport aircraft, including parameters such as design payload, design range, etc. However, the method by which some of these parameters, specifically the design payload, are determined is very different. For a civil transport aircraft, the payload is primarily passengers which are regularly arranged in a fuselage, often of cylindrical cross-section to minimise weight, whereas the payload for a military airlifter varies in both shape and weight. Items such as helicopters are relatively light, yet require a large volume of space in the hold. This can lead to 'bulking-out' of the cargo hold where the volumetric space has been used, yet the payload is far below the maximum that can be carried by the aircraft. This occurred frequently with the C-141A Starlifter and led to the development of the stretched version (C-141B) with a substantially increased hold volume.

Another loading difficulty can occur with very dense items, such as tanks and artillery vehicles. These often have very high floor loadings (often expressed as kg/m^2 of contact area) and this requires strengthened floors. The C-17 achieves this with minimum weight penalty by strengthening the floor ramps between the three strengthened fuselage frames that take the main wing spar connections and undercarriage attachment points. This configuration, however, limits the carriage of heavy items to the centre of the aircraft, disrupting the length of the cargo hold available for other items. This position helps to maintain an adequate centre of gravity position for the aircraft. As a replacement for the C-5 Galaxy it would be expected to carry several heavy items at once leaving lighter items to the C-17 and C-141.

Although the C-17 is a relatively new design, the performance requirements placed constraints on the size of the hold such that it cannot accommodate many items in the US inventory. The other US airlifters were designed in the 1960s when many items now in the US inventory did not exist. Some of these items are too bulky to be carried by even the C-5 Galaxy. Also, it must be remembered, particularly after the Gulf War, that rapid reaction forces may require deployment of outsize items from other NATO/Coalition forces as well as from the US inventory. This rapidly expands the possible options that must be considered in sizing the fuselage for the aircraft.

The requirement to carry heavy wheeled/tracked items, implies a high-wing configuration, so that vehicles may be driven on/off the main cargo deck. This requires a large rear cargo door to enable quick loading/unloading of cargo with minimum ground support. Special consideration must be given to the provision of a hinged nose door to facilitate simultaneous loading and unloading of cargo (as provided on the C-5 and An-124 aircraft).

Design payload

To aid in the determination of the design payload and the cargo hold shape, data is required on the possible types of vehicles that are in the US and NATO forces inventory. Table 17.1 shows a summary of the data collated with key types listed for each sub-category.

After some consideration, the design payload was set at 150 000 kg. This gives enhanced capability compared with the C-5 Galaxy and enables the carriage of a greater number of medium to heavy items.

In comparison with current military aircraft typical design range appears to be in the range 2400–3400 nm (Table 17.2). In general, the older aircraft have shorter ranges, with the exception of the An-225 whose low range is possibly attributable to the carriage of external cargo. A design range of 3300 nm was selected.

The design range may be extended with the use of in-flight refuelling. However, this is expensive and will lead to some loss of flexibility if tanker support is required on a regular basis.

The cruise performance of military airlifters has often been inferior to that of civil transport aircraft of comparable technology due to the higher drag configuration and the requirement for good field performance. Better field performance normally demands a large wing area which directly affects the cruise performance. However, with improvements in the drag reduction of wing high-lift aerodynamics and the use of super-critical wing technology it should be possible to offer similar cruise performance to that of civil transport aircraft. This will also make the aircraft more compatible with operations on civil air-traffic routes. In view of this, the cruise speed was set at Mach 0.80 with an initial cruise altitude of 33 000 ft.

Finally the field performance must be specified. For a conventional civil transport aircraft, this is normally defined as a single distance for both take-off and landing. However, the operation of a military airlifter is quite different. A typical mission may be to take-off at maximum weight from a relatively long paved runway and land at a short semi-prepared strip and then return unloaded. This mission implies a landing field length much shorter than that for take-off. Based on this mission requirement, the take-off distance was set at 2000 m (6560 ft) with a landing field length of 1250 m (4100 ft).

Table 17.1 Dimensions and masses of typical loads

	Overall length (m)	Width (m)	Height (m)	Mass unloaded/ loaded (kg)
Helicopters				
Bell AH-1W Super Cobra	13.87	3.23	4.11	4627/6690
McD AH-64A Apache	14.63	5.23	3.84	5165/9525
Eurocopter Tiger	14.00	2.52–4.46	3.81	3300/6000
Boeing/Sikorsky RAH-66A Comanche	13.22	2.31	3.39	4167/7790
Westland Lynx	10.62	3.75	3.50	3072/4535
EH Industries EH101	18.90	4.52	5.22	9000/14 288
Sikorsky UH-60A Black Hawk	12.60	3.56	3.76	5216/9979

	Overall length (m)	Width (m)	Height (m)	Mass unloaded/ loaded (kg)	Hull length (m)	Maximum ground loading (kg/m²)
Main battle tanks						
Challenger 1	9.80	3.25	2.95	–/62 000	6.98	0.97
Leopard 2	8.49	3.70	2.79	–/55 150	7.72	0.83
M1A1 Abrams	9.03	3.66	2.89	–/57 524	7.92	0.96

	Overall length (m)	Width (m)	Height (m)	Mass unloaded/ loaded (kg)	Maximum ground loading (kg/m²)
Mechanised infantry combat vehicles					
AAV7A1	7.94	3.27	3.26	18 663/23 991	0.56
M2 Bradley	6.45	3.20	2.97	20 053/22 590	0.54
M3 Bradley	6.45	3.20	2.99	19 732/22 443	0.83
Marder	6.79	3.24	2.99	28 200/29 207	0.83
Warrior	6.34	3.03	2.79	–/24 500	0.65
Armoured recovery vehicles					
Challenger 1	9.64	3.55	3.91	62 000/67 500	1.06
M88A1	8.27	3.43	3.23	–/50 803	0.76
Warrior	6.60	3.09	2.88	–/29 500	0.78
Miscellaneous					
LAV-25	7.56	2.83	2.31	–/10 300	–
M998 Hummer	4.57	2.15	1.75	2295/3870	–
MLRS	6.97	2.97	2.62	20 189/25 191	0.54
Tracked Rapier	6.40	2.80	2.50	–/14 010	–

The completed specification is now summarised:

Design payload	150 000 kg	Design range	3300 nm
Cruise speed	Mach 0.80	Initial cruise altitude	33 000 ft
Take-off field length	2000 m	Landing field length	1250 m

Table 17.2 Current military transports

Aircraft	Design payload (kg)	Design range (nm)
An-22 Cock	80 000	2700
An-124 Condor	150 000	2430
An-225 Cossack	250 000	1350
C-5 Galaxy	118 388	2985
C-17 Globemaster III	78 110	2430
C-141A Starlifter	32 136	3450
C-141B Starlifter	41 222	2550
IL-76A Candid	40 000	2700
IL-76B Candid	48 000	–
Shorts Belfast	36 285	870

Cargo-hold sizing

Apart from the performance specification, the cargo-hold sizing of large capacity military transport aircraft is of paramount importance. As mentioned earlier the actual dimensions required are dictated by the equipment and materials that are to be transported. During the definition of the specification, reference was made to the possible items that might be air-lifted to provide rapid deployment. The geometry of such items is now considered for sizing of the cargo hold. To aid this process a 50 m by 8 m grid was drawn representing the possible cargo hold. Onto this grid the silhouettes of various items can be positioned to determine the number of items that can be transported and the volume/mass used.

This model was used to evaluate possible payload combinations and at the same time provide a graphical representation of the physical size of such combinations in terms of length, width, height and total weight.

It is worth noting that the cargo cross-sections of current military transports are not rectangular. Key dimensions are the fuselage floor width and the maximum height. To ensure that the cargo hold does not 'bulk out' when low density items are carried sufficient height and width must be used to provide the appropriate volume without an unacceptably long cargo hold.

After consideration of several payload combinations the following (length, width, height) cargo hold dimensions were chosen to be:

45.0 m (147.7 ft) × 5.9 m (19.3 ft) × 4.2 m (13.8 ft)

These dimensions are greater than those of the C-5 Galaxy and enable this aircraft to carry the widest range of items within the US and NATO inventories. It is important to note that the total length of the cargo hold includes the rear-loading ramp which is designed to be load bearing and forms part of the cargo volume. This is similar practice to that used on the C-5 Galaxy and C-17 Globemaster III aircraft. These dimensions are now used to determine the overall fuselage layout and size.

Fuselage layout

A major requirement for any military transport aircraft is that of being able to load and unload any possible payload quickly, preferably with little or no extra handling systems to those available on the aircraft. An obvious necessity is to have the floor of the cargo hold as close to the ground as possible. This is achieved through choice of wing position and undercarriage design. A high-wing with undercarriage-mounted to the fuselage sides will, in general, provide the best solution. An examination of the airlifters listed in Table 17.2 shows that they all adopt such a configuration.

To prevent the wing carry-through section from obstructing the cargo hold, it is positioned to pass above the cargo hold. To provide an aerodynamic shape this is then blended with the fuselage to form an upper deck above the cargo hold. This deck provides four abreast seating (shown in Fig. 17.1) in two sections, one forward and one aft of the wing carry-through sections. The forward section may be used for crew rest quarters for long endurance flights. This upper deck feature is also common to the C-5 Galaxy and the Antonov An-124 aircraft.

Cargo hold access is of particular importance. Most civil freighters use side cargo doors which are not suitable for use on military transport aircraft. Access is provided through longitudinal loading using the nose and/or rear. All the existing aircraft provide rear-loading ability. The larger aircraft all have

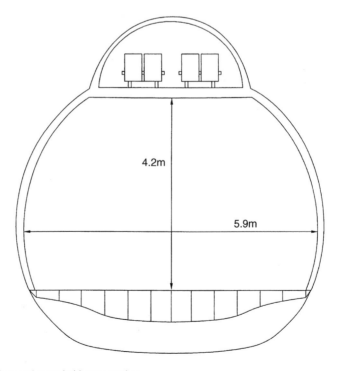

Fig. 17.1 Fuselage and cargo hold cross-section.

nose-loading providing simultaneous loading/unloading capability. This may be an issue, particularly close to the front-line where it is desirable to expose the aircraft to threat for the shortest possible time. Both nose and rear access produce weight penalties in terms of fuselage mass. For initial design estimates this may be included in the design calculations by increments added to the basic fuselage mass estimated using the methods of Chapter 7. Typical values for rear-loading ramps are 7–9% of fuselage mass. Nose-loading might be expected to incur a slightly lower penalty of 6–8%. More precise values can only be refined using detailed analysis once the specific geometry and loads are known.

For this aircraft the provision of an upper deck and high level flight deck lends itself to nose-loading, similar to that on the C-5 Galaxy, An-124 and freighter variants of the Boeing 747. It is then questionable whether a rear-loading ramp is necessary. This feature, however, allows the air-drop of certain cargo and troops. This was a design requirement for the C-17 Globemaster.

With the basic fuselage configuration determined and the cargo hold dimensions known, the external fuselage size can be set out through the development of engineering drawings and comparisons to existing military transport aircraft. For our design the overall fuselage dimensions are:

length: 70.0 m (229.7 ft)
maximum width: 7.3 m (23.9 ft)
maximum height: 7.5 m (24.6 ft)

Initial estimates

To determine an initial estimate of the maximum take-off mass, values are required for the operating empty mass ratio and the fuel mass ratio for the design mission. For an initial value of the operating empty mass ratio comparison was made with existing aircraft. For the aircraft listed in Table 17.2, the computed values (not shown) range from 0.423 to 0.463 with a mean value of 0.435. Little variation is observed with aircraft size in contrast to that shown for regional aircraft in Chapter 16.

The fuel mass ratio is estimated using the rearranged Breguet Range equation shown below:

$$\frac{M_F}{M_{TO}} = 1 - e^{\left[\frac{-Rc}{V\frac{L}{D}}\right]}$$

Using the chart and expression provided in Chapter 10 for a design range of 3300 nm, the equivalent still air range is found to be approximately 4050 nm.

It is anticipated that the aircraft will need a 64 000 lb (285.7 kN) class high bypass ratio turbofan engine such as the Trent 700 series. Cruise SFC for this engine is approximately 0.553 lb/hr/lb.

For the baseline design an initial cruise altitude of 10 000 m (33 000 ft) and cruise speed of Mach 0.80 were chosen. This corresponds to a cruise speed of 239.6 ms^{-1} (465.7 kt). Little public information is provided on lift/drag ratios for military transport aircraft; however, they are known to be lower than similar civil

Table 17.3 Sensitivity of take-off mass to payload/range variation

Payload (kg) range (nm)	2000	2500	3000	3300	3500
100 000	254 150	275 274	299 239	315 222	326 639
120 000	304 981	330 329	359 087	378 266	391 967
150 000	381 226	412 911	448 858	472 833	489 958
200 000	508 301	550 548	598 478	630 444	653 278

transport. The Boeing 747 L/D ratio is about 17 and this suggests that the C-5 Galaxy cruise L/D may be around 15. Since more modern civil transports are now achieving L/D values of 18–20 it may be appropriate for the initial estimates to choose a lift/drag ratio of 17 since the configuration is not too dissimilar to civil aircraft.

Applying these values to the Breguet Range equation results in a fuel mass ratio of 0.2478. With a design payload of 150 000 kg, the initial estimate for the take-off mass is then:

$$M_{TO} = \frac{M_P}{1 - 0.2478 - 0.435}$$
$$= 472\,833\,kg$$

This value is comparable to that of the Antonov An-124 (405 tonnes) which has similar payload capability but shorter range.

Take-off mass for various payload/range values is shown in Table 17.3. A payload of 120 tonnes and 2500 nm range corresponds to the C-5 Galaxy. Note how the 330 tonnes estimate compares with the Galaxy take-off mass of 380 tonnes. Altering the assumed L/D to 15 and SFC to 0.63 (typical of 1960s technology) gives a take-off mass of 383 tonnes, i.e. very close to the C-5 Galaxy. The reduction to 330 tonnes shows the improvements in airframe and engine technology since the C-5 design was designed.

Take-off masses above 600 tonnes would probably require four engines, each with a static thrust in excess of 100 000 lb (454 kN) or alternatively six engines (the configuration used by Antonov for the An-225) if such large engines were not available.

Aircraft geometry

With an initial estimate of the maximum take-off mass, the remaining aircraft geometry can be sized. As for the regional jet in Chapter 16, this is principally done by comparison with existing aircraft in the first instance to complete the baseline design. Parameters are then changed later in the process as a result of parametric analysis and trade studies.

The next parameter to be chosen is the wing loading. This will determine the wing reference area and has an effect on the aerodynamics, performance and size of other airframe components. Analysis of the transports listed in Table 17.2 shows that wing loadings vary between 450 and 750 kgm^{-2}. Higher values tend to be associated with more modern and larger transports without a rough/short field take-off requirement. Based on this information a wing area of 700 m^2 was chosen, corresponding to a wing loading of 672 m^2. Wing aspect ratio was chosen to be 8.0, typical of aircraft of this size. Higher values will increase wing mass and also wing span which is an issue for large military transports where a shorter span is preferable to reduce congestion at military bases. A parametric trade study can be done later in the design process to identify the sensitivity of aspect ratio on various design parameters (see later in this chapter).

A super-critical wing section was adopted for the aircraft as used on more recent aircraft. This should substantially improve cruise performance relative to aircraft such as the C-5 Galaxy. For a cruise speed of Mach 0.8, Fig. 6.6 suggests that a quarter chord sweep angle of 30 is sufficient. This compares with 35 for the C-5 Galaxy. The reduction in wing sweep is brought about through improved wing sectional aerodynamics, and this should reduce wing mass and improve high-lift device efficiency.

High-lift devices are critical to good field performance by military transport aircraft, although significant advances have been made in aerodynamics since the introduction of the C-5 Galaxy. The main benefits will be associated with the reduction of mass and drag, rather than increases in lift generation. The aircraft is assumed to have full span leading edge slats and double-slotted Fowler flaps. Based on the method presented in Chapter 8, maximum landing lift coefficient is estimated to be 3.05 and maximum take-off lift coefficient around 2.7. The take-off value is quite high and will result in a low L/D ratio for second-segment climb. This, however, should not present a problem for a four-engined transport which requires lower engine-out and second-segment climb capability.

A number of options are available for the tail layout. The classical layout for military transport aircraft with rear-loading ramps is a T-tail configuration. Recently, the An-124 demonstrated that it is possible to incorporate a low-set horizontal tail behind the rear-loading ramp. McDonnell–Douglas studied both configurations for the C-17 and determined that a T-tail provides improved ride quality at low level in turbulence where a low-set tail position suffers since it is in the wake from the wing. Based on this information a T-tail arrangement was chosen.

At the initial stages of the design process tail sizing is made on the basis of tail area ratio comparisons with existing aircraft. Tail volume coefficients may be evaluated later in the process once the tail arm dimensions have been determined. The horizontal tail area ratios are seen to vary between 0.117 and 0.167. An initial estimate of area is made of 100 m^2 corresponding to a ratio of 0.143. This is towards the lower limit of values but reflects the introduction of new technologies such as improved aerodynamics and relaxed static stability. A similar strategy was taken for sizing the vertical tail. Area ratios vary between 0.129 and 0.205, the low value corresponding to the C-141 and the higher value the C-17. This may seem counter to the trends of technology, but is partly due to the short tail arm on the C-17. Based on this data an area of 97.3 m^2 was chosen corresponding to an area ratio of 0.139. Once the configuration was complete the tail arms and volume coefficients were determined to further check the tail geometry.

To complete the layout four Trent 700 class high bypass ratio turbofan engines are mounted beneath the wing at relative spanwise positions of 0.266 and 0.50 respectively. Static thrust for the baseline design was assumed to be 64 000 lb (285.7 kN).

Spreadsheet model

Design methods similar to those described in the preceding chapters were coded into a spreadsheet aircraft design model. With some minor changes the model can be adapted from the one used for the regional aircraft of Chapter 16.

Using the model the baseline design was found to have a $MTOW$ of 445.6 tonnes, somewhat lower than the initial prediction. This reduction is mainly due to a higher than expected lift/drag ratio of 17.55. The empty weight ratio is also slightly lower than expected (0.423), compared with the 0.435 value assumed for the initial estimate. This slight reduction may be attributable to advances in propulsion technology and materials.

Aircraft performance analysis shows that equipped with 64 000 lb (284.7 kN) turbofans the baseline design meets all the specification criteria. Take-off field length is 1818 m and landing field length is 1060 m. Note these are unfactored distances which are normally quoted for military transport aircraft. The second-segment climb gradient is 0.04 which comfortably meets the minimum requirement for four-engined aircraft of 0.03. Sensitivity analysis shows that aspect ratio has a significant effect on the second-segment climb gradient. Assuming other variables remain constant Table 17.4 illustrates the effect of changes in aspect ratio on second-segment climb gradient.

The effect is particularly powerful for this design because of the high lift coefficient assumed and hence the resulting high induced drag. Reducing the take-off lift coefficient to 2.5 would increase the gradient from 0.027 to 0.033 for an aspect ratio of 7.0, but reduction in C_L would, however, have a detrimental effect on take-off performance.

Undercarriage design

Once a more accurate estimate of the maximum take-off mass has been achieved a more detailed analysis of the undercarriage may be made. Although the mass estimation method presented in Chapter 7 does not require detailed information

Table 17.4 Sensitivity of aspect ratio of second segment climb gradient

Aspect ratio	7.0	7.5	8.0
Net thrust (kN)	683 363	683 363	683 363
Drag (N)	567 356	536 297	509 120
Climb gradient	0.027	0.034	0.040

about the undercarriage configuration it is important for this class of aircraft to determine the number of wheels required and the tyre size for a given pavement loading. Using the references provided in Chapter 3 tyre sizes and allowable loads may be used to select an appropriate tyre. For this aircraft a large low pressure tyre was chosen with the following characteristics:

diameter: $= 1.63\,\mathrm{m}$
width: $= 0.572\,\mathrm{m}$
inner diameter: $= 0.66\,\mathrm{m}$

At a pressure of $90\,\mathrm{lb/in^2}$ ($620\,\mathrm{kN/m^2}$) the tyre can accept a maximum static load of approximately $25\,000\,\mathrm{kg}$. For this load the pavement load classification number (LCN) is approximately 50. This corresponds with an LCN of 44 for the C-5 Galaxy.

A simple configuration with 10 undercarriage struts (five per side) and two wheels per strut is assumed, similar to that used by Antonov for the An-124/225. A check must now be made that 10 struts and 20 wheels provides an equivalent single wheel load ($ESWL$) below $25\,000\,\mathrm{kg}$.

Assuming, for simplicity, that the aircraft static load is applied to the main undercarriage as though it were a tricycle layout, approximately 8% of $MTOW$ is applied to the nose gear and 92% of $MTOW$ to the main gear. Now dividing by 10 gives the total load per strut. Using the final design $MTOW$, this gives a load per strut of $40\,945\,\mathrm{kg}$.

The $ESWL$ must now be determined. For a strut with two wheels or a bogie with four or more wheels this is not simply the strut load divided by the number of wheels. This is because the wheels are relatively close together and so the $ESWL$ will be higher. For a twin-wheel strut the reduction factor is dependent on the spacing between the wheels. The range is typically 1.65–1.8. For this case the wheels are closely spaced to minimise the volume required when retracted into the fuselage sides and thus a lower value of 1.65 is used. This produces an $ESWL$ of $24\,815\,\mathrm{kg}$, just inside the allowable value of $25\,000\,\mathrm{kg}$.

If the allowable load had not been achieved, the spacing would need to be increased or, if this was not possible, a greater number of struts/wheels or a more capable tyre would be needed. All these options would increase weight and drag. This provides a classical trade-off scenario where, at the detailed design stage, the effect on aircraft mass of LCN limits can be determined.

Optimisation

The baseline design whilst satisfying all design criteria is far from optimum. Field and cruise performance both exceed specification with some margin. This suggests that the efficiency of the design can be improved. The margin on landing field length would allow the wing area to be reduced and this should improve cruise performance. Table 17.5 shows a simple trade-off analysis on wing area for the baseline design.

Reducing wing area from the baseline value of $700\,\mathrm{m^2}$ reduces $MTOW$ slightly but at the expense of a slight increase in cruise power setting. Based on the above

Table 17.5 Wing area trade-off study

Gross wing area (m^2)	MTOW (kg)	Cruise power setting (%)	L/D	W/S (kg/m^2)
660	442 654	100.26	17.35	670.61
680	443 979	99.95	17.46	652.91
700	445 384	99.71	17.55	636.26
710	446 114	99.62	17.60	628.33

data, the wing area was reduced to 680 m^2. To help reduce cruise thrust, which is now the critical requirement for sizing the powerplant, a trade analysis was conducted for aspect ratio with the new wing area. An increase in aspect ratio to 8.5 increases cruise L/D to 17.76 (from 17.46) and reduces the cruise power setting. This enables engine static thrust to be reduced to 62 000 lb (275.8 kN). A small weight penalty is incurred increasing gross weight close to that of the baseline design.

Table 17.6 Final design mass breakdown

Mass breakdown	
Wing group (kg)	59 264
Fuselage group (kg)	44 820
Tail group (kg)	4955
Undercarriage group (kg)	12 736
Surface controls group (kg)	3438
Nacell group (kg)	12 140
Airframe mass (kg)	137 353
Propulsion group (kg)	24 497
Hydraulic and electrical systems group (kg)	4295
Cabin and safety systems (kg)	1573
Instrumentation (kg)	2743
Systems (kg)	8611
Furnishings (kg)	2550
Aircraft empty mass (kg)	173 010
Operational items group (kg)	15 550
Flight crew (kg)	186
Cabin crew	279
Operational empty mass (kg)	189 025
Payload (kg)	150 000
Zero fuel mass (kg)	339 025
Fuel mass (kg)	106 025
Maximum take-off mass (kg)	445 051

This gives a brief indication of the trade and optimisation studies possible once the baseline design is defined. Although wing area and aspect ratio were treated independently here, they are highly dependent parameters and a better design may have been found had the two parameters been analysed as a matrix of aircraft with different wing areas and aspect ratios.

The changes to the design have increased the landing field length to 1080 m and the take-off field length to 1927 m, whilst second-segment climb gradient remains unchanged. All parameters remain within specification.

Final design

A mass breakdown for the final design is given in Table 17.6 and three-view general arrangement in Fig. 17.2. The actual numbers may differ slightly from those estimated directly from the equations given in Chapter 7 due to the assumptions made for the configuration and advanced materials used.

Conclusions

This case study has expanded the concepts introduced in Chapter 16. For this aircraft, development of the specification required detailed analysis of the possible payload options. Development of the baseline design was then followed with more detailed analysis of the undercarriage including pavement load classification number assessment. The final design provides a useful basis for students to continue through to the detailed design phase. In particular, detailed design of the rear-loading ramp and undercarriage extension/retraction geometry will need more careful consideration.

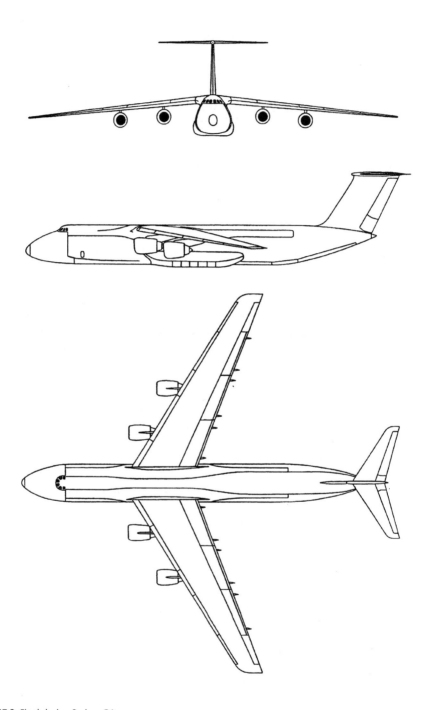

Fig. 17.2 Final design 3-view GA.

18

Unconventional designs

Summary

Unconventional designs are often difficult to analyse using traditional design programs that are inflexible. In contrast the modular layout of a spreadsheet aircraft design program and the rapid development time, allow the model to be tailored to the requirements of the unconventional design study.

This study was made possible due to the availability of a wing mass formula for multi-hull aircraft. This was incorporated into the spreadsheet model developed for the regional airliner and updated for application to large, long-haul aircraft. The revised model was first validated on large aircraft such as the Boeing 747–400 before being applied to three aircraft with different fuselage configurations.

Optimisation analysis demonstrates the structural benefits of distributing fuselage mass along the wing span. The significance of aspect ratio in design optimisation is illustrated by determining the aspect ratio for minimum gross weight, minimum direct operating cost and minimum fuel weight for each of the three designs.

Introduction

One of the limitations of some of the established aircraft design programs is the difficulty of assessing aircraft of widely different configurations from the conventional layouts on which the software was based. Modifications often require substantial redesign, recoding, recompiling and debugging of the program. With spreadsheet methods the analysis is structured for the particular problem under investigation; therefore the inflexibility of traditional programs is avoided.

This study was used to assess the difficulties associated with designing unconventional layouts using spreadsheet methods. Some earlier work on ultra-high capacity civil transport aircraft had identified serious technical and operational problems associated with aircraft size. To overcome some of these problems multi-hull layouts were investigated. This offered the opportunity of assessing the flexibility of spreadsheet methods applied to unusual layouts.

Multi-hull aircraft have been considered in several aircraft projects in the past.[1,2,3] The main advantage claimed for the configuration is the increased structural efficiency compared to conventional single-hull layouts. This increase in structural efficiency results from the reduced wing bending moment due to the outboard position of the fuselage mass. This in turn, produces a lighter wing and enables higher aspect ratios to be used. There are, however, many problems surrounding the configuration which have resulted in the configuration rarely being adopted. The most notable exception is the Voyager aircraft.[4]

The design point for the study was set at 1500 seats (i.e. an ultra-ultra high capacity aircraft) in a three-class layout, flying 4000 nm with usual reserves at Mach 0.82 and operating from 3000 m runways. The range is shorter than that proposed for the new large aircraft as it was felt that the payload/range trade-off would offer more flexibility than for conventional designs.

The multi-hull layout necessitated developing the wing mass formula within the overall mass model to account for the novel configuration. The wing mass formula for twin fuselage aircraft developed by Udin et al.[5] was used. The method may be applied to single-, twin- and triple-hull aircraft configurations. In the latter case the aircraft is modelled as a single fuselage aircraft with the two outer fuselages considered as concentrated loads. In this way all three aircraft can be modelled using a consistent analytical method, enabling accurate comparisons to be made between configurations.

The implementation of the new wing mass formula confirmed that existing spreadsheet modules could be modified quickly. Unlike traditional procedural languages, the spreadsheet code does not require recompiling after each alteration, and the formula may be debugged faster because all the calculated values in the spreadsheet are visible. Since no multi-hull transport aircraft exist the spreadsheet was applied to aircraft designs proposed by Lockheed in its multi-body aircraft study[6] to validate the analytical methods used.

Configuration

Fuselage design was an area of specific interest. Each design required a different fuselage cross-section to produce a near optimum fuselage shape. To achieve this a range of fuselage cross-sections was analysed for all three configurations.

The single-fuselage design proved the most difficult since this requires 24–28 seats abreast to maintain acceptable fuselage length. Such a number of seats immediately dictates a multi-deck aircraft. The critical question here is either a twin-aisle triple-deck aircraft or a triple-aisle twin-deck configuration. Neither of these options is likely to be viable in terms of passenger evacuation; however the twin-deck layout was chosen as being most representative for this class of aircraft. After considering double bubble and ovoid designs a circular section was adopted as mechanically the simplest shape. Figure 18.1 shows that the main deck has 14 seats arranged 3–4–4–3, whilst the upper deck has 12 seats in 2–4–4–2 configuration. In terms of scale the upper deck floor width is the same as a Boeing 747 main deck. The under floor cargo hold can accommodate three LD3 containers abreast. For the number of seats abreast this represents a reduction in cargo hold

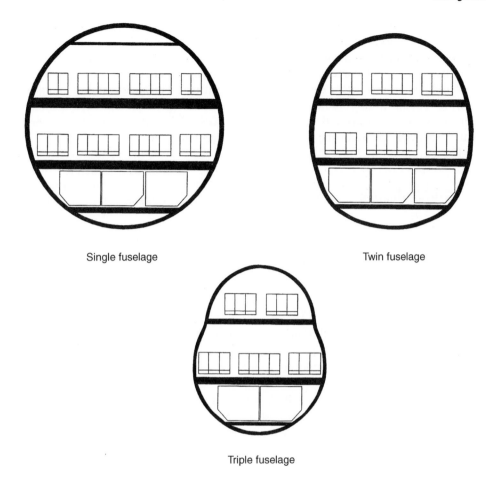

Single fuselage

Twin fuselage

Triple fuselage

Fig. 18.1 Fuselage cross-sections.

volume compared to current single-deck civil transport aircraft. This will be a serious disadvantage to airline profitability potential.

The twin-fuselage design requires 750 passengers per fuselage. The three-class layout was split equally between the fuselages. The implications for passenger loading are not considered here. Circular and double-bubble designs were analysed and the final choice is based on a double-bubble shape with flattened top and bottom. The upper deck is 10 abreast in 3–4–3 configuration. The main deck is 11 abreast with a 3–5–3 seating arrangement. The underfloor cargo hold can accommodate three LD3 containers abreast. This equates to 7pax/LD3 compared with 8.7pax/LD3 for the single-fuselage aircraft. The twin-fuselage aircraft will, however, have two wing carry-through sections and cargo hold volume should be similar to that of the single-fuselage aircraft.

The triple-fuselage aircraft requires 500 passengers per fuselage. Again the three-class layout has been split between the fuselages. A number of cross-sections were analysed, yet it was concluded that the Boeing 747 cross-section with 3–3 seating on the upper deck and 3–4–3 seating on the lower would be suitable. The under floor cargo contains two LD3 containers abreast. The cross-section equates to 8pax/LD3, compared with 10pax/LD3 for the twin-fuselage design.

Statistical data of current large transport aircraft was then used to determine the baseline parameters for each of the configurations. This included determination of wing loadings, tail volume coefficients, high-lift device type and capability, etc. Engines were based around scaled variants of the Rolls Royce Trent 800 series which was well defined at the time this study was conducted.

Optimisation

The aircraft were then optimised using spreadsheet methods. Figure 18.2 (reproduced directly from the spreadsheet output) shows the variation of gross weight with aspect ratio for each of the designs.

The minimum gross weight for single-fuselage aircraft is seen to occur for an aspect ratio of 6.75. This is very close to the values used on the first generation wide-body transports as shown in Table 18.1.

This illustrates two points. First, it confirms that early wide-body transports were optimised around minimum gross weight. This is mainly because computer power was not available for more advanced optimisation and also because lower fuel prices meant weight and initial aircraft cost were more important than direct operating cost.

Secondly, the approximate aspect ratio for minimum gross weight has not changed significantly despite the advances in propulsion, aerodynamics and structures. It is quite likely that the value found is more closely related to the basic (single-fuselage) configuration.

The twin-fuselage configuration shows a considerable improvement compared with

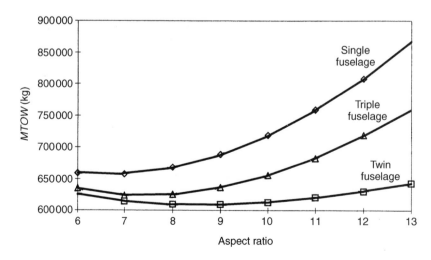

Fig. 18.2 Effect of aspect ratio on gross weight.

Table 18.1 Aspect ratios of early wide body transports

Aircraft	Aspect ratio
Boeing 747–100/200	6.96
Lockheed L1011 TriStar	6.97
McDonnell Douglas DC10	6.91
Single-fuselage design	6.75

the single-fuselage aircraft. For the aspect ratio range 6–12 it is at least 6% lighter than the single-fuselage configuration. The aspect ratio for minimum gross weight is approximately 8.75, a considerable increase over the single-fuselage design due to the relief of wing load associated with the outboard location of the fuselages.

The triple-fuselage design does not move all the fuselage and payload mass outboard and so is not as efficient (structurally) as the twin-fuselage design. It still, however, proves to be better than a single-fuselage design. A reduction in gross weight of at least 5% is evident for a given aspect ratio relative to the single-fuselage design.

DOC optimisation

The aspect ratio of the final designs was selected on the basis of DOC optimisations. A selection of fuel prices were used to determine the optimum aspect ratio for each design and the sensitivity to changes in fuel price. Figures 18.3(a)–(c) show the DOC as a function of aspect ratio for different fuel prices for the three aircraft configurations respectively. For a given fuel price, optimum aspect ratios are higher for the twin- and triple-fuselage designs relative to the single-fuselage design. In all cases the optimum aspect ratio is higher than that for minimum gross weight. Optimising for minimum DOC is now common practice and a feature of modern designs such as the Boeing 777.

Overall, at current fuel prices the DOC for the triple-fuselage design is about

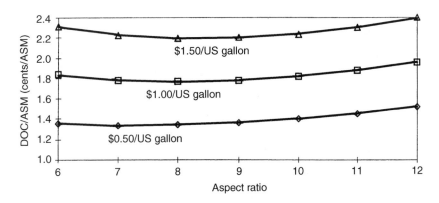

Fig. 18.3(a) Single fuselage DOC vs aspect ratio.

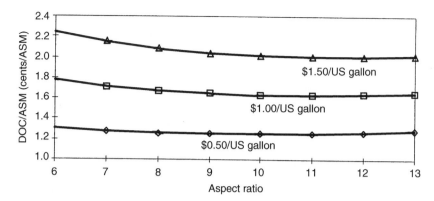

Fig. 18.3(b) Twin fuselage DOC vs aspect ratio.

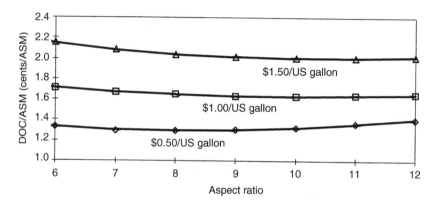

Fig. 18.3(c) Triple fuselage DOC vs aspect ratio.

3% lower that of the single-fuselage design. The greater efficiency of the twin-fuselage design increases this margin to 6%. Such gains are relatively small considering the technological risk associated with multi-hull configurations.

Design summary

The general arrangement drawings of the final designs are shown in Fig. 18.4. A comparison of the main parameters for the three configurations is shown in Table 18.2.

Study conclusions

The study has analysed unconventional configurations using spreadsheet methods. Three conceptual designs with different configurations have been developed and optimised using direct operating cost estimates.

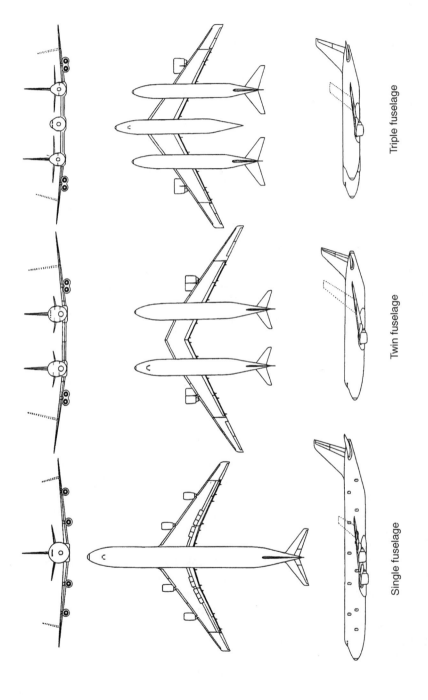

Single fuselage Twin fuselage Triple fuselage

Fig. 18.4 General arrangement of multi-hull designs.

Table 18.2 Multi-hull final design summary

	Single hull	Twin hull	Triple hull
FUSELAGE (twin-deck)			
Length (m)	102.5	67.5	63.0
Maximum width (m)	9.75	8.1	6.5
Maximum height (m)	9.75	9.2	8.1
Wetted area (m^2)	2847	3037	3605
Number of passengers per hull	1500	750	500
WING			
Reference area (m^2)	900	900	950
Span (m)	87.5	101.7	92.5
Aspect ratio	8.5	11.5	9.0
MASS (tonnes)			
Operational empty mass	207.2	179.3	182.8
Fuel mass	184.9	165.7	193.8
Maximum take-off mass	672.6	626.9	651.4
ENGINE			
Static thrust (kN/engine)	435.0	360.0	460.0
PERFORMANCE			
Cruise Mach number	0.82	0.82	0.82
Initial cruise altitude (ft)	32000	32000	32000
Design range (nm)	4000	4000	4000
FAR take-off field length (m)	2460	2580	2340
FAR landing field length (m)	1900	1920	1990
Cruise L/D ratio	17.9	18.8	16.3
ECONOMICS (1992 prices)			
Aircraft price ($M)	244.0	225.0	232.0
Aircraft direct operating cost (cents)	1.34	1.27	1.33

Comparing the designs, it can be seen that the twin-fuselage design is most efficient with a gross mass 6.8% below that of the single-fuselage design. Because the triple-fuselage design does not move all the fuselage mass outboard it is less efficient with a gross mass 3.2% lower than the single-fuselage design. These mass savings also help to reduce aircraft total cost and direct operating costs.

There are many reasons why such designs have not been developed for commercial transport aircraft. First is the technological risk associated with the multi-hull configuration. Advances in technology may, in time, solve many of these issues. There are, however, other operational problems which remain to be overcome. The most important of these relate to integration with existing airport and terminal facilities and also the potential difficulties for twin-fuselage aircraft of the pilot being located away from the aircraft centre-line. Although none of these issues has been considered here, it is important to consider operational factors when analysing unconventional configurations in greater detail.

References

1. Houbolt, J. C., 'Why Twin Fuselage Aircraft?', *Astronautics & Aeronautics*, April 1982, pp. 26–38.
2. *Jane's All The World's Aircraft*, 1987–1988, pp. 529–530.
3. Jenkinson, L. R. and Rhodes, D. P., 'Beyond Future Large Transport Aircraft', AIAA Paper 93-4791, August 1993.
4. Lockheed-Georgia Co., Marietta, 'Multibody Aircraft Study, Final Contractor Report', September 1981. Published May 1982.
5. Maglieri, D. J. and Dollyhigh, S. M., 'We Have Just Begun to Create Efficient Transport Aircraft', *Astronautics & Aeronautics*, February 1982, pp. 26–37.
6. Udin, S. V. and Anderson, W. J., 'Wing Mass Formula for Twin Fuselage Aircraft', *Journal of Aircraft*, September–October 1992, pp. 907–914.

19

Economic analysis

Summary

This study investigates the effect of aircraft size on aircraft economics. Two separate design specifications (300- and 600-seat aircraft) are taken through the conceptual design process and optimised to produce final designs. Comparison on key parameters shows that the 600-seat aircraft is structurally more efficient than two 300-seat aircraft. Total fuel and gross mass are also seen to be lower compared to two 300-seat aircraft.

Assuming equal production runs (1000 aircraft), the 600-seat design purchase cost is seen to be less expensive compared to two 300-seat aircraft. However, if only half as many 600-seat aircraft are produced (500), the 600-seat design is seen to be more expensive. Estimates show that for equal production runs (1000 aircraft), the 600-seat design direct operating cost (DOC) is seen to be 10.4% below that of the 300-seat aircraft. If only five hundred 600-seat aircraft are produced compared to one thousand 300-seat aircraft, then the DOC advantage for the 600-seat aircraft reduces slightly to 9.0%.

Introduction

Having shown the ability of spreadsheet methods to handle conventional and unconventional layouts from a technical/engineering standpoint, this study investigates the use of such methods for economic/cost analysis.

Analysis of past aircraft shows that aircraft direct operating costs have decreased with increasing aircraft size.[1] However, such aircraft have been designed at different technology levels making comparisons difficult. In order to make fair comparisons between two designs, the technology level must be similar.

Two aircraft of different size (300- and 600-seat) were designed to a common mission specification and optimised for minimum direct operating cost. The mission specification was based on the proposed new large aircraft (NLA) design.

- Single-class layout with 34-inch seat pitch. A single-class layout was adopted to simplify the design of the fuselages and allow more accurate comparisons to be made.

- 7200 nm range with full passengers and baggage.
- 3000 m field length.
- Cruise at M0.82 to M0.85 with initial altitude of at least 31 000 ft.

The spreadsheet model that was developed for this study incorporates detailed aircraft cost and direct operating cost methods so that the effects of production run/rate could be investigated. The cost model by J. Wayne Burns[2] was used to estimate aircraft development and purchase costs. Direct operating costs were estimated to European rules using Association of European Airlines (AEA) methods.[3]

Configuration

Similar to the designs presented in Chapter 18, the fuselage layout is central to the initial design for this study. As mentioned above, single-class seating arrangements were adopted. Galley space and other amenities were then set according to AEA recommendations for long-haul aircraft.

300-seat design

The fuselage design is highly dependent on the choice of cross-section. Obviously for a given number of passengers the cabin will be longer if the number of seats across is reduced. For a 300-seat aircraft, a cross-section incorporating 7, 8, 9 or 10 abreast seating is possible. This will produce fuselages of different length which will also affect the carriage of baggage and cargo. Within this variation the designer must be aware of operational issues such as cargo hold container sizes and passenger preferences, e.g. to be seated no more than one seat from an aisle.

To assess the effects of seating layout, several fuselage designs were configured. This enabled the effects of the fuselage on the overall design to be taken into account.

Table 19.1 shows the primary characteristics for four different fuselage cross-sections. Fuselage length, wetted area and floor area all decrease with increasing number of seats abreast. Cargo hold volume per passenger is also seen to decrease as the cross-section increases from eight to 10 abreast. This is mainly attributable to the shorter fuselage and hence fewer LD3 containers may be accommodated.

Table 19.2 shows the effects of different fuselage configurations on the baseline designs. The fuselage group mass is seen to decrease as the number of seats abreast is increased. This is mainly due to the lower finesse ratio of the shorter fuselage along with the reduction in wetted area. The higher drag associated with the shorter fuselage, however, offsets the reduction in fuselage mass. This makes the total mass, cost and DOC greater for nine and 10 abreast configurations compared with the baseline eight abreast seating configuration. Based on this analysis the eight abreast seating layout was adopted for the 300-seat aircraft.

Table 19.1 300-seat fuselage characteristics

No. pax	Seats abreast	Fuselage length (m)	Wetted area (m^2)	Floor area (m^2/pax)	Cargo volume (m^3/pax)
300	7	61.33	871.68	0.787	0.2856*
300	8	57.58	868.30	0.763	0.3874
300	9	52.88	860.20	0.729	0.3576
300	10	50.56	853.27	0.682	0.2980

* The low cargo hold volume per passenger for the seven abreast aircraft is due to a change of container from LD3 to LD2. Because of the poor cargo hold volume this configuration was not considered for further analysis.

Table 19.2 Effect of fuselage configuration on baseline design

300-seat aircraft cross-section	8 abreast	9 abreast	10 abreast
Fuselage group mass (kg)	Baseline	−1.38%	−2.20%
Fuselage wetted area (m$^{2)}$	Baseline	−0.93%	−1.73%
Aircraft cost	Baseline	+0.69%	+0.27%
DOC/ASM	Baseline	+0.65%	+0.50%

600-seat design

The fuselage configuration for the 600-seat design is more complex than for the 300-seat design. Above 500 seats the option of a double-deck seating arrangement becomes attractive. This prevents the fuselage from being unacceptably long whilst providing acceptable interior seating with a traditional twin-aisle seating arrangement.

A number of single-deck and twin-deck fuselage designs were analysed. The number of seats abreast ranged from nine to 19 with values greater than 10 implying a double-deck seating arrangement.

Table 19.3 shows 600-seat fuselage characteristics as a function of number of seats abreast. Again similar trends are seen as the number of seats abreast is increased. Fuselage length and wetted area reduce with double-deck designs being

Table 19.3 600-seat fuselage characteristics

No. pax	Seats abreast	Fuselage length (m)	Wetted area (m^2)	Floor area (m^2/pax)	Cargo volume (m^3/pax)
600	9	95.21	1632.63	0.737	0.5066
600	10	85.49	1536.26	0.729	0.4321
600	14 (9 + 5)	69.40	1340.68	0.978	0.2533
600	16 (10 + 6)	65.04	1241.27	0.977	0.2384
600	19 (11 + 8)	58.18	1197.90	0.907	0.2352

Table 19.4 Effect of fuselage configuration on baseline design

600-seat aircraft	14 abreast	16 abreast	19 abreast
Fuselage group mass (kg)	Base	−6.81%	−7.97%
Fuselage wetted area (m²)	Base	−7.41%	−10.65%
Price	Base	−1.35%	+1.04%
DOC/ASM	Base	−0.94%	+1.20%

significantly more efficient. Floor area is seen to reduce for either single- or double-deck designs, although double-deck seating appears to offer much greater floor area per passenger than for single-deck seating arrangements.

This is actually the additional floor space required to allow for the curved fuselage sides on the upper deck. This may be used for carry-on baggage but it does not necessarily imply increased passenger comfort. Hold volume per passenger is significantly higher for the nine and 10 abreast seating arrangements compared with the eight abreast 300-seat aircraft. However, the volume per passenger then drops dramatically for double-deck designs to around $0.25\,\mathrm{m}^3$ per passenger compared with $0.35\,\mathrm{m}^3$ per passenger for the 300-seat aircraft. This reduction is entirely due to the double-deck seating arrangement and may help to improve the direct operating cost per air seat mile. This may be a disadvantage as airlines would like to use extra cargo space for passenger and crew rest accommodation.

Noting that the single-deck arrangements produce unacceptably long fuselages, the three twin-deck options were analysed using the baseline aircraft parameters. Again fuselage group mass reduces for 16 and 19 abreast seating arrangements compared with the 14 abreast seating design. This is shown in Table 19.4. Further analysis shows that the 16 abreast seating arrangement produces the minimum gross weight and minimum cost design. Based on the analysis, the 16 abreast seating arrangement was selected for the 600-seat aircraft.

Optimisation

The aircraft were optimised using DOC/ASM as an objective function to produce the final designs. The effect of wing area and aspect ratio on DOC/ASM for the 300-seat aircraft is shown in Fig. 19.1 (three-dimensional plotting facilities are automatically available within the spreadsheet). It can be seen that aspect ratio has a pronounced effect on the DOC/ASM. The optimum aspect ratio for current fuel prices is seen to be around 9.0. This is in agreement with current designs. It can also be seen that a smaller wing improves the DOC/ASM, for a long-range aircraft, as expected. The minimum wing area was set by fuel volume requirements. The buffet boundary constraint was not considered at this initial project stage.

The 600-seat aircraft was optimised in a similar manner to the 300-seat aircraft. The optimum aspect ratio was found to be around 8.5 for this aircraft. This is slightly lower than that for the 300-seat aircraft which may be due to the increased overall size of the aircraft. Allowable wing fuel volume is acceptable for this aircraft, partly due to the lower wing aspect ratio, enabling a higher wing loading

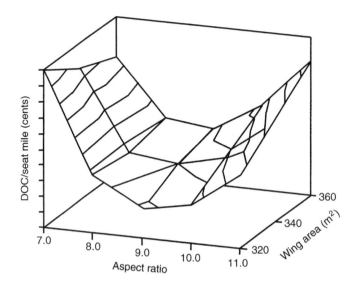

Fig. 19.1 Carpet plot for DOC/ASM.

to be achieved than in the 300-seat design. The general arrangements of the final 300- and 600-seat aircraft are shown in Figs 19.2 and 19.3.

A comparison of the two designs is shown in Table 19.5. The 600-seat design is seen to be structurally more efficient with a lower empty weight ratio than the 300-seat designs. Cruise lift/drag ratio is slightly lower for the larger aircraft; however, mission fuel is slightly below that of two 300-seat aircraft resulting in a lower gross weight for the 600-seat aircraft relative to two 300-seat aircraft.

Economic analysis

For the initial economic analysis, the production volume and build rate were assumed to be similar for both aircraft. However, if two 300-seat aircraft were to perform the role of a single 600-seat aircraft, the production of the 300-seat design would be double that of the 600-seat design. Also, the 300-seat design would be sold in its original 300-seat market. Based on these assumptions a production run of 500 at four per month was set for the 600-seat design and 1000 at eight per month for the 300-seat design.

Table 19.6 shows the results for both production scenarios. Assuming the same production run and rate for both designs, the results, as expected, show the 600-seat design to be less expensive than two 300-seat designs. This trend is reflected in a reduction in DOC/ASM. The 600-seat aircraft shows a 7.4% saving in aircraft cost and a 10.4% saving in DOC/ASM. For the second scenario, which reflects

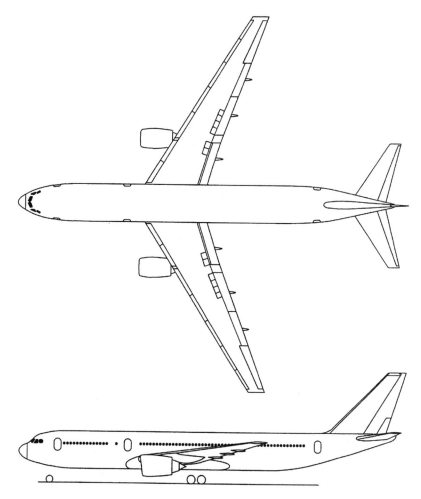

Fig. 19.2 General arrangement of 300-seat aircraft.

different production volumes and rates, the 600-seat aircraft is seen to be 9.0% more expensive than two 300-seat designs and the DOC/ASM advantage is reduced 4.3%.

Study conclusions

The study has produced conceptual aircraft designs for two specifications differing only in passenger capacity. The spreadsheet model has been used to determine the aircraft mass and drag characteristics along with performance estimates. The two baseline designs have been optimised for wing area and aspect ratio to produce minimum DOC designs. Optimum aspect ratio was found to be 9.0 and 8.5 for the 300-seat and 600-seat aircraft respectively, suggesting a reduction in optimum aspect ratio with increasing aircraft size.

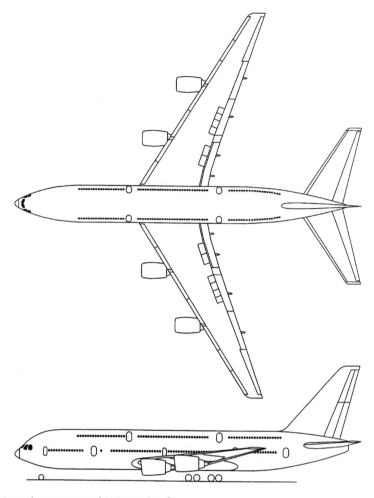

Fig. 19.3 General arrangement of 600-seat aircraft.

Structural efficiency of the 600-seat aircraft is also better than the 300-seat aircraft. The empty weight ratio is seen to be 0.464 compared with 0.467 for the 300-seat aircraft.

The economic analysis has shown that the 600-seat aircraft is superior in terms of aircraft acquisition cost and direct operating cost per passenger provided that the same number of each aircraft are made. However, since the 600-seat aircraft market is smaller an alternative scenario shows that the 600-seat aircraft may be more expensive than two 300-seat aircraft. Direct operating cost for the larger aircraft is still below that of the 300-seat aircraft illustrating the effects of aircraft size.

There are many issues not covered by this simple study. These include the greater flexibility offered to airlines by the smaller aircraft with respect to matching traffic variations, the use of the smaller aircraft on less dense routes. Also not considered

Table 19.5 Final design data

	300-seat aircraft	600-seat aircraft
FUSELAGE		
Length (m)	57.58	65.04
Maximum width (m)	5.52	6.90
Maximum height (m)	5.52	7.50
WING		
Reference area (m^2)	345	620
Span (m)	55.72	72.59
Aspect ratio	9.0	8.5
MASS (tonnes)		
Operational empty mass	114.3	221.1
Fuel mass	99.8	196.8
Maximum take-off mass	243.2	476.1
Operating empty mass ratio	0.469	0.464
ENGINE		
Number of engines	2	4
Static thrust (kN/engine)	370	349
PERFORMANCE		
Wing loading (kg/m^2)	705.0	784.0
FAR take-off field length (m)	2668	2456
FAR landing field length (m)	2361	2456
Cruise L/D ratio	18.22	18.05

Table 19.6 Final design data

Economics: scenario 1	Two 300-seat	One 600-seat
Production run	1000	1000
Development cost $B 1994	3.35	5.94
Purchase cost $M 1994	178.84	165.53
Direct operating cost	5.39	4.83
Cash operating cost	3.49	3.07

Economics: scenario 2	Two 300-seat	One 600-seat
Production run	1000	500
Development $B 1994	3.35	5.94
Purchase cost $M 1994	178.84	195.01
Direct operating cost	5.39	5.16
Cash operating cost	3.49	3.10

is the difficulty of operating large aircraft in congested airports. Such issues may counter the slight cost advantage shown for the large aircraft by this purely technical study. It is important to understand and take into account all the criteria that are superficial when conducting a full project study.

References

1. Association of European Airlines 'Long-range aircraft requirements', December 1989.
2. Cleveland, F. A., 'Size effects in conventional aircraft design', *Journal of Aircraft*, **7**, No. 6, November–December 1970.
3. Wayne Burns, J. W., 'Aircraft Cost Estimation Methodology and Value Of A Pound Derivation For Preliminary Design Development Applications', 53rd Conference of Society of Allied Weight Engineers, Long Beach CA, 23–25 May 1994.

Index